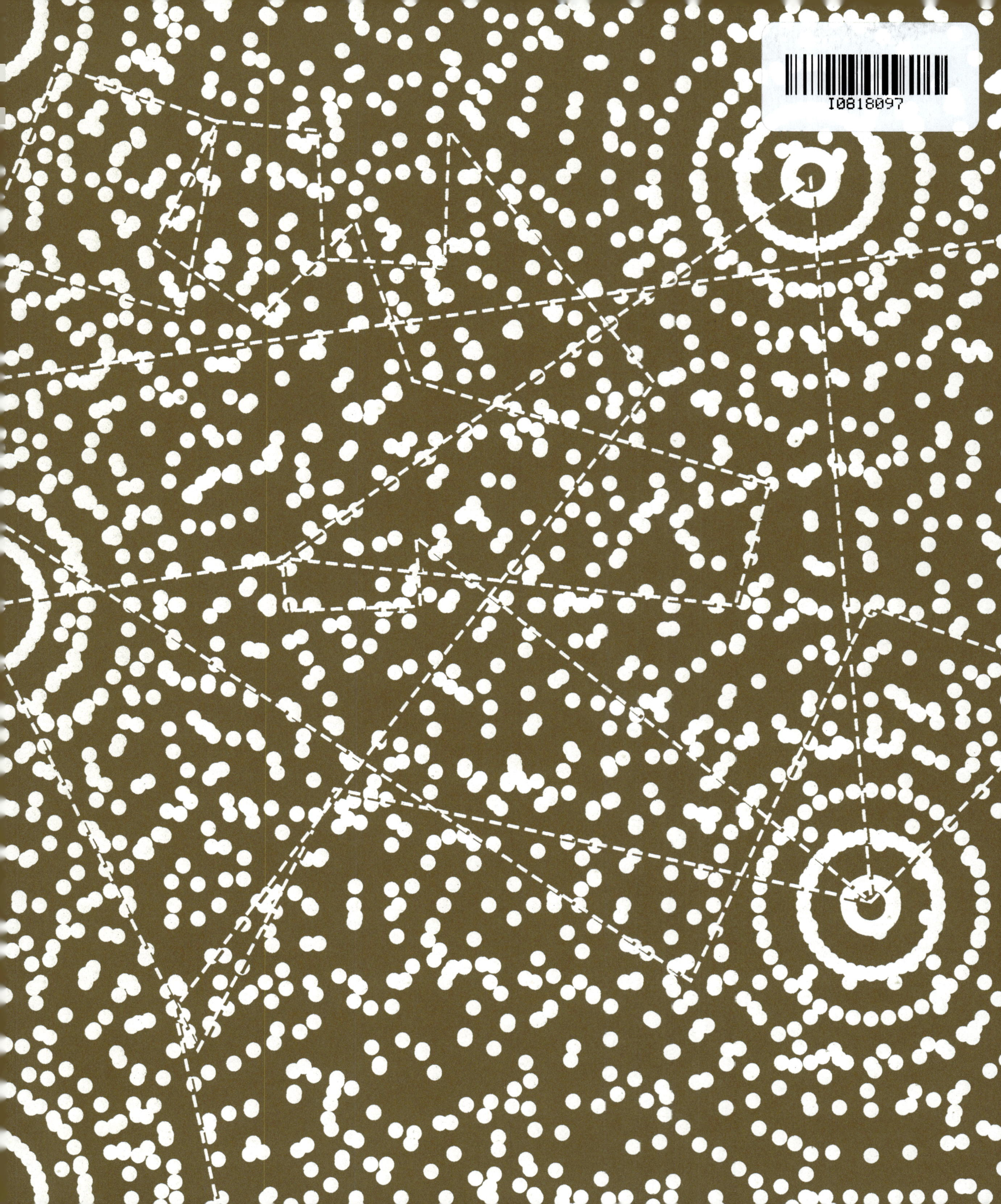
I0818097

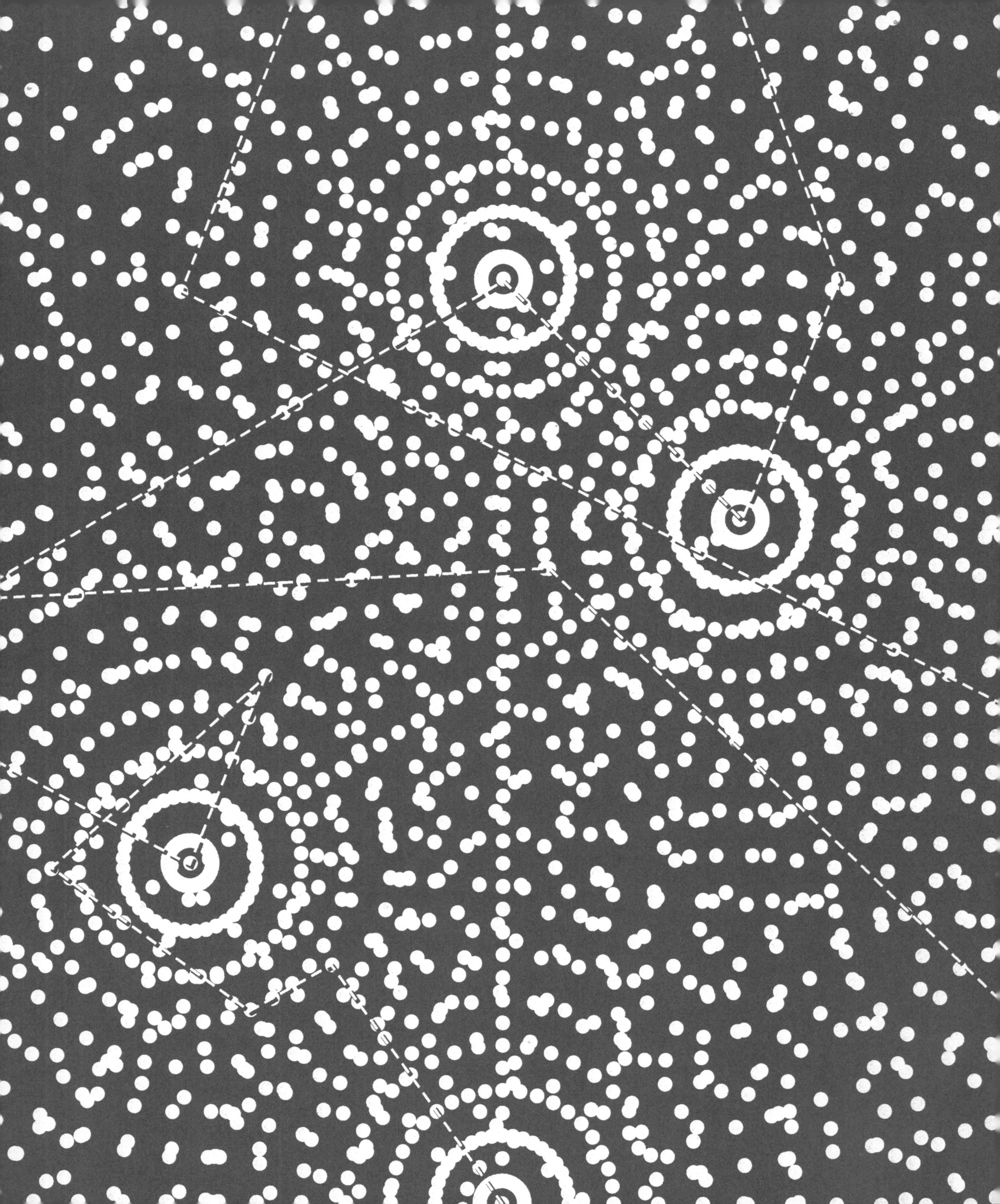

Other Networks

Other

A RADICAL
TECHNOLOGY SOURCEBOOK

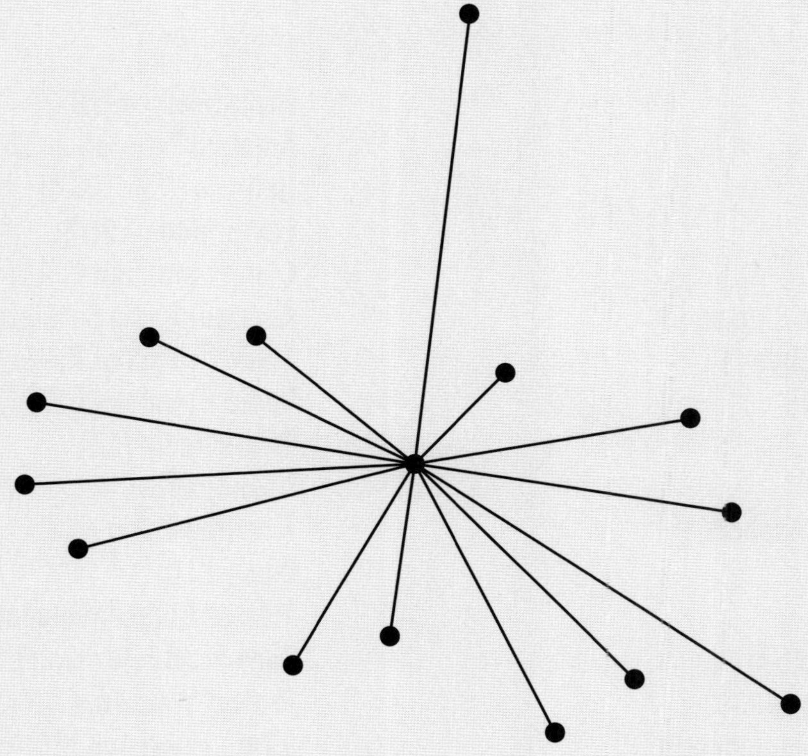

LORI EMERSON

Networks

ANTHOLOGY EDITIONS
NEW YORK

WIRELESS NETWORKS

WIRED NETWORKS

HYBRID NETWORKS

IMAGINARY NETWORKS

Chronological List of Networks

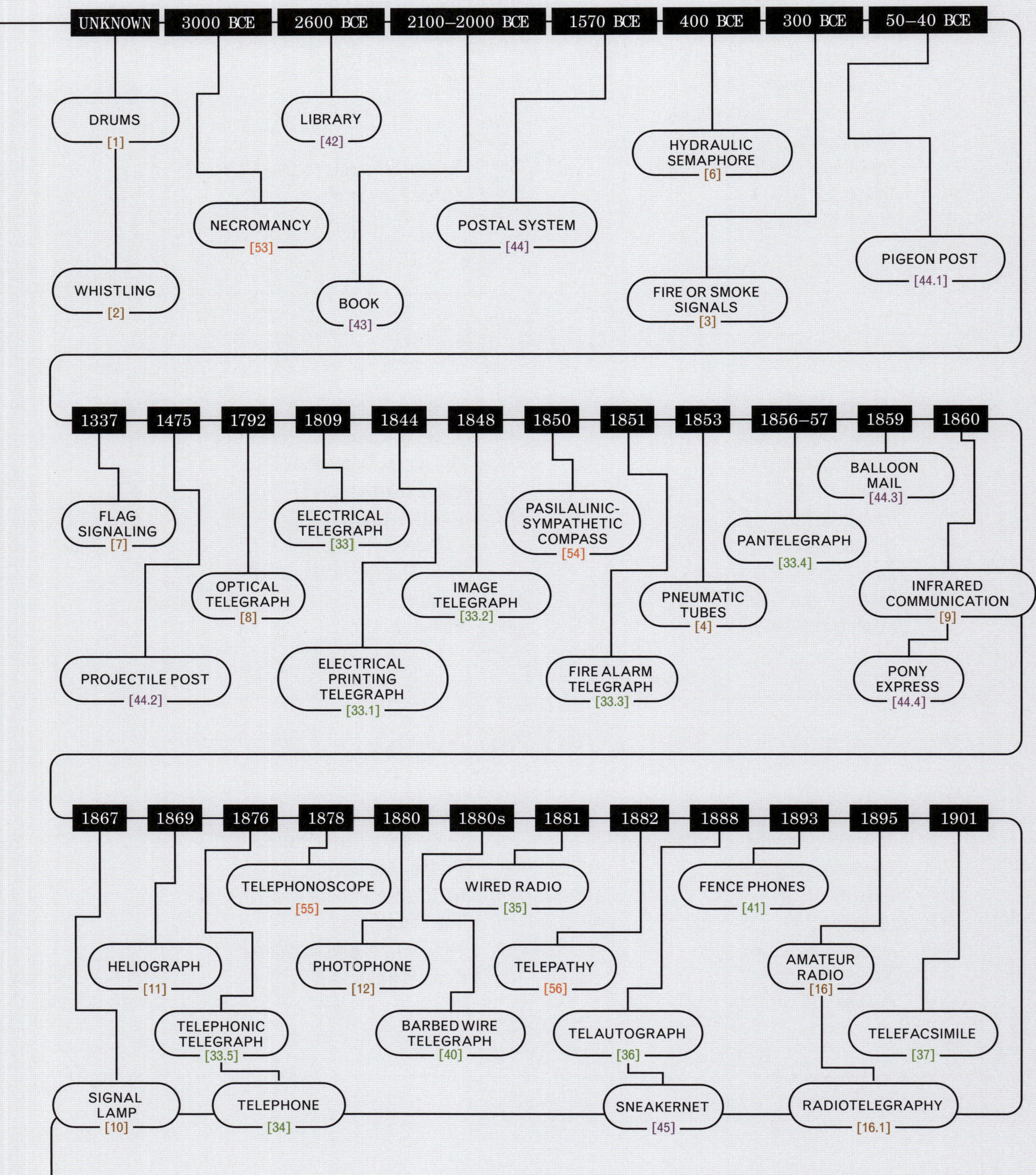

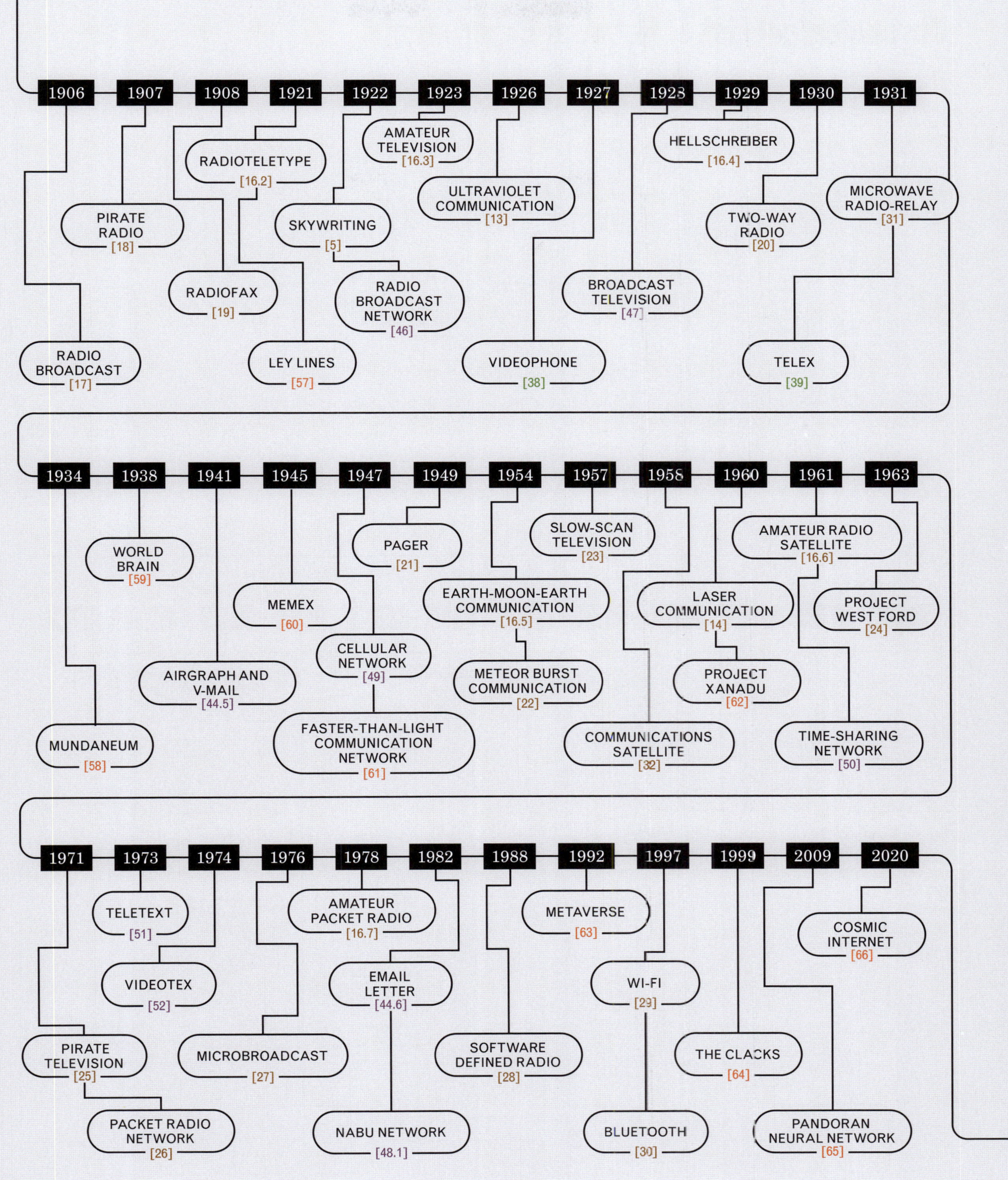

1906
1907
1908
1921
1922
1923
1926
1927
1928
1929
1930
1931
RADIO BROADCAST [17]
PIRATE RADIO [18]
RADIOFAX [19]
RADIOTELETYPE [16.2]
LEY LINES [57]
SKYWRITING [5]
RADIO BROADCAST NETWORK [46]
AMATEUR TELEVISION [16.3]
ULTRAVIOLET COMMUNICATION [13]
VIDEOPHONE [38]
BROADCAST TELEVISION [47]
HELLSCHREIBER [16.4]
TWO-WAY RADIO [20]
MICROWAVE RADIO-RELAY [31]
TELEX [39]
1934
1938
1941
1945
1947
1949
1954
1957
1958
1960
1961
1963
MUNDANEUM [58]
WORLD BRAIN [59]
AIRGRAPH AND V-MAIL [44.5]
MEMEX [60]
CELLULAR NETWORK [49]
FASTER-THAN-LIGHT COMMUNICATION NETWORK [61]
PAGER [21]
EARTH-MOON-EARTH COMMUNICATION [16.5]
METEOR BURST COMMUNICATION [22]
SLOW-SCAN TELEVISION [23]
COMMUNICATIONS SATELLITE [32]
LASER COMMUNICATION [14]
PROJECT XANADU [62]
AMATEUR RADIO SATELLITE [16.6]
TIME-SHARING NETWORK [50]
PROJECT WEST FORD [24]
1971
1973
1974
1976
1978
1982
1988
1992
1997
1999
2009
2020
PIRATE TELEVISION [25]
PACKET RADIO NETWORK [26]
TELETEXT [51]
VIDEOTEX [52]
MICROBROADCAST [27]
AMATEUR PACKET RADIO [16.7]
EMAIL LETTER [44.6]
NABU NETWORK [48.1]
SOFTWARE DEFINED RADIO [28]
METAVERSE [63]
WI-FI [29]
BLUETOOTH [30]
THE CLACKS [64]
PANDORAN NEURAL NETWORK [65]
COSMIC INTERNET [66]

Chronological List of Network Experiments

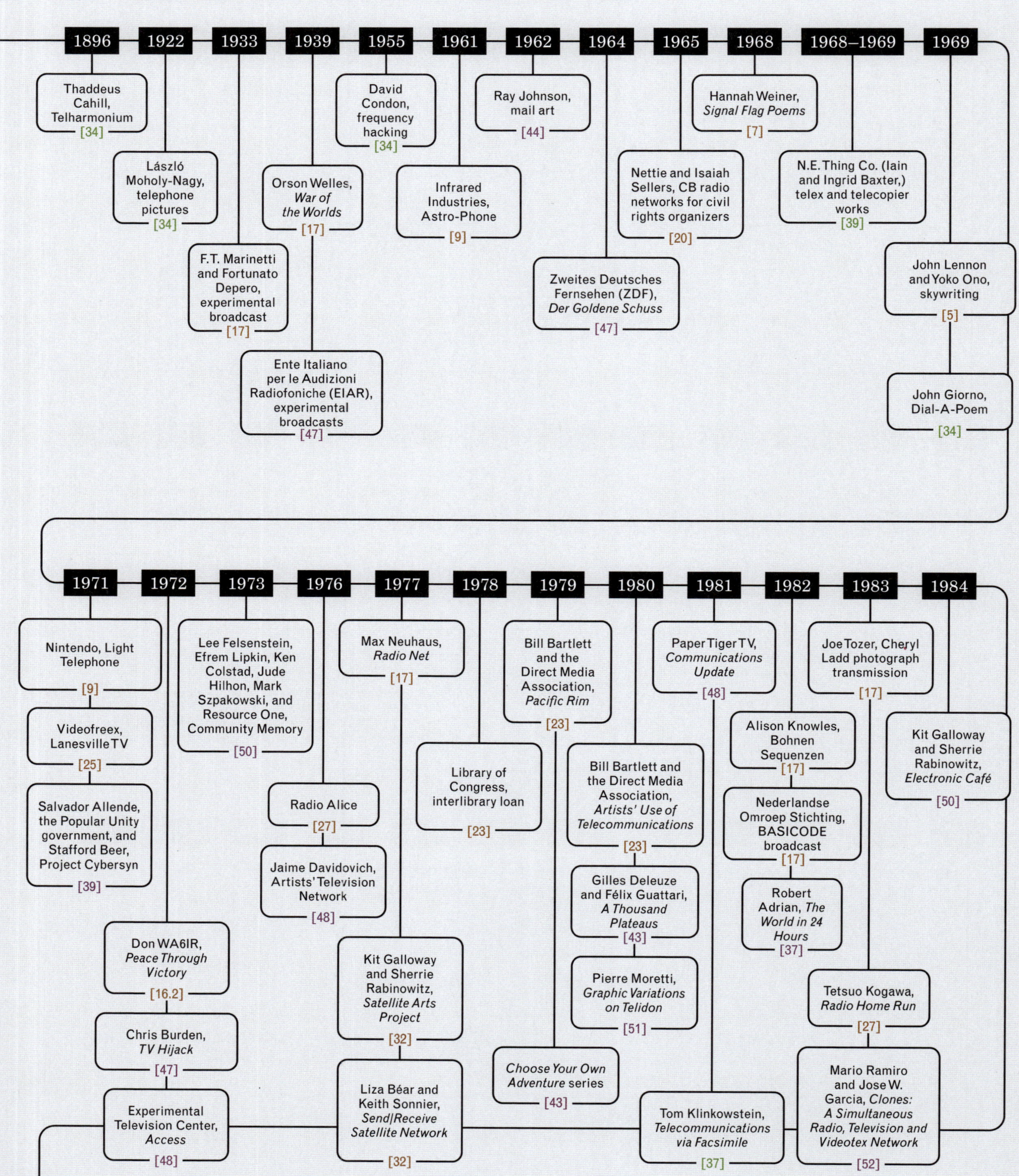

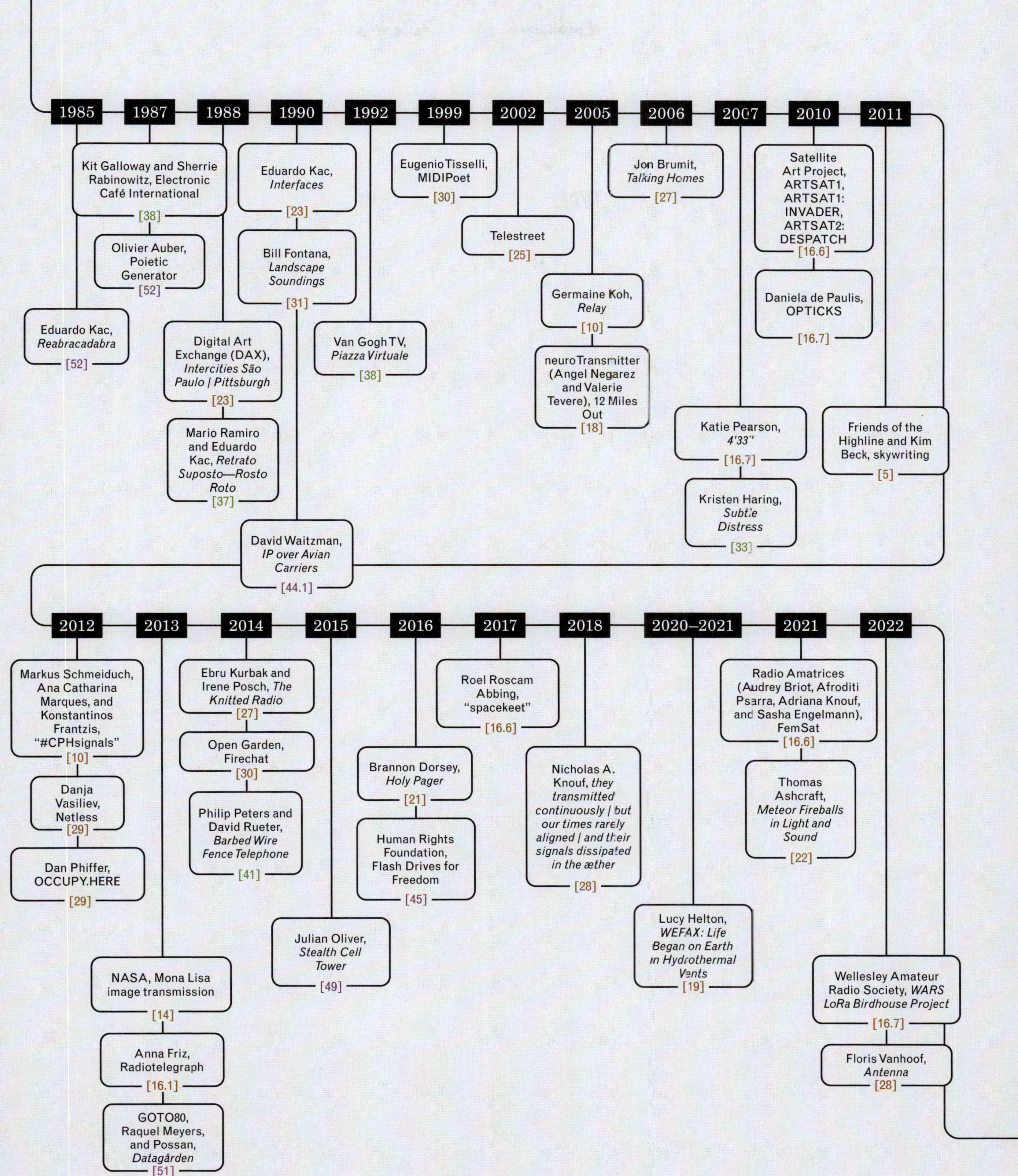

1985
1987
1988
1990
1992
1999
2002
2005
2006
2007
2010
2011
Kit Galloway and Sherrie Rabinowitz, Electronic Café International [38]
Olivier Auber, Poietic Generator [52]
Eduardo Kac, Reabracadabra [52]
Eduardo Kac, Interfaces [23]
Bill Fontana, Landscape Soundings [31]
Digital Art Exchange (DAX), Intercities São Paulo | Pittsburgh [23]
Mario Ramiro and Eduardo Kac, Retrato Suposto—Rosto Roto [37]
Van Gogh TV, Piazza Virtuale [38]
David Waitzman, IP over Avian Carriers [44.1]
Eugenio Tisselli, MIDIPoet [30]
Telestreet [25]
Germaine Koh, Relay [10]
neuroTransmitter (Angel Negarez and Valerie Tevere), 12 Miles Out [18]
Jon Brumit, Talking Homes [27]
Katie Pearson, 4'33" [16.7]
Kristen Haring, Subtle Distress [33]
Satellite Art Project, ARTSAT1, ARTSAT1: INVADER, ARTSAT2: DESPATCH [16.6]
Daniela de Paulis, OPTICKS [16.7]
Friends of the Highline and Kim Beck, skywriting [5]
2012
2013
2014
2015
2016
2017
2018
2020–2021
2021
2022
Markus Schmeiduch, Ana Catharina Marques, and Konstantinos Frantzis, "#CPHsignals" [10]
Danja Vasiliev, Netless [29]
Dan Phiffer, OCCUPY.HERE [29]
NASA, Mona Lisa image transmission [14]
Anna Friz, Radiotelegraph [16.1]
GOTO80, Raquel Meyers, and Possan, Datagården [51]
Ebru Kurbak and Irene Posch, The Knitted Radio [27]
Open Garden, Firechat [30]
Philip Peters and David Rueter, Barbed Wire Fence Telephone [41]
Julian Oliver, Stealth Cell Tower [49]
Brannon Dorsey, Holy Pager [21]
Human Rights Foundation, Flash Drives for Freedom [45]
Roel Roscam Abbing, "spacekeet" [16.6]
Nicholas A. Knouf, they transmitted continuously | but our times rarely aligned | and their signals dissipated in the æther [28]
Lucy Helton, WEFAX: Life Began on Earth in Hydrothermal Vents [19]
Radio Amatrices (Audrey Briot, Afroditi Psarra, Adriana Knouf, and Sasha Engelmann), FemSat [16.6]
Thomas Ashcraft, Meteor Fireballs in Light and Sound [22]
Wellesley Amateur Radio Society, WARS LoRa Birdhouse Project [16.7]
Floris Vanhoof, Antenna [28]

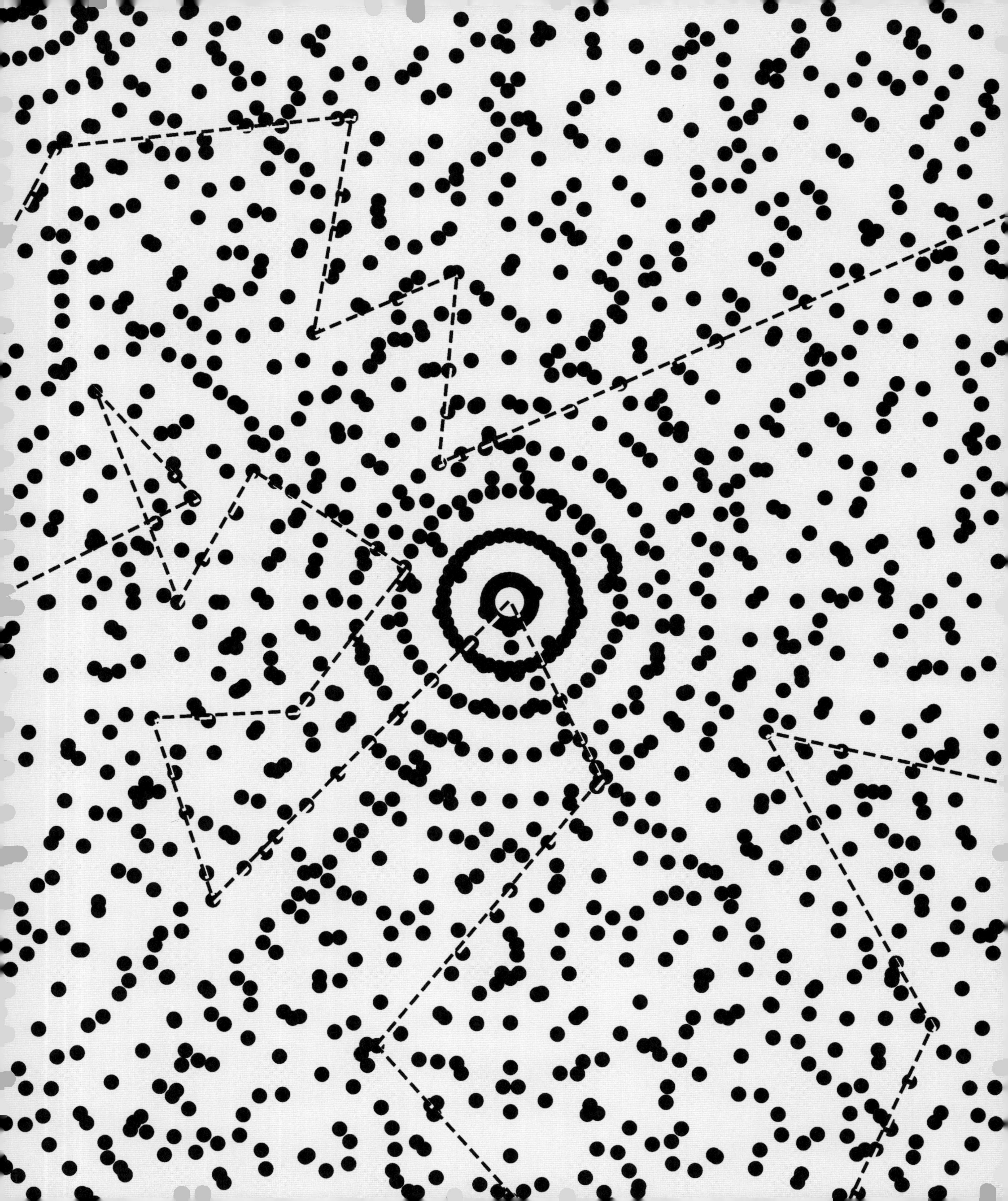

FOREWORD
Joanne McNeill

You don't need to know much about a network—what it is, how it works, or where to find it—to receive its transmissions.

Say you are in a small town in Denmark. It's 1958 and you have turned the radio dial all the way to the right where you catch a crackling racket of rock music over pirate radio from an anchored vessel somewhere in the Öresund strait. In mainland China, in 1989, someone has faxed you a news digest about the Tiananmen Square student protests. They don't know who you are; they found your fax number in the yellow pages. Then you are in the Amazon rainforest. It's the nineteenth century and you are listening through the cambarysu drum in the ground to messages vibrating through the earth sent from someone in a neighboring village. Aboard a ship in England in the late sixteenth century, the naval officer at the dock is waving flags to signal your departure, a code devised by Queen Elizabeth I's secretaries. It's 1973, and you cut across campus at Berkeley to visit Leopold's Records, where you see a flyer for a band's upcoming performance on the analog bulletin board and read about the same upcoming performance on the digital message board at the Community Memory terminal. Then you are in Brandon, Wisconsin in 1883, waiting for a telegraph from Milwaukee sent through a barbed wire fence. It doesn't work and the message remains unsent, but twenty years later you will hear about telegraph operators outside of Boston who attempt the same experiment with success. Or it's Rome in ancient times and there's a pigeon at your window with a scroll tied to its leg with string. In the Mesopotamia of 2100 BCE, you are reading a poem carved in cuneiform on a clay tablet. Maybe it's the late nineties and you are connecting your laptop to Wi-Fi for the very first time. On each of these occasions, you enter a network.

Usually when people hear the word "network," in a technology and telecommunications context, they think of one network in particular. But there are other internets, *other networks*: histories, alternate presents, and potential futures that have long existed in the shadow of the capital-I "Internet." To explore the full expansive possibilities of networks, both past and future, both real and imagined, I can think of no better introduction than this book in your hands, and no better guide, as expert and inviting, than Lori Emerson.

As founding director of the Media Archaeology Lab (MAL), which has since 2009 offered visitors hands-on encounters with hardware and software from the past, Emerson's vital work brings to life the idea that technology does not follow a linear path. Progress is an illusion, as one can see for themselves in the lab: projects with enormous potential and purpose may have failed in the marketplace, while successful innovations often emerged from unexpected places. The archival gadgets and machines on hand at the MAL are fully functional, as testament to the tremendous care that has been extended to these technologies long after they have been hacked together or shipped from a factory. It's instructive and emotionally salient to experience old technology as a user of it: there are quirks, both seductive and frustrating, that reveal themselves and offer unexpected moments of friction. These quirks shape how we communicate through various networks.

Lately, technology prioritizes frictionless connectivity. This apparent ease, however, can be yet another illusion. Because networks don't just happen. Networks are built and developed intentionally. *Other Networks* breaks down what the intentions of each network have been, and identifies examples by origin and ingredients, including often surprising materials and infrastructures. These details are crucial to know. Each network features ways you can and cannot communicate and connect; some networks are better suited for one style of communication than another. In familiarizing ourselves with examples of networks from the past—including dreams, projections, and failures—we revisit missed opportunities and ideas that remain relevant and imbued with potential, even in the present.

Other Networks is a galvanizing call for experimentation in materials, radical change on the protocol level, and wild reimagining of infrastructure. Reviewing the networks highlighted in this book, both unrealized futures and real histories, the reader discovers there is one fundamental commonality: each network features a message that is sent and received. As the subtitle says, it is a "sourcebook"—a sourcebook for new dreams. People are going to connect and continue to want to connect unencumbered by time, space, and other limits of nature. So where can we go from here?

This book is a message to you to explore other networks: better networks, other futures, and new ways forward. Now that you have received this message, how will you respond?

have to come to grips with is how the nightmares that many people are forced to endure are the underside of elite fantasies about efficiency, profit, and social control." Benjamin then reminds us that racism, not unlike a technology (as we learned from Lisa Nakamura), is an axis of domination that "helps produce this fragmented imagination, misery for some, monopoly for others." One powerful alternative to this fragmented imagination is "radical imagination"—one that inspires us "to push beyond the constraints of what we think, and are told, is politically possible."

In the course of any attempt to activate radical imagination by pluralizing networks, it's clear we also need to ask hard questions: Who imagines networks for whom? Why and how does access take place? How is access to the network made difficult or impossible? Likewise, these questions need to be accompanied by pragmatic considerations: in any push to decentralize and miniaturize networks, it may not be realistic or even desirable to also push for leaving the internet altogether in favor of setting up legions of microbroadcasting stations or small mesh networks. Legacy Russell reminds us that the internet "still provides opportunity for queer propositions for new modalities of being and newlyproposed worlds." The internet remains, for many, an aid to survival. I am wary, then, to suggest that *Other Networks* provides us with a blueprint to leave the internet behind altogether. Instead, I hope it opens up the possibility for choice—it is, after all, a sourcebook—a means of reawakening our sense of possibility by the excavation of networks from the past.

SOURCES Félix Guattari (trans. David Sweet), "Popular Free Radio," in *Radiotext(e)* (Autonomedia Press, 1993); Jenny Hval (trans. Marjam Idriss), *Girls Against God* (Verso, 2020); Paul Baran, "On Distributed Communications: I. Introduction to Distributed Communications Networks" (RAND Corporation, 1964); Stewart Brand, "SPACEWAR: Fanatic Life and Symbolic Death Among the Computer Bums," *Rolling Stone* (December 7, 1972); Anil Dash, "The Internet Is About to Get Weird Again," *Rolling Stone* (December 30, 2023); Kevin Driscoll, *The Modem World: A Prehistory of Social Media* (Yale University Press, 2022); Doron Galili, *Seeing by Electricity: The Emergence of Television, 1878–1939* (Duke University Press, 2020); Judy Malloy, ed., *Social Media Archaeology and Poetics* (MIT Press, 2016); Charlton McIlwain, *Black Software: The Internet & Racial Justice, from the AfroNet to Black Lives Matter* (Oxford University Press, 2019); Cait McKinney, *Information Activism: A Queer History of Lesbian Media Technologies* (Duke University Press, 2020); Joy Lisi Rankin, *A People's History of Computing in the United States* (Harvard University Press, 2018); Michel Foucault, *The Order of Things: An Archaeology of the Human Sciences* (Vintage, 1994); Eric Kluitenberg, *The Book of Imaginary Media* (NAi Publishers, 2006); Ruha Benjamin, "The New Jim Code? Race, Carceral Technoscience, and Liberatory Imagination," online talk for the Othering & Belonging Institute (October 17, 2019); Lisa Nakamura, "Race and/as Technology, or How to Do Things to Race," in *Race After the Internet* (Routledge, 2012); Ruha Benjamin, *Imagination: A Manifesto* (W.W. Norton, 2024); Legacy Russell, *Glitch Feminism: A Manifesto* (Verso, 2020)

—Lori Emerson

vacuum that transports capsules from point to point. Newspapers had a particular use for tubes: sending copy to be laid out and then printed. Since the final destination for the material being sent via pneumatic tube sometimes is beyond a delivery location, pneumatic tube companies were often paired with telegraph companies or wagon services. Given the speed and efficiency of the pneumatic tube system, according to Holly Kruse, the hope was that eventually all houses would be connected via a pneumatic system of communication, just like contemporary fiber optic cables. They are still in use today in hospitals, banks, stores, and libraries, and transmission speeds can reach up to forty miles per hour.

SOURCES "Clark, Josiah Latimer," *Encyclopædia Britannica* 6 (1911); Molly Wright Steenson, "Interfacing with the Subterranean," *Cabinet* 41 (2011); Tom Standage, *The Victorian Internet* (Bloomsbury, 2007); Jason Farman, *Delayed: The Art of Waiting from the Ancient to the Instant World* (Yale University Press, 2018); Holly Kruse, "Pipeline as Network: Pneumatic Tubes and the Social Order," in *The Long History of New Media* (Peter Lang, 2011); Shannon Mattern, "Puffs of Air: Communicating by Vacuum," *Alphabet City* 15 (MIT Press, 2010); Sara Wykes, "Gone with the Wind: Tubes Are Whisking Samples Across Hospital," Stanford Medicine News Center (January 11, 2010)

Skywriting [5]

COUNTRY OF ORIGIN Great Britain and USA
CREATOR(S) John C. Savage
EARLIEST KNOWN USE 1922

BASIC INFRASTRUCTURE/MATERIALS Two to five aircraft, paraffin oil

RELATED Fire or smoke signals [3]

DESCRIPTION Skywriting uses a viscous oil combined with hot exhaust from a plane in midair to produce (with the help of highly skilled pilots who can easily fly upside down) cloud-like messages in the sky. The origins of skywriting are disputed, but most sources agree that British pilot John C. Savage was the first to skywrite "Daily Mail" over England and then "Hello USA" over New York City in 1922. The same year witnessed the birth of skywriting as a form of advertising, as a pilot spelled out "Hello USA Call Vanderbilt 7200" over Times Square, reportedly producing 47,000 phone calls to the Vanderbilt Hotel. Letters that appear in the sky using this method usually have to be produced between 7,000 and 17,000 feet above

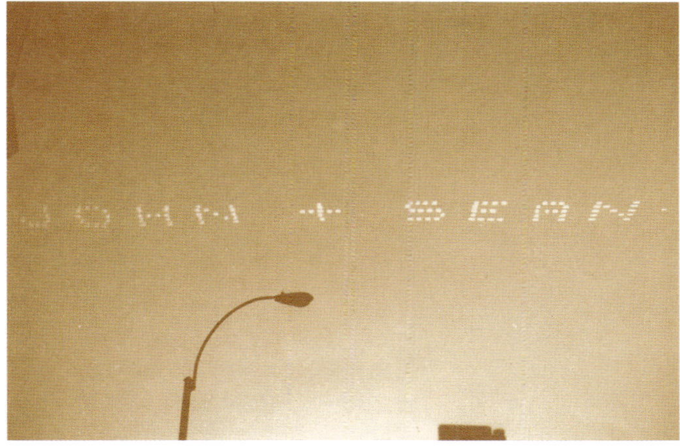

Dot-matrix style skytyped message over New York City in 1980, commissioned by Yoko Ono.

ground and characters can be as high as one mile and as long as fifteen miles. While skywriting fell out of favor with the broad adoption of radio and television, especially as it was too vulnerable to weather, it is still used along with a faster, more legible (and thus more costly) form known as "skytyping." According to the Skywriting Corporation, this method involves five airplanes flying side-by-side and linked by a computer such that they produce a dot-matrix style of messaging in the sky. It usually takes one to two minutes to communicate a character.

EXPERIMENTS In 1969, John Lennon and Yoko Ono commissioned a pilot to fly over Toronto, Canada, and deliver the longest skywritten message that had been written up to that point: "War is over, if you want it. Happy Xmas from John and Yoko." Shortly after the death of John Lennon and the birth of Sean Lennon in 1980, Yoko Ono once again commissioned a pilot to write "HAPPY BIRTHDAY JOHN + SEAN • LOVE YOKO" nine times across the New York City skyline. More recently, in 2011, the New York–based nonprofit Friends of the High Line commissioned artist Kim Beck to choose messages from advertising billboards for skywriting. The messages included phrases such as "LAST CHANCE," "LOST OUR LEASE," and "NOW OPEN."

SOURCES "How Do Skywriting and Skytyping Work?" Library of Congress (November 19, 2019); Adrienne LaFrance, "What Happened to Skywriting?," *The Atlantic* (April 16, 2014); Barrie Paddock, "'Last Chance' Skywritten Messages over NYC Not Warning but an Art Project," *New York Daily News* (October 9, 2011)

[4] A stereograph of the Pneumatic Tube Station located in Sears Roebuck, shown here sometime between 1893 and 1920, reportedly could send 135,900 messages a day.

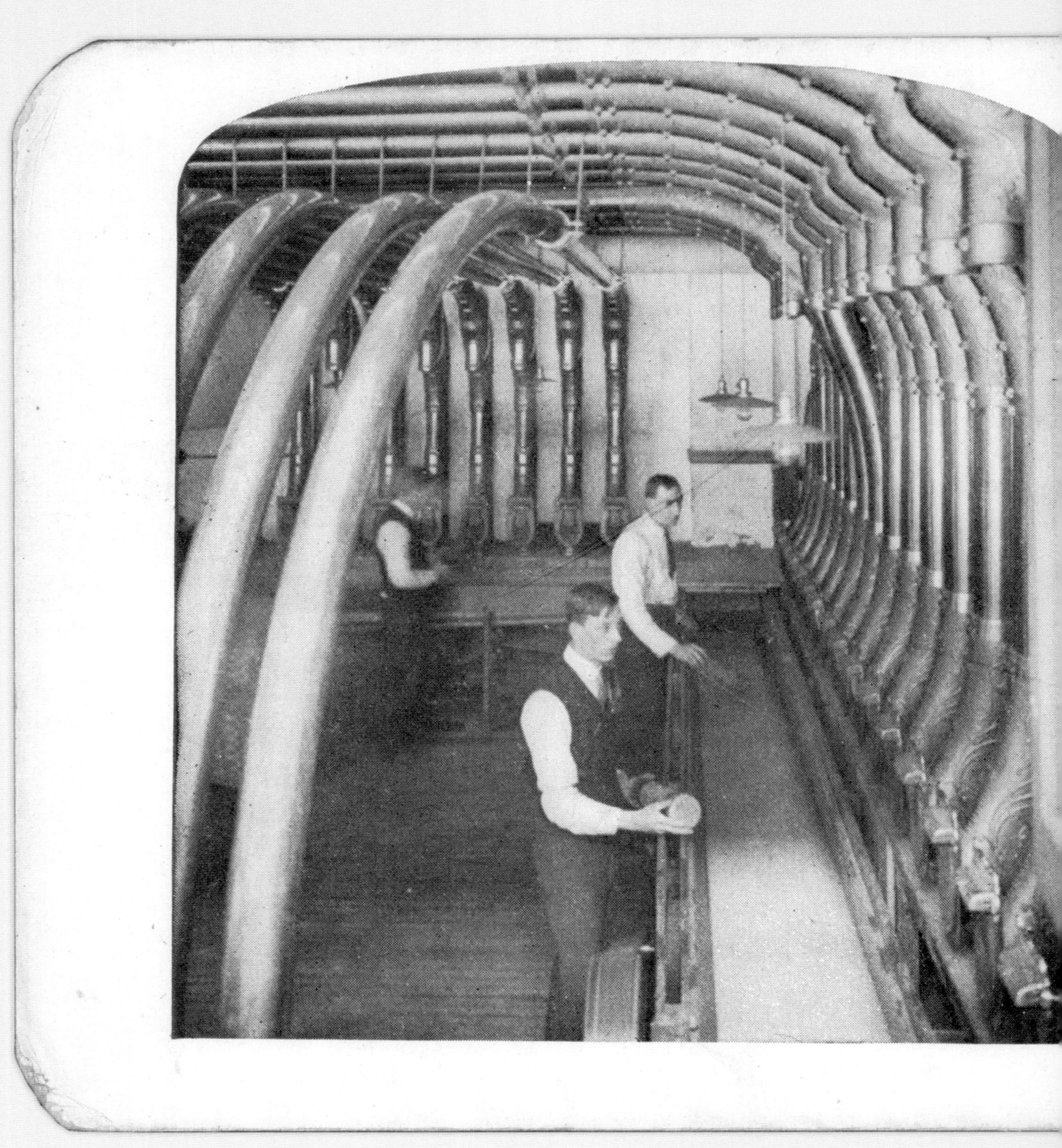

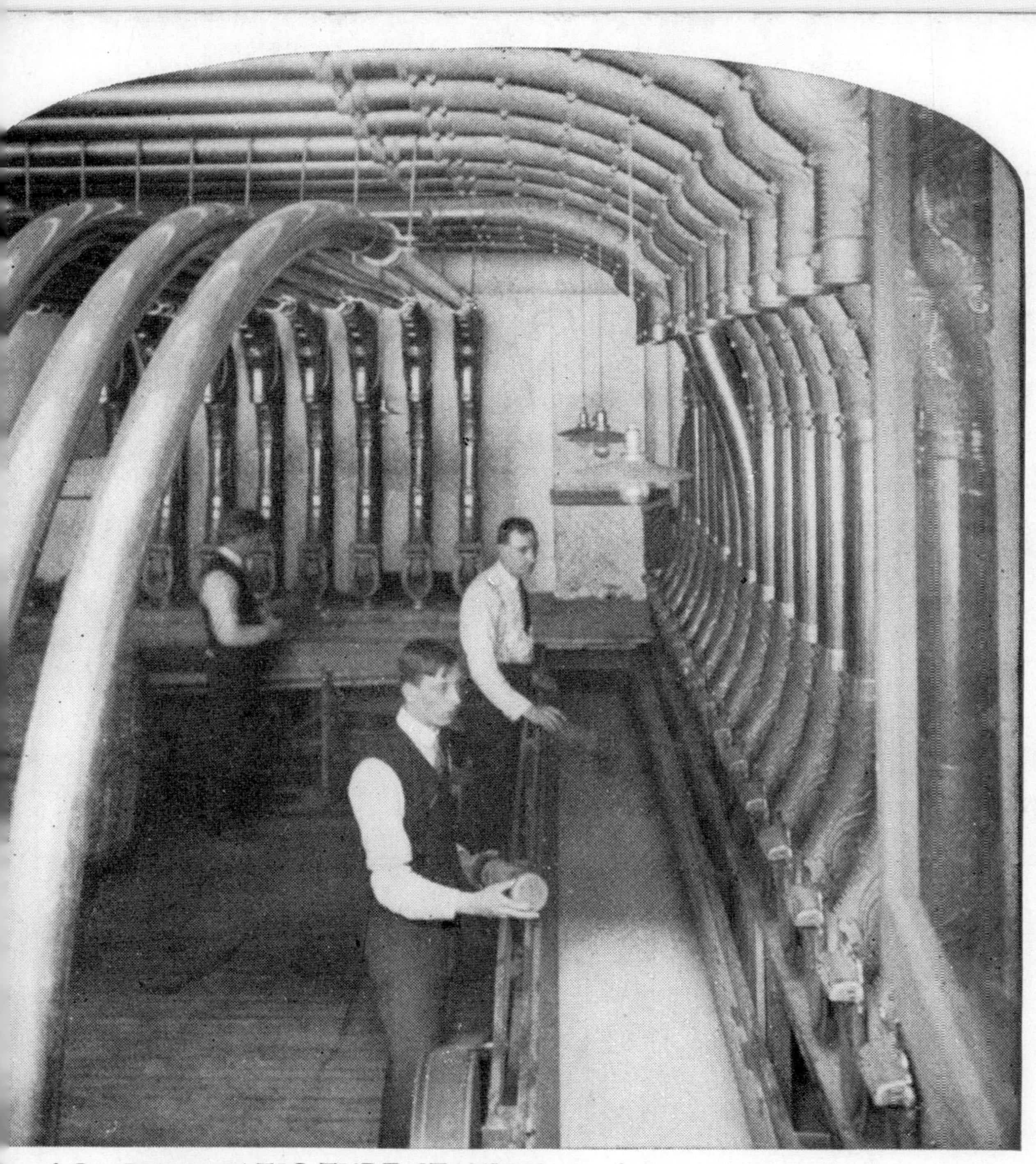

o. 16 PNEUMATIC TUBE STATION.
SEARS, ROEBUCK & CO., Chicago, Ill.

[6] Illustration of a hydraulic semaphore by Max de Nansouty from 1911.

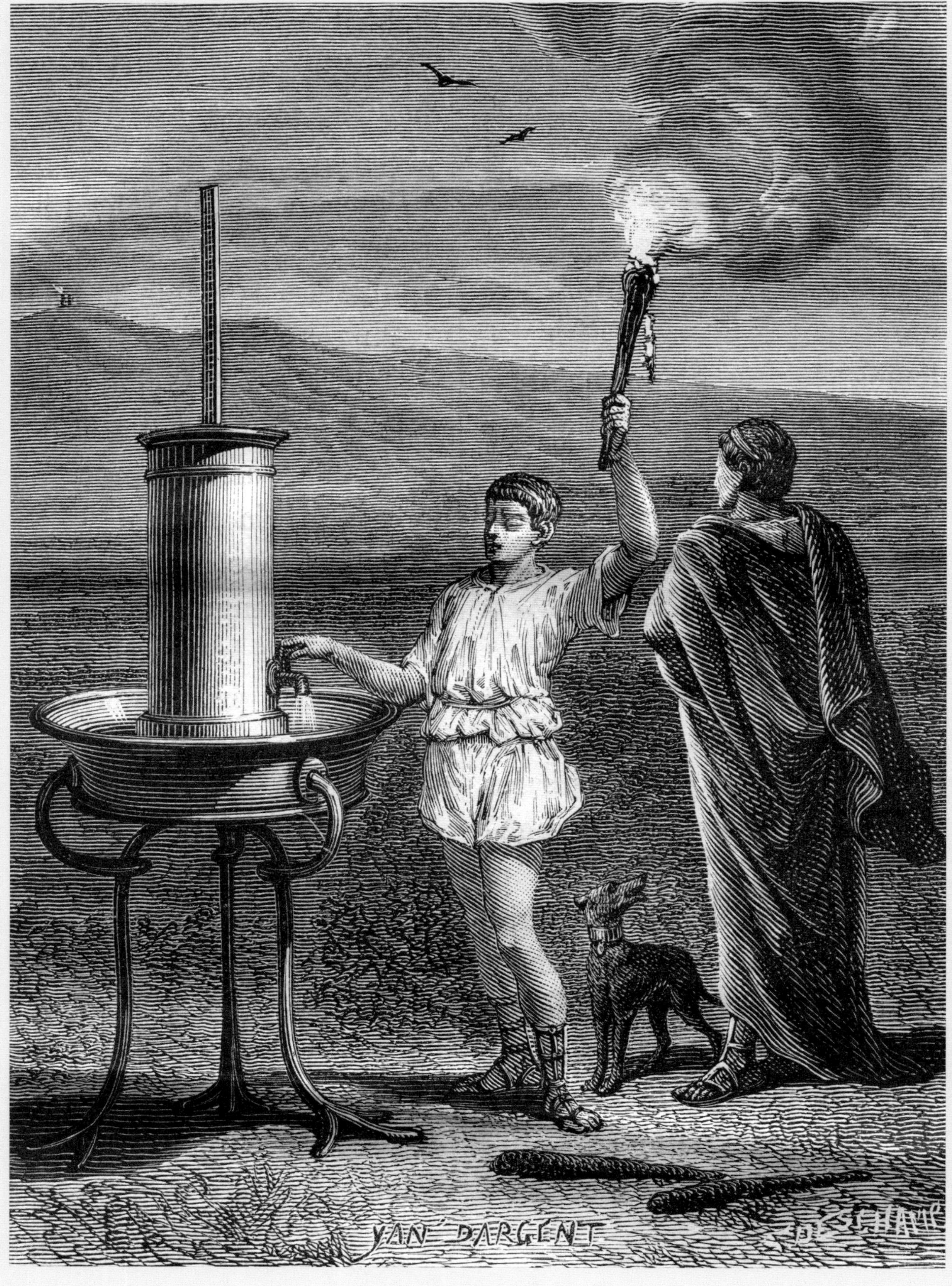

WATER NETWORKS

While there are numerous methods for tracking and mapping animate and inanimate objects underwater (sonar, for example), hydraulic semaphore [6] is one of the only methods for human communication that uses water and does not necessarily need to operate underwater.

Hydraulic Semaphore [6]

COUNTRY OF ORIGIN Ancient Greece
CREATOR(S) Aeneas Tacticus
EARLIEST KNOWN USE Roughly 400 BCE

BASIC INFRASTRUCTURE/MATERIALS Cylindrical container, faucet, float/buoy

RELATED Heliograph [11], radiotelegraphy [16.1], electrical telegraph [33], The Clacks [64]

DESCRIPTION The ancient Greeks were the first to construct a hydraulic semaphore system for military use. First proposed by Aeneas Tacticus in the fourth century BCE and later described more thoroughly by the Greek historian Polybius, the system would employ individuals to use a cylindrical container, faucet, and buoy-like object to transmit a statement corresponding to a notch on the container. According to Cesare Rossi and Flavio Russo, each container was "Meticulously identical in volume and type of faucet . . . filled with water up to the top, awaiting use. To begin transmission a metal mirror was used to send a flash of light to the receiver. Once receipt was confirmed by a return flash, a third flash ordered the temporary opening of the faucets. Water began to flow from the containers, causing a simultaneous synchronous descent of the graduated float in both, notch by notch. When the numbered notch in the transmitter touched the upper border of the cylinder, a final flash ordered the closing of the faucets, allowing the receiver to read the same number as the one transmitted, that is, the message." Since the only message being transmitted from one container to another related to the opening or closing of the faucet, the hydraulic semaphore was a secure means of communication. The next record we have of the development of the hydraulic semaphore is from British civil engineer Francis Whishaw's 1837 proposal for what he called a "hydraulic telegraph." *The Times* reported that Whishaw's proposal included a network of station houses, the distance between them depending on the topography of the land to be traversed by water; lead pipes filled with water, which would be laid five feet underground; glass pipes that would be placed perpendicularly at the terminus of the network with markings on them and/or floats that would correspond to specific words or messages; and faucets that would be used to add or remove water to send a particular message. Whishaw's proposal was never commercially realized. The time it took to transmit a message using hydraulic semaphore was highly variable depending on the size of container and type of faucet; Rossi and Russo estimate that a container thirty centimeters in diameter and one meter high with ten notches would take roughly eighty seconds to transmit one notch.

SOURCES Cesare Rossi and Flavio Russo, *Ancient Engineers' Inventions: History of Mechanism and Machine Science* (Springer, 2017); "Hydraulic Telegraph," *The Times* (December 11, 1837)

OPTICAL NETWORKS

Optical networks are a form of wireless network that use waves on the electromagnetic spectrum of the optical spectrum; these waves include visible, ultraviolet, and infrared. While optical fiber is often classified by industry professionals as a form of wireless optical communication, in this book fiber optic cables are folded into the section on electrical wire networks, given that optical fiber requires cables. Furthermore, while fire and smoke signals could also be considered optical networks, these have been classified as air networks because of their reliance on nitrogen and oxygen to feed the fire and enable the suspension of characters in the atmosphere. The first signal lamp network using limelight dates to 1867. From that point on, wireless networks have been constructed using both visible and invisible light, mostly over fairly short distances. Most of these optical networks are today referred to as Li-Fi, and most of them are still in the early stages of development.

Flag Signaling [7]

COUNTRY OF ORIGIN Great Britain
CREATOR(S) Unknown
EARLIEST KNOWN USE 1337

BASIC INFRASTRUCTURE/MATERIALS Flag(s), short poles (optional), flagpole (optional)

RELATED Fire or smoke signals [3], hydraulic semaphore [6], heliograph [11], electrical telegraph [33]

DESCRIPTION Flag signaling, also known as maritime or marine flag signaling and which includes flag semaphore, is the use of one or two flags to signal a code over a distance of up to two miles (3.2 kilometers), depending on weather conditions and visibility. The earliest reference in print to flag signaling appears in a 1337 version of *The Black Book of the Admiralty*, a compilation of English maritime law dating from 1160. According to Robert W. Burns, only two signals using a banner atop a ship's flagpole are mentioned in *The Black Book*: one requesting nearby captains to attend a meeting aboard the admiral's flagship and one reporting an enemy sighting. By 1596, Queen Elizabeth I's secretaries had devised "the first regular sets of signals and orders to the commanders of the English fleet," and in 1647 the English parliament introduced the first complete maritime flag signaling code. In 1857, the British Board of Trade introduced the *Commercial Code of Signals* as a way to standardize competing codes; the latter became what is now known as the *International Code of Signals*, completed in 1897 and revised numerous times thereafter. The text included instructions on the use of "semaphore

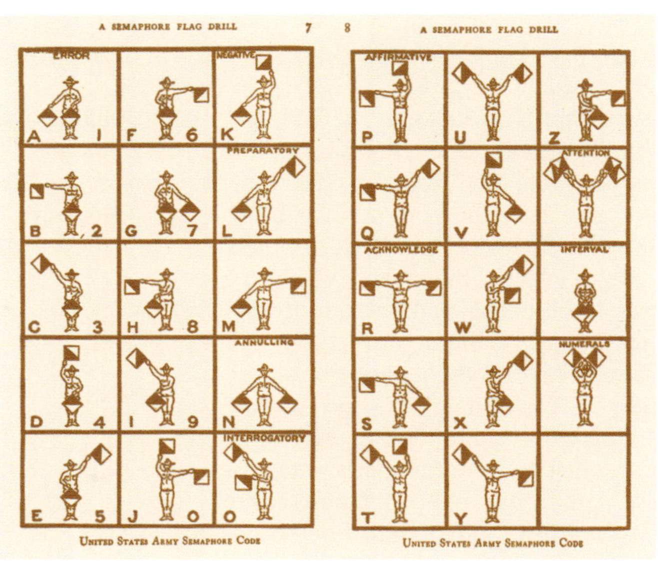

An illustration of semaphore flag signaling for the letters *A* through *Z* and numbers 0 through 9 from a 1918 handbook *A Semaphore Flag Drill*.

flags," or two handheld flags placed in various positions to express individual letters; semaphore flags were considered a faster and more secure method of communication than signaling with light. While the system is no longer used as an official means of communication, NATO had established a sending standard of fifteen words per minute at the height of its use.

Notably, the now universally recognized symbol for peace was designed by Gerald Holtom in 1958 by overlaying the semaphore flag symbols for *N* (nuclear) and *D* (disarmament) for an antiwar protest. Currently, the beach patrol in Ocean City, Maryland, still uses flag signaling (or beach semaphore) to communicate

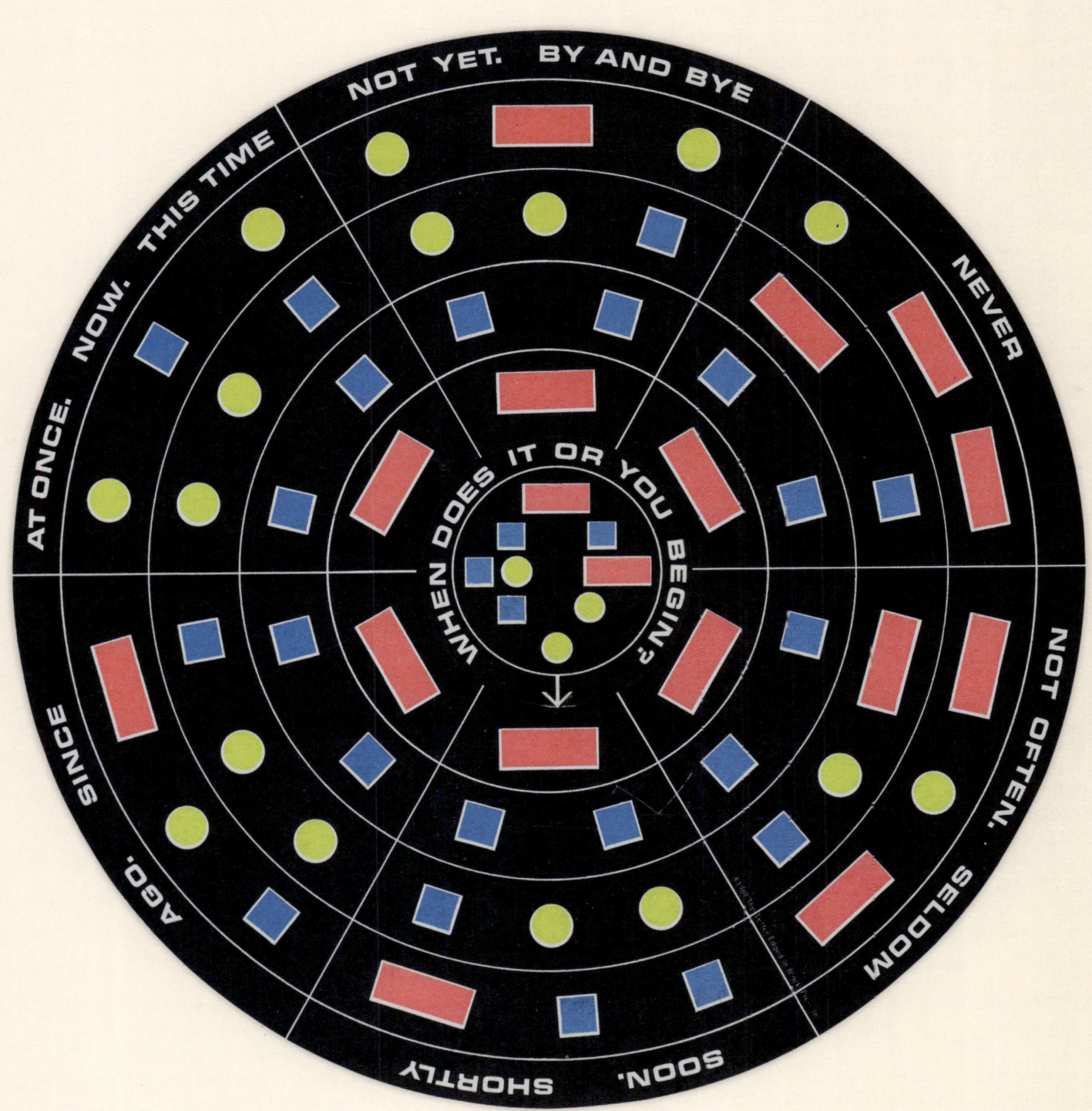

[7] Volvelle or paper wheel chart that appears in Hannah Weiner's 1968 book of visual poems, *Signal Flag Poems*.

messages from lifeguard station to station along the ten miles (sixteen kilometers) of beach.

EXPERIMENTS In 1968, American poet Hannah Weiner published *Signal Flag Poems*, a book of visual poems using flag signals as outlined by the *International Code of Signals*. The book includes a volvelle, or a paper wheel chart of signals, that the reader must assemble. In 1969, Weiner wrote, "I am interested in exploring methods of communication that will be understood face to face or at any distance, regardless of language, country or planet or origin, by all sending and receiving." In 1982, she published another related collection of code poems and performance pieces titled *Code Poems: From the International Code of Signals for the Use of All Nations*.

SOURCES Russell W. Burns, *Communications: An International History of the Formative Years* (Institution of Electrical Engineers, 2004); Howard M. Chapin, "Notes on the Early Development of the Designs in Marine Signal Flags," *U.S. Naval Institute Proceedings* 53:11 (November 1927); Intergovernmental Maritime Consultative Organization, *International Code of Signals: As Adopted by the Fourth Assembly of the Inter-Governmental Maritime Consultative Organization in 1965 for Visual Sound and Radio Communications* (Starpath Publications, 2009); Jacopo Prisco, "Three Lines and a Circle: A Brief History of the Peace Symbol," *CNN* (February 8, 2019); "Semaphore," Town of Ocean City, Maryland, website (2023); Hannah Weiner, *Signal Flag Poems* (Letter Edged in Black Press, 1968); Hannah Weiner, *Trans-Space Communication* (July 1969); Hannah Weiner, *Code Poems: From the International Code of Signals for the Use of All Nations* (Open Book, 1982)

Optical Telegraph [8]

COUNTRY OF ORIGIN France
CREATOR(S) Claude Chappe
EARLIEST KNOWN USE 1792

BASIC INFRASTRUCTURE/MATERIALS Large display face, regulators and indicators, elevated structure (i.e., clock or church tower), counterweights, cranks, ropes, telescopes

RELATED Fire or smoke signals [3], hydraulic semaphore [6], heliograph [11], electrical telegraph [33], The Clacks [64]

DESCRIPTION Originally dubbed a "tachygraphe" in French (from the Greek for "rapid writer"), optical

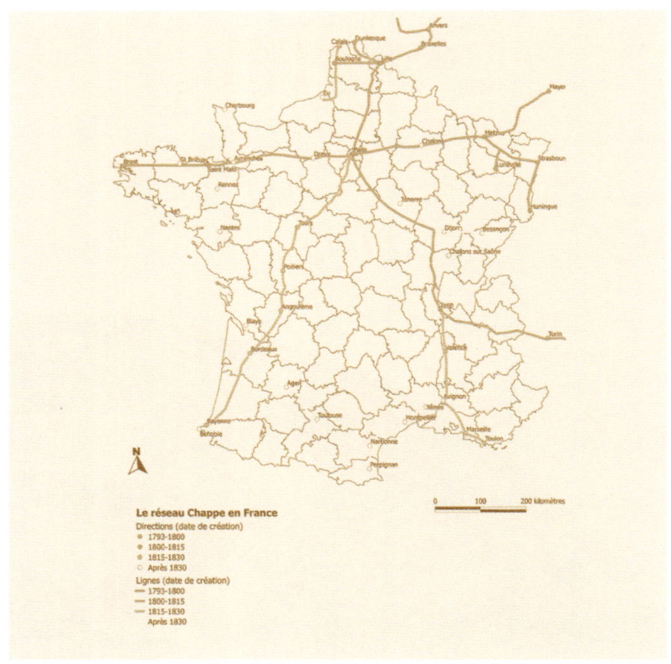

Chappe optical telegraph network in France from 1793 to 1830.

telegraph was first developed when Claude Chappe and his brother René began experimenting with sending and receiving coded signals throughout 1790–1791. The first, less successful version involved two tachygraphes that appeared as if they were large clocks, except the dials featured a set of previously agreed-upon symbols; one tachygraphe would point to a particular symbol and the second tachygraphe (located within a visible distance of the first) would simply repeat the symbol. In 1792, Claude Chappe designed a new model that allowed for more elaborate communication over longer distances. This device was soon renamed a telegraph and, as Anton Huurdeman describes it, consisted of "a regulator, approximately 4.5 m long and 0.35 m wide, to which two indicators were attached, each approximately 2 m long and 0.33 m wide. The indicators were made like a window shutter, with alternating slats and apertures, half the slats being set to the right and half to the left, to lessen wind resistance and increase visibility. The indicators were balanced by thin iron counterweights. Generally, the regulator and indicators were painted black, but to improve visibility where necessary, blue triangles were painted on the white regulator and indicators . . . The positions of the regulator and indicators could be changed via three cranks and wire ropes." After trial and error, Chappe established ninety-six signal positions that made possible a total of 25,392 different messages. In an attempt to create rapid communications for the military at the height of the French Revolution, the first government-approved line was constructed between Paris and the town of Lille, located about 210 kilometers north. According to Huurdeman, the transmission of a signal from Paris to Lille took about two minutes. Optical telegraphs of varying designs were soon adopted by countries such as Sweden, the UK, Ireland, Canada, Spain, Portugal, Malta, Australia, Germany, Russia, and the US. Most of these networks were restricted to government use, but occasionally lines (for example in Germany and the US) were constructed for commercial use.

SOURCES Anton A. Huurdeman, *The Worldwide History of Telecommunications* (IEEE Explore and Wiley, 2005); Gerard J. Holzmann and Björn Pehrson, *The Early History of Data Networks* (Wiley, 2003); Tom Standage, *The Victorian Internet* (Bloomsbury, 2014)

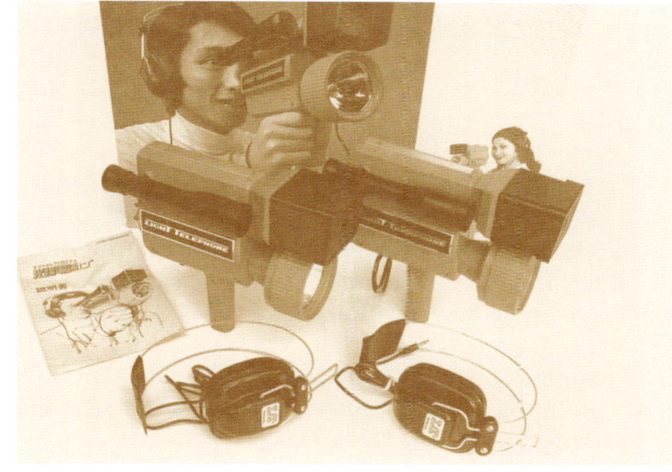

The Nintendo Light Telephone (1971).

Infrared Communication [9]

COUNTRY OF ORIGIN Great Britain
CREATOR(S) Sir William Herschel
EARLIEST KNOWN USE 1860

BASIC INFRASTRUCTURE/MATERIALS Transmitter (or IR light source, for example lightbulbs, sunlight, fire, LED, laser diode), receiver (IR enabled device with silicon photo diodes), modulator, output interpretation device

RELATED Signal lamp [10]

DESCRIPTION British astronomer Sir William Herschel is considered the first person to thoroughly observe and prove the existence of infrared radiation (IR). IR communication involves the use of light undetectable by the human eye and whose wavelength is longer than that of visible light [15]. It is a form of electromagnetic radiation and thus it is emitted by many sources of heat (such as the sun, fire, and light bulbs). While IR's ubiquity is an advantage, its ubiquity also increases the difficulty of eliminating sources of interference when attempting to harness IR for communication. Thus, IR communication usually can only take place over short distances and within a line of sight. That said, there are a number of advantages to using IR: transceivers (combination transmitters and receivers) are quite inexpensive, it is very effective for short-range communication, it offers almost unlimited and unregulated bandwidth, and it is a fairly secure way to transmit information, as it is largely immune to eavesdropping. The simplest method for transmitting information over IR involves quickly turning an LED on and off, typically at a frequency of 38,000 hertz; the pulsing makes it possible for the receiver to differentiate between signal and noise coming from objects in the surrounding environment emitting IR radiation. IR communication can generally send ten to fifteen megabits per second.

EXPERIMENTS In 1961, a US company called Infrared Industries began producing a toy called the Astro-Phone, which used a flashlight bulb and a filter that allowed only IR light to pass through; users would each insert an earphone connected to the phone, aim their devices at each other, pull a trigger, and speak to each other through a microphone that transmitted sound as IR light. The Astro-Phone would, according to the advertising of the time, send "your voice over hundreds of yards on a beam of invisible light." Ten years later, Nintendo released a very similar device, only in Japan, called the Light Telephone. The device could reportedly transmit messages over a distance of ten to thirty meters.

SOURCES Bob Berman, *Zapped: From Infrared to X-Rays, the Curious History of Invisible Light* (Little, Brown, 2017); Joseph M. Kahn and John R. Barry, "Wireless Infrared Communications," *Proceedings of the IEEE* 85:2 (February 1997); Melvin H. Friedman and Ronald G. Driggers, *Introduction to Infrared and Electro-Optical Systems* (Artech House, 2012); "Nintendo Light Telephone," *BeforeMario* blog (June 18, 2011)

Signal Lamp [10]

COUNTRY OF ORIGIN Great Britain
CREATOR(S) Philip Howard Colomb
EARLIEST KNOWN USE 1867

BASIC INFRASTRUCTURE/MATERIALS Shutters or concave mirror, light source (lamp, lantern, electric bulb, etc.), lens (optional), binoculars (optional), infrared viewing device (optional)

RELATED Fire or smoke signals [3], optical telegraph [8], heliograph [11], photophone [12], radiotelegraphy [16.1], The Clacks [64]

DESCRIPTION The first documented use of a lantern (at least in the Global North) to broadcast code via flashes of limelight was reportedly developed by Captain Philip Howard Colomb of the British Navy in 1867. By 1885, kerosene oil lamps (or begbie lamps) were also used. According to amateur historian Jerry Proc, as time went on the Royal Navy, Royal Canadian Navy, and US Navy used slightly different versions—for example, the US Navy "used a shutter to key the light beam while those in the [Royal Canadian Navy] and [Royal Navy] used a tilting mirror." Communication could be one-way or two-way; in the case of the former, infrared communication [9] was sometimes used at night; in the case of the latter, the receiver would broadcast a short flash to indicate receipt of each word. A system of flashing colored lights is still used today for the control of aircraft, ground vehicles, equipment, and personnel not equipped with radio or in cases where radio transmission has been interrupted. The US Navy also switched from the use of signal lamps to a flashing light to text converter in 2016; this system uses a camera and a digital tablet to send Morse code back and forth between ships via LED lights. The reception speed of a signal lamp transmission depends on whether the decoding is manual or automatic; manual decoding ranges from ten to twenty words per minute.

EXPERIMENTS In 2005, Canadian artist Germaine Koh installed *Relay* in the UK's Baltic Centre for Contemporary Art; her project involved the use of "lights around the Level 4 outdoor viewing platform which flashed in Morse code—broadcasting messages received via SMS (text message) . . . Visitors were able to text their own messages to Relay for the duration of the exhibition." In 2012, three students at the Copenhagen Institute of Interaction Design (Markus Schmeiduch, Ana Catharina Marques, and Konstantinos Frantzis) created "#CPHsignals," which connected two sides of the inner Copenhagen harbor using a graphical user interface to allow users to type their message. An Arduino (a set of inexpensive, open-source electrical components) translates the message into Morse code and transmits the signals using old maritime lamps. The same interface is placed on the receiving side, which serves to decode and display the messages.

SOURCES Christopher H. Sterling, *Military Communications: From Ancient Times to the 21st Century* (ABC-CLIO, 2007); Jerry Proc, "Visual Signalling in the RCN," *Light Signalling* (July 2021); Federal Aviation Administration, "Aeronautical Information Manual" (May 19, 2022); Sam LaGrone, "Office of Naval Research Set to Upgrade the 200-Year-Old Signal Lamp for Modern Stealth Communication," *USNI News* (July 17, 2017); Germaine Koh, *Relay*, Baltic Centre for Contemporary Art (2005); Tom Schofield, "Null by Morse: Historical Optical Communication to Smartphones," *Leonardo* 46:4 (2013)

Heliograph [11]

COUNTRY OF ORIGIN Present-day Pakistan
CREATOR(S) Henry Christopher Mance
EARLIEST KNOWN USE 1869

BASIC INFRASTRUCTURE/MATERIALS Tripod, signaling mirror, U-frame, plate, tangent screw, vertical screw rod, spring, sighting vane

RELATED Fire or smoke signals [3], hydraulic semaphore [6], optical telegraph [8], electrical telegraph [33]

DESCRIPTION While Charles Babbage briefly discusses instances of the use of "night signals," "sun signals," and "zenith-light signals" in his 1864 *Passages from the Life of a Philosopher*, the first reliable documentation of the use of heliography (or the practice of visual communication at a distance by flashing reflected rays of the sun to transmit semaphore) dates to 1869. That year, Henry Christopher Mance, a member of the British Government Persian Gulf Telegraph Department, developed the first widely used heliograph while he was stationed in Karachi, Pakistan. Mance's model, which he also called a "sun telegraph," weighed only seven pounds and could be operated by a single person transmitting over a distance of about thirty-five miles, depending on weather conditions, the opacity of the air, the absence of obstacles between sender and receiver, and the diameter of the mirror. Since Morse code was well-known at the time, Mance's system (adopted by nearly all heliograph users thereafter) used a short flash to represent a dot and a long flash to represent a dash. In 1890, as part of its bloody war on Native Americans, the US Army demonstrated it was possible to create a network of heliographs that extended over roughly 2,000 miles. In 1894, a group of US signal sergeants successfully used a single heliograph with larger mirrors than between Paris and the town of Lille, located about 210 kilometers north. According to Huurdeman, the transmission of a signal from Paris to Lille took about two minutes. Optical telegraphs of varying designs were soon adopted by countries such as Sweden, the UK, Ireland, Canada, Spain, Portugal, Malta, Australia, Germany, Russia, and the US. Most of these networks were restricted to government use, but occasionally lines (for example in Germany and the US) were constructed for commercial use.

SOURCES Anton A. Huurdeman, *The Worldwide History of Telecommunications* (IEEE Explore and Wiley, 2005); Gerard J. Holzmann and Björn Pehrson, *The Early History of Data Networks* (Wiley, 2003); Tom Standage, *The Victorian Internet* (Bloomsbury, 2014)

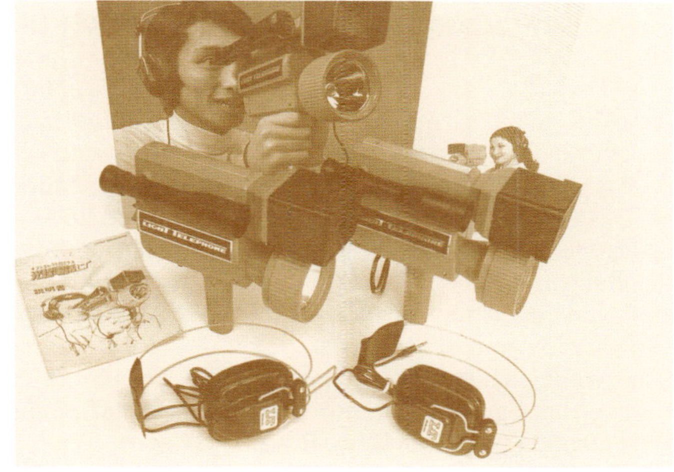

The Nintendo Light Telephone (1971).

Infrared Communication [9]

COUNTRY OF ORIGIN Great Britain
CREATOR(S) Sir William Herschel
EARLIEST KNOWN USE 1860

BASIC INFRASTRUCTURE/MATERIALS Transmitter (or IR light source, for example lightbulbs, sunlight, fire, LED, laser diode), receiver (IR enabled device with silicon photo diodes), modulator, output interpretation device

RELATED Signal lamp [10]

DESCRIPTION British astronomer Sir William Herschel is considered the first person to thoroughly observe and prove the existence of infrared radiation (IR). IR communication involves the use of light undetectable by the human eye and whose wavelength is longer than that of visible light [15]. It is a form of electromagnetic radiation and thus it is emitted by many sources of heat (such as the sun, fire, and light bulbs). While IR's ubiquity is an advantage, its ubiquity also increases the difficulty of eliminating sources of interference when attempting to harness IR for communication. Thus, IR communication usually can only take place over short distances and within a line of sight. That said, there are a number of advantages to using IR: transceivers (combination transmitters and receivers) are quite inexpensive, it is very effective for short-range communication, it offers almost unlimited and unregulated bandwidth, and it is a fairly secure way to transmit information, as it is largely immune to eavesdropping. The simplest method for transmitting information over IR involves quickly turning an LED on and off, typically at a frequency of 38,000 hertz; the pulsing makes it possible for the receiver to differentiate between signal and noise coming from objects in the surrounding environment emitting IR radiation. IR communication can generally send ten to fifteen megabits per second.

EXPERIMENTS In 1961, a US company called Infrared Industries began producing a toy called the Astro-Phone, which used a flashlight bulb and a filter that allowed only IR light to pass through; users would each insert an earphone connected to the phone, aim their devices at each other, pull a trigger, and speak to each other through a microphone that transmitted sound as IR light. The Astro-Phone would, according to the advertising of the time, send "your voice over hundreds of yards on a beam of invisible light." Ten years later, Nintendo released a very similar device, only in Japan, called the Light Telephone. The device could reportedly transmit messages over a distance of ten to thirty meters.

SOURCES Bob Berman, *Zapped: From Infrared to X-Rays, the Curious History of Invisible Light* (Little, Brown, 2017); Joseph M. Kahn and John R. Barry, "Wireless Infrared Communications," *Proceedings of the IEEE* 85:2 (February 1997); Melvin H. Friedman and Ronald G. Driggers, *Introduction to Infrared and Electro-Optical Systems* (Artech House, 2012); "Nintendo Light Telephone," *BeforeMario* blog (June 18, 2011)

Signal Lamp [10]

COUNTRY OF ORIGIN Great Britain
CREATOR(S) Philip Howard Colomb
EARLIEST KNOWN USE 1867

BASIC INFRASTRUCTURE/MATERIALS Shutters or concave mirror, light source (lamp, lantern, electric bulb, etc.), lens (optional), binoculars (optional), infrared viewing device (optional)

RELATED Fire or smoke signals [3], optical telegraph [8], heliograph [11], photophone [12], radiotelegraphy [16.1], The Clacks [64]

DESCRIPTION The first documented use of a lantern (at least in the Global North) to broadcast code via flashes of limelight was reportedly developed by Captain Philip Howard Colomb of the British Navy in 1867. By 1885, kerosene oil lamps (or begbie lamps) were also used. According to amateur historian Jerry Proc, as time went on the Royal Navy, Royal Canadian Navy, and US Navy used slightly different versions—for example, the US Navy "used a shutter to key the light beam while those in the [Royal Canadian Navy] and [Royal Navy] used a tilting mirror." Communication could be one-way or two-way; in the case of the former, infrared communication [9] was sometimes used at night; in the case of the latter, the receiver would broadcast a short flash to indicate receipt of each word. A system of flashing colored lights is still used today for the control of aircraft, ground vehicles, equipment, and personnel not equipped with radio or in cases where radio transmission has been interrupted. The US Navy also switched from the use of signal lamps to a flashing light to text converter in 2016; this system uses a camera and a digital tablet to send Morse code back and forth between ships via LED lights. The reception speed of a signal lamp transmission depends on whether the decoding is manual or automatic; manual decoding ranges from ten to twenty words per minute.

EXPERIMENTS In 2005, Canadian artist Germaine Koh installed *Relay* in the UK's Baltic Centre for Contemporary Art; her project involved the use of "lights around the Level 4 outdoor viewing platform which flashed in Morse code—broadcasting messages received via SMS (text message) . . . Visitors were able to text their own messages to Relay for the duration of the exhibition." In 2012, three students at the Copenhagen Institute of Interaction Design (Markus Schmeiduch, Ana Catharina Marques, and Konstantinos Frantzis) created "#CPHsignals," which connected two sides of the inner Copenhagen harbor using a graphical user interface to allow users to type their message. An Arduino (a set of inexpensive, open-source electrical components) translates the message into Morse code and transmits the signals using old maritime lamps. The same interface is placed on the receiving side, which serves to decode and display the messages.

SOURCES Christopher H. Sterling, *Military Communications: From Ancient Times to the 21st Century* (ABC-CLIO, 2007); Jerry Proc, "Visual Signalling in the RCN," *Light Signalling* (July 2021); Federal Aviation Administration, "Aeronautical Information Manual" (May 19, 2022); Sam LaGrone, "Office of Naval Research Set to Upgrade the 200-Year-Old Signal Lamp for Modern Stealth Communication," *USNI News* (July 17, 2017); Germaine Koh, *Relay*, Baltic Centre for Contemporary Art (2005); Tom Schofield, "Null by Morse: Historical Optical Communication to Smartphones," *Leonardo* 46:4 (2013)

Heliograph [11]

COUNTRY OF ORIGIN Present-day Pakistan
CREATOR(S) Henry Christopher Mance
EARLIEST KNOWN USE 1869

BASIC INFRASTRUCTURE/MATERIALS Tripod, signaling mirror, U-frame, plate, tangent screw, vertical screw rod, spring, sighting vane

RELATED Fire or smoke signals [3], hydraulic semaphore [6], optical telegraph [8], electrical telegraph [33]

DESCRIPTION While Charles Babbage briefly discusses instances of the use of "night signals," "sun signals," and "zenith-light signals" in his 1864 *Passages from the Life of a Philosopher*, the first reliable documentation of the use of heliography (or the practice of visual communication at a distance by flashing reflected rays of the sun to transmit semaphore) dates to 1869. That year, Henry Christopher Mance, a member of the British Government Persian Gulf Telegraph Department, developed the first widely used heliograph while he was stationed in Karachi, Pakistan. Mance's model, which he also called a "sun telegraph," weighed only seven pounds and could be operated by a single person transmitting over a distance of about thirty-five miles, depending on weather conditions, the opacity of the air, the absence of obstacles between sender and receiver, and the diameter of the mirror. Since Morse code was well-known at the time, Mance's system (adopted by nearly all heliograph users thereafter) used a short flash to represent a dot and a long flash to represent a dash. In 1890, as part of its bloody war on Native Americans, the US Army demonstrated it was possible to create a network of heliographs that extended over roughly 2,000 miles. In 1894, a group of US signal sergeants successfully used a single heliograph with larger mirrors than

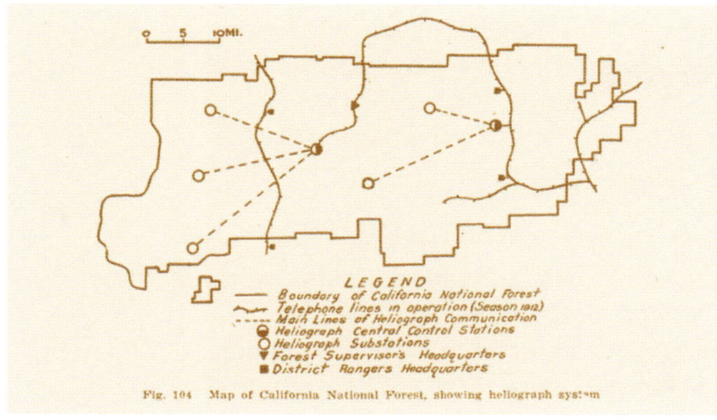

A network of heliographs used to protect national forests in California, from *Methods of Communication Adapted to Forest Protection* (1920).

Mance's model over a record-breaking distance of 183 miles. Heliographs were then used throughout the world (often in war and sometimes for forestry protection across the US and Canada) because of their portability, low energy consumption, and security, especially if the beam of light was narrowed to fifty feet per mile. Heliographs are still included in most aircraft and wilderness survival kits. The reception speed of a heliograph transmission depends on the decoder's skill level; the average decoder speed is roughly ten to twenty words per minute.

SOURCES Philip and Emily Morrison, eds., *Charles Babbage and His Calculating Engines* (Dover Publications, 1961); David Herres, "Morse Code, the First Serial Communication Protocol," *Test & Measurement Tips* (October 4, 2019); Henry Mance, "The Heliograph or Sun Telegraph," *Journal of the United Service Institution of India* (February 10, 1872); W. N. Millar, *Methods of Communication Adapted to Forest Protection* (Canadian Forestry Service, 1920)

Photophone [12]

COUNTRY OF ORIGIN USA
CREATOR(S) Alexander Graham Bell and Charles Sumner Tainter
EARLIEST KNOWN USE 1880

BASIC INFRASTRUCTURE/MATERIALS Light source, mirrors, selenium cell photodetector, telephone, battery, electromagnetic earphone

RELATED Heliograph [11]

DESCRIPTION While Alexander Graham Bell claimed that A.C. Brown was the first to demonstrate the transmission of sound by light in 1878, Bell and his assistant Charles Sumner Tainter were the first to fully develop the technology and file four patents for what they called "a photophone" based on their experiments in 1880. In its early form, Tainter and Bell based their nonelectronic design on the photoacoustic effect, or the way in which sound waves form as a result of varying light intensity that is absorbed in a material such as carbon black (then called "lamp-black"). In order to achieve a varying light intensity, the pair realized that "the simplest form of apparatus for producing the effect consists of a plane mirror of flexible material against the back of which the speaker's voice is directed. Under the action of the voice the mirror becomes alternately convex and concave and thus alternately scatters and condenses the light." By 1881, the photophone had transformed into an electrical device. It used a flexible mirror, attached to the end of a speaking tube, that changed shape in response to corresponding changes in sound pressure, thereby modulating the intensity of the reflected light. The receiver was a selenium cell mounted on another mirror and paired to a battery and telephone receiver. Both designs, however, were limited by the fact that transmissions were vulnerable to anything that interferes with the reception of sunlight, such as cloud, fog, rain, snow, etc. Still, the photophone is widely recognized as the primary influence on the creation of modern fiber optic telecommunications. In 1937, Germany also devised a battery-powered photophone called the Lichtsprechgerät 80. Communication between these devices could take place over about a three-mile distance.

SOURCES Alexander Graham Bell, "On the Production and Reproduction of Speech by Light," *American Journal of Science* (October 1880); Alexander Graham Bell, "Production of Sound by Radiant Energy," *Journal of the Franklin Institute* (June 1881);

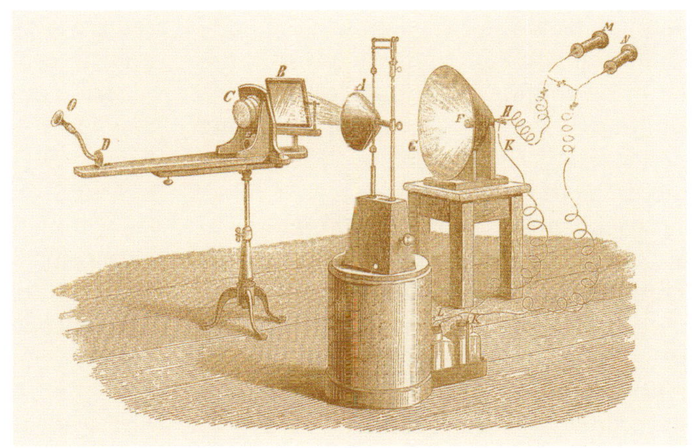

An illustration of Bell and Taintor's photophone from 1880.

J. Clark, "An Introduction to Communications with Optical Carriers," *IEEE Students' Quarterly Journal* (June 1966); Dirk W. Rollema, "Optical Communications—1935 Style," *Electronics and Wireless World* (August 1985)

Ultraviolet Communication [13]

COUNTRY OF ORIGIN USA
CREATOR(S) E. O. Hulburt / US Naval Research Laboratory
EARLIEST KNOWN USE 1926

BASIC INFRASTRUCTURE/MATERIALS Transmitter (signal generator, current sources, modulation circuits, and light sources that could include deep ultraviolet LEDs, deep ultraviolet lasers, or mercury lamps), receiver (solar-blind filter, UV detectors, or avalanche photodiode, signal amplifier, demodulator, oscilloscope), output interpretation device

RELATED Signal lamp [10], electrical telegraph [33]

DESCRIPTION Ultraviolet Communication (UVC) uses UV radiation to transmit signals over fairly short distances. Some of the earliest experiments with UVC date from 1926 to 1933 at the US Naval Research Laboratory (NRL). The NRL, under the direction of E. O. Hulburt, developed an ultraviolet signaling system using a searchlight, ultraviolet filter, and telescope equipped with a fluorescent screen; two unlit ships used the system to communicate at night over a distance of four miles. Contemporary UVC networks can transmit during the daytime or night at a broad range of speeds and over varying distances. Generally, according to Guo et al., these networks take electrical signals from the signal generator, code them with a modulation circuit, and then convert them into high-frequency currents. "Then high-frequency currents are amplified by current sources and supplied to light sources to generate light signals." Next, "the light signal detected by the UV detector [is] converted back into the electrical signal. In order to increase the SNR [signal-to-noise-ratio], a signal amplifier is required to increase signal power. After demodulation and reconstruction, the signal [is] restored." While there is an increased risk of certain cancers in humans from continuous exposure to UV light, UVC has been shown to work well over short distances on land, underwater, between vehicles, and between machines. Further, UVC is considered secure because it is not subject to interference from technologies operating in lower radio frequencies. The transmission speed of UVC varies depending on transmission distance, light source, detector, and wavelength; generally, it ranges from 2 megabits per second to 2.4 gigabytes per second.

SOURCES G. Harvey, *A Survey of Ultraviolet Communication Systems* (Naval Research Laboratory, 1964); L. Guo, Y. N. Guo, J. X. Wang, and T. B. Wei, "Ultraviolet communication technique and its application," *Journal of Semiconductors* 42:8 (2021); Z. Xu and B. M. Sadler, "Ultraviolet Communications: Potential and State-of-the-Art," *IEEE Communications Magazine* 46:5 (May 2008)

Laser Communication [14]

COUNTRY OF ORIGIN USA
CREATOR(S) Gordon Gould, Charles Townes, Arthur Schawlow, Theodore H. Maiman
EARLIEST KNOWN USE 1960

BASIC INFRASTRUCTURE/MATERIALS Transmitter (consisting of a laser, modulator, laser drive, telescope), receiver (consisting of a telescope, detector, and decoder), output interpretation device

RELATED Amateur radio [16], amateur radio satellite [16.6], Radio broadcast [17], communications satellite [32]

DESCRIPTION Laser communication is considered a type of free space optical communication, or any optical communication technology that uses light in "free space" to wirelessly transmit data; "free space" includes the earth's atmosphere, outer space, or a vacuum. Free space optical links are often used in conjunction with other types of networks such as radio or optical fiber as a way to close connectivity gaps in urban areas or as a way to maintain communication in outer space. While lasers were first conceived by Albert Einstein in 1917 as well as physicists Charles Townes and Arthur Schawlow in the 1940s and 1950s, the first successful construction of a laser took place in 1960 at the Hughes Research Laboratories in Malibu, California, by Theodore H. Maiman. An acronym for light amplification by stimulated emission of radiation, lasers are artificially created, narrow, and bright beams of light that have the potential for exceptionally high bandwidth. Some advantages of laser communication are that, at present, there are no licenses or frequency allocations required; lasers are also portable, lightweight, and have fairly low power requirements. Some disadvantages are that weather or other forms of "optical turbulence" can interfere with performance; also, a clear line of sight is required between the transmitter and receiver. As Arun K. Majumdar puts it, the transmitter for laser communication consists basically of "a laser, a modulator . . . a laser driver . . . and a telescope. Depending on the modulation format, the modulator converts bits of information into an electrical signal and modulates the laser to generate an optical signal. The telescope expands the optical beam to reduce diffractive effects that spread the beam out." The receiver, on the other hand, consists of "a telescope, a detector, and a decoder to get the signal out. The telescope collects the optical signal and contracts the beam size small enough to fit onto the detector. The detector converts the optical signal back to an electrical signal." Generally, lasers can send multiple gigabits per second.

EXPERIMENTS In 2013, scientists at NASA's Goddard Space Flight Center in Greenbelt, Maryland, successfully used laser communication to transmit a digital image of the *Mona Lisa* to a satellite, orbiting the moon, roughly 240,000 miles away. Lasers are normally used to track the position of the satellite, and in this instance the digital image accompanied these pulses of light. According to the NASA press release, scientists "divided the Mona Lisa image into an array of 152 pixels by 200 pixels. Every pixel was converted into a shade of gray, represented by a number between zero and 4,095. Each pixel was transmitted by a laser pulse, with the pulse being fired in one of 4,096 possible time slots during a brief time window allotted for laser tracking. The complete image was transmitted at a data rate of about 300 bits per second." The team used error detection codes to clean the image upon its arrival on earth.

SOURCES Arun K. Majumdar, *Optical Wireless Communications for Broadband Global Internet Connectivity* (Elsevier, 2019); "December 1958: Invention of the Laser," *American Physical Society News* 12:11 (December 2003); Arun K. Majumdar and Jennifer C. Ricklin, *Free Space Laser Communications: Principles and Advances* (Springer Science+Business Media, 2008); E. Leitgeb, M. Gebhart, and U. Birnbacher, "Optical Networks, Last Mile Access and Applications," *Journal of Optical and Fiber Communications Reports* 2 (2005); Megan Gambino, "Mona Lisa Travels by Laser, to Space and Back Again," *Smithsonian Magazine* (January 25, 2013)

Visible Light Communication [15]

COUNTRY OF ORIGIN Japan
CREATOR(S) Nakagawa Laboratory (Keio University)
EARLIEST KNOWN USE 2003

BASIC INFRASTRUCTURE/MATERIALS Transmitter (LEDs, fluorescent lamps, or lasers), receiver (optical filter, lens concentrator, photodetector, preamplifier, postamplifier), modulator, output interpretation device

RELATED Photophone [12], Wi-Fi [29], faster-than-light communication networks [61]

DESCRIPTION While experiments with visible light communication (VLC) took place before 2003 (for instance, the photophone [12] is often referred to as a VLC device), the term itself dates to 2003, when the Nakagawa Laboratory at Keio University in Japan successfully used LEDs and photodiodes to communicate via visible light existing between 430 and 790 terahertz on the electromagnetic spectrum. VLC is also often referred to as wireless light communication (Wi-Li). According to Sklavos et al., VLC "uses modulated light, emitted and received by LEDs for downlink and IR LEDs for uplink path. Both uplink and downlink could be provided sufficiently." Given the centrality of LEDs (light emitting diodes, which are known for their low power consumption) for VLC, this network has the potential to be extremely energy efficient. VLC is also highly effective for indoor and outdoor line-of-sight communications (including underwater communication); as a result of its reliance on line of sight, it is also considered secure. The transmission speed of VLC varies depending on transmission distance and light source; generally it ranges from ten kilobits per second to ten gigabytes per second.

SOURCES Nikolas Sklavos, Michael Hübner, Diana Goehringer, and Paris Kitsos, eds., *System-Level Design Methodologies for Telecommunication* (Springer, 2014); Latif Ullah Khan, "Visible Light Communication: Applications, Architecture, Standardization and Research Challenges," *Digital Communications and Networks* 3:2 (May 2017)

RADIO NETWORKS

The wireless radio networks listed in this section range from ultra-local transmissions to international broadcasts of audio as well as televised content. It may surprise some to find television included in this section. However, TV was originally broadcast via radio waves between roughly 52 and 600 megahertz on the VHF and UHF bands; in addition, TV antenna use continues to grow worldwide, as hundreds of millions of viewers use antennas instead of or in addition to cable and/or satellite TV (particularly in most parts of Asia, Europe, Russia, and sub-Saharan Africa). The radio wave portion of the electromagnetic spectrum is also incredibly flexible and complex; as Kimberley Peters puts it, "Radio or electromagnetic waves do not require a medium to move through. They travel, buoyed by their own energy, through a vacuum of space. Radio waves are both nowhere and yet everywhere. They are all around us." Unlike electricity, radio waves are omnipresent; coupled with the fact that they are also considered a limited "natural" resource, the electromagnetic spectrum is carefully regulated by national and international bodies. One also cannot separate these regulations from the technologies that have been developed to harness radio waves; the one influences the other and vice versa. Wireless radio is a deeply entangled, profoundly variable assemblage which means, first, that the list of wireless radio networks below is far from complete, and second, since many radio-based networks function on several different frequency ranges and since different national regulatory bodies have different rules for which bands may be used for which kinds of radio communications, readers will note that few fit neatly into any one type or category. Finally, I have attempted to focus on wireless radio technologies rather than services (for example, those designated in the US as Family Radio Services or Maritime Mobile Service); however, once again, the boundaries between services and technologies are significantly blurred. As such, I have resorted to having one large category for "amateur radio" and additional categories for networks that do not require a license and for networks that are used by government or commercial entities.

Amateur Radio [16]

COUNTRY OF ORIGIN UK
CREATOR(S) Guglielmo Marconi
EARLIEST KNOWN USE 1895

BASIC INFRASTRUCTURE/MATERIALS Transmitter (including power supply, oscillator, amplifier, and modulator), receiver, antennas, telegraph key (optional), teleprinters (optional), computer (optional), microphone (optional), earphones (optional)

RELATED Radiotelegraphy [16.1], radioteletype [16.2], Amateur television [16.3], Hellschreiber [16.4], earth-moon-earth communication [16.5], slow-scan television [23], amateur radio satellite [16.6], amateur packet radio [16.7], radio broadcast [17], pirate radio [18], two-way radio [20], packet radio network [26], microbroadcast [27], software defined radio [28]

DESCRIPTION The International Telecommunications Union (ITU) defines amateur radio as "A radiocommunication service for the purpose of self-training, intercommunication and technical investigations carried out by amateurs, that is, by duly authorized interested in radio technique solely with a personal aim and without pecuniary interest." Since radio waves are ubiquitous and their propagation can easily cause interference, most amateur radio operators need to obtain a license (along with a unique identifying call sign that must be used with nearly all transmissions) issued by the appropriate regulatory body in their home country. That said, amateur radio users do not need a license for listening to FM, emergency weather, or for ultra-local transmissions. It is also worth noting that, even though turn-of-the-century radio pioneers predated licensing practices and the establishment of the ITU, they could still be considered amateur radio operators.

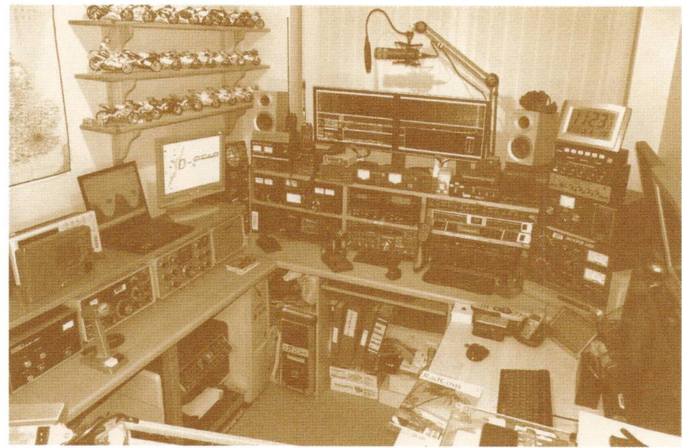

An amateur radio "shack," 2012.

Now, amateur radio users may take advantage of only small frequency bands located across the radio spectrum; however, the transmissions modes are quite variable as they include voice (such as amplitude modulation or AM and frequency modulation or FM), text (such as continuous wave or Morse code, radioteletype [16.2] or RTTY, and amateur teleprinting over radio or AMTOR), image (such as slow-scan television [23], fast scan television, and facsimile), and data communications (such as packet radio).

The first amateur radio transmission was Guglielmo Marconi's successful experiment with transmitting a radiotelegraphic signal over a distance of about two kilometers in 1895. The first amateur radio transmission of audible speech and music was Canadian-born Reginald Fessenden's transmission in December 1906 from Brant Rock, Massachusetts, to ships in the Atlantic. That same year, the UK began requiring licenses for amateur radiotelegraphy [16.1], and in 1912 the US also began requiring licenses specifically for amateur radio enthusiasts. While the International Amateur Radio Union has ceased collecting data on the number of amateur radio users around the world, as of 2021 the US had issued roughly 780,000 amateur radio licenses, Japan roughly 381,000, and Sweden roughly 13,000. According to these same statistics, the vast majority of these users are male.

SOURCES International Association for Amateur Radio, "ITU Radio Regulations" (2019); "Radioland Fee Is Not Necessity," *Los Angeles Times* (January 6, 1924); Christopher H. Sterling, ed., *The Concise Encyclopedia of American Radio* (Routledge, 2010); International Telecommunications Union, http://itu.int; H. Ward Silver, *Ham Radio* (John Wiley & Sons, 2021); *ARRL Ham Radio License Manual*, 5th edition (ARRL, 2022); Clinton B. DeSoto, *Two Hundred Meters Down: The Story of Amateur Radio* (ARRL, 1936)

Radiotelegraphy [16.1]

COUNTRY OF ORIGIN UK
CREATOR(S) Guglielmo Marconi
EARLIEST KNOWN USE 1895

BASIC INFRASTRUCTURE/MATERIALS Transmitter (including power supply, oscillator, amplifier, and modulator), receiver, antennas, telegraph key (optional), teleprinters (optional), computer (optional), microphone (optional), earphones (optional)

RELATED Amateur radio [16], radioteletype [16.2], amateur television [16.3], Hellschreiber [16.4], earth-moon-earth communication [16.5], slow-scan television [23], amateur radio satellite [16.6], amateur packet radio [16.7], radio broadcast [17], pirate radio [18], two-way radio [20], packet radio network [26], microbroadcast [27], software defined radio [28], barbed wire telegraph [40]

DESCRIPTION Radiotelegraphy, historically also referred to as "wireless telegraphy" and today known as "continuous wave" (CW), is the process whereby information is transmitted by pulses of different lengths of radio waves produced by modulating their amplitude, frequency, or phase. At the receiving end, the information carried by the waves is extracted and decoded either by a manual key, automatic key, or, contemporarily, a computer. Morse code is the most commonly used communication system, whereby a short radio wave is indicated by a dot and a longer radio wave is indicated by a dash. It is worth noting that even though "radiotelegraphy" and "wireless telegraphy" are used interchangeably, historically "radiotelegraphy" specifically referred to the use of radio waves to transmit information via telegraphic device. Similarly, the term "wireless telegraphy" does not necessarily mean no wires are involved; instead, it refers to a wide range of manually operated optical signaling devices as well as to early experiments from the 1840s through the 1880s that used water, land, or air to conduct electrical signals. In these cases, wires were not used for the impulses of the current, but may still have been present either as antennas or as power sources.

The beginning of telegraphy using radio waves may be marked by Sir David Lodge's 1895 demonstration at a meeting of the British Association when he sent a signal from the Clarendon Laboratory to the University Museum, a distance of 150 yards. However, as I.V. Lindell points out, "since the receiver was a mirror galvanometer instead of a Morse inker, it has been argued that this was not a real demonstration of the wireless telegraph and thus, Lodge

[16.1]

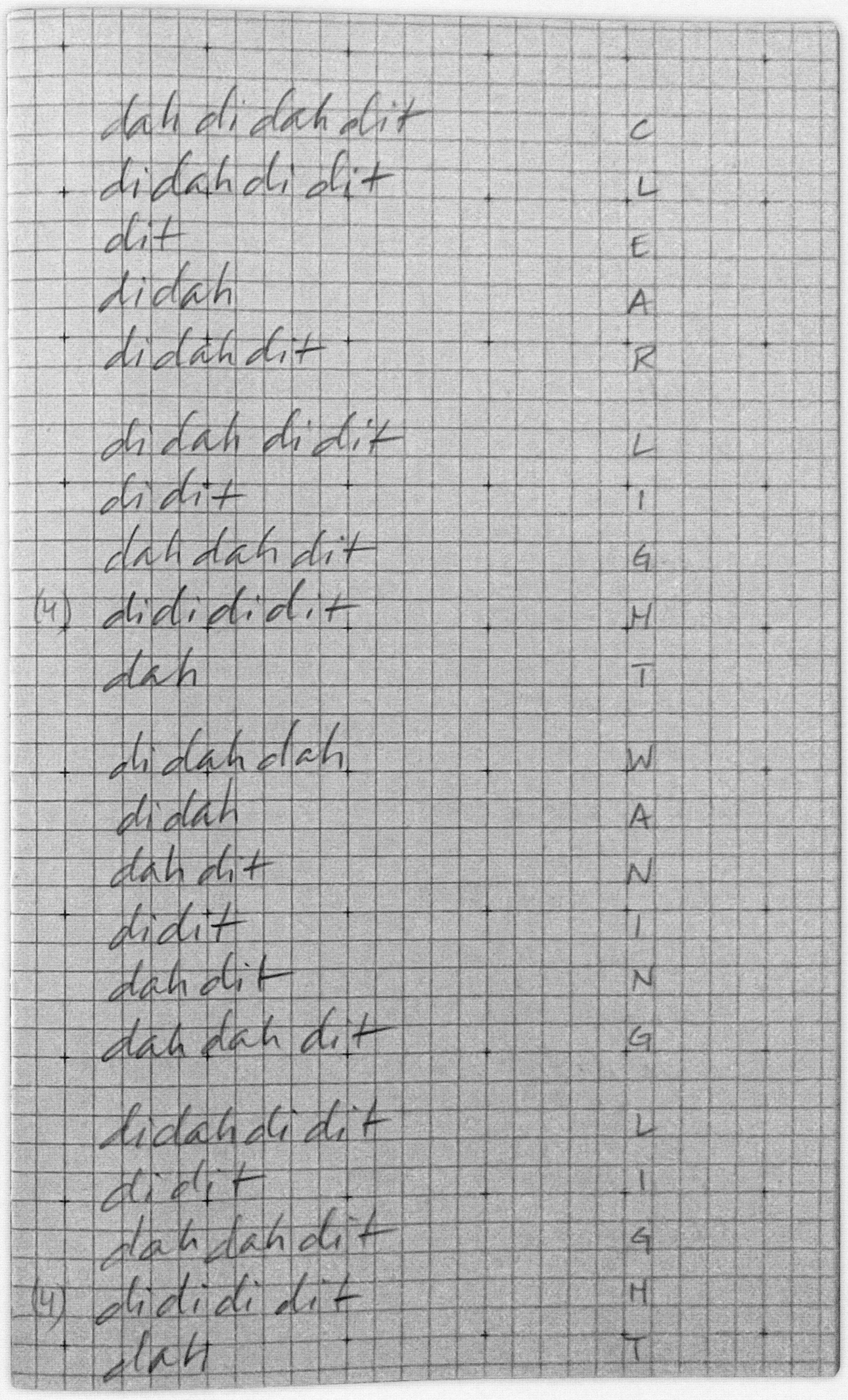

Excerpt of verbal Morse code text from Anna Friz's *Radiotelegraph* (2013).

cannot be considered as the inventor of the radio." Radiotelegraphy became what it is now understood to be after Guglielmo Marconi's experiments in the same year, 1895. Thereafter—particularly in response to the need for ship-to-ship communication at distances that exceeded the capabilities of various forms of optical signaling and also the need for communication in wartime that did not rely upon cables—it became a crucial means of person-to-person communication in military, maritime, and commercial contexts. Numerous networks of international chains of radiotelegraphy stations were also created as a way to support communication between countries in, for example, the British Empire. Manual radiotelegraphy works by having an operator use a device called a "telegraph key" to tap out a message, which then turns a transmitter on and off in order to produce the pulses of radio waves; at the receiving end, the pulses can be heard as long and short beeps, which are in turn translated into Morse code. Automatic telegraphy uses teleprinters at the sending and receiving ends to produce typed text.

Radiotelegraphy is still used as part of national maritime services, weather bulletins, time signals, and among amateur radio [16] enthusiasts. The reception speed of a radiotelegraphy transmission depends on whether the decoding is manual or automatic; manual decoding ranges from ten to twenty words per minute and automatic decoding is roughly two hundred words per minute.

EXPERIMENTS Innumerable experiments have taken place since the late nineteenth century with radiotelegraphy. One of the most recent was undertaken by artist Anna Friz in October 2013. *Radiotelegraph* features Friz performing spoken Morse code and simultaneously broadcasting this code via private low-watt transmitter at two locations—Seyðisfjörður in Iceland, on 107.1 FM, and Chicago, on 88.9 FM—over a period of five days. As Friz put it in an interview, the verbal Morse code "refers to the daily loss of light after the solstice, and is a little inspired by reading translations of Icelandic sagas, where a lot of collective action, for good or for ill, takes place after nightfall." Friz continued: "I was also aware of the text needing to speak to two very different geographic circumstances, that of this northern village on the edge of the Arctic Circle and the huge metropolis of Chicago further south, but for it not to be a warning. The beacon tells that long nights are coming, but we will not be alone. It's the basic promise and premise of a signal, however faulty, asking and declaring: who's there? I am here."

SOURCES Arthur Vaughan Abbott, William Mayer, Thomas Commerford Martin, and William Mott Steuart, *Telephones and Telegraphs: 1902* (US Government Printing Office, 1906); I.V. Lindell, "Wireless Before Marconi," in Tapan K. Sarkar et al., *History of Wireless* (John Wiley & Sons, 2006); Brian Winston, *Media, Technology, and Society: A History, from the Telegraph to the Internet* (Routledge, 1998); J.J. Fahie, *A History of Wireless Telegraphy 1838–1899* (W. Blackwood and Sons, 1900); J. Erskine-Murray, *A Handbook of Wireless Telegraphy* (D. Van Nostrand, 1907); "Here for the Night," interview with Anna Friz, *Desperado Philosophy* (October 21, 2013); Anna Friz, *Radiotelegraph* (2013)

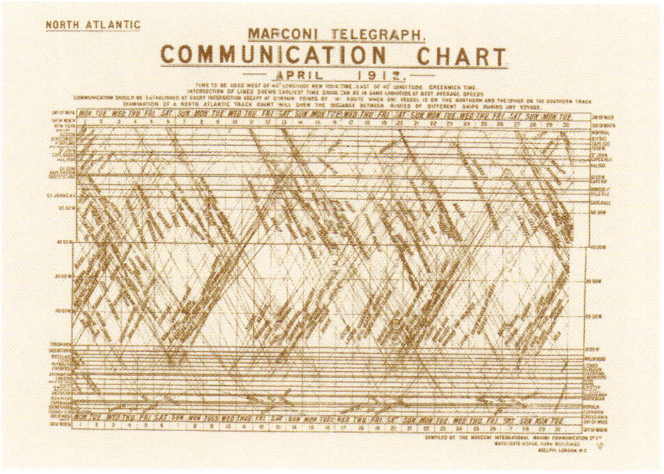

"Marconi Telegraph Communication Chart" from April 1912 depicting an international network of ships, including the *Titanic*, traversing the North Atlantic Ocean. The chart specifies when and where each ship ought to be on any given day of the month; intersecting ship routes indicate where and when messages should be relayed from one ship to the next. It was thanks to this same network that passengers were rescued when the *Titanic* sunk on April 14, 1912.

Radioteletype [16.2]

COUNTRY OF ORIGIN USA
CREATOR(S) John B. Brady
EARLIEST KNOWN USE 1921

BASIC INFRASTRUCTURE/MATERIALS Teletypewriter or teleprinter, transmitter (including power supply, oscillator, amplifier, and modulator), receiver, antennas, computer (optional), modem (optional)

RELATED Amateur radio [16], radiotelegraphy [16.1], amateur television [16.3], Hellschreiber [16.4], earth-moon-earth communication [16.5], slow-scan television [23], amateur radio satellite [16.6], amateur packet radio [16.7], telautograph [36], radio

[16.2] Example of RTTY that appeared in "RTTY Art Made Easy" in *73 Magazine* (January 1972).

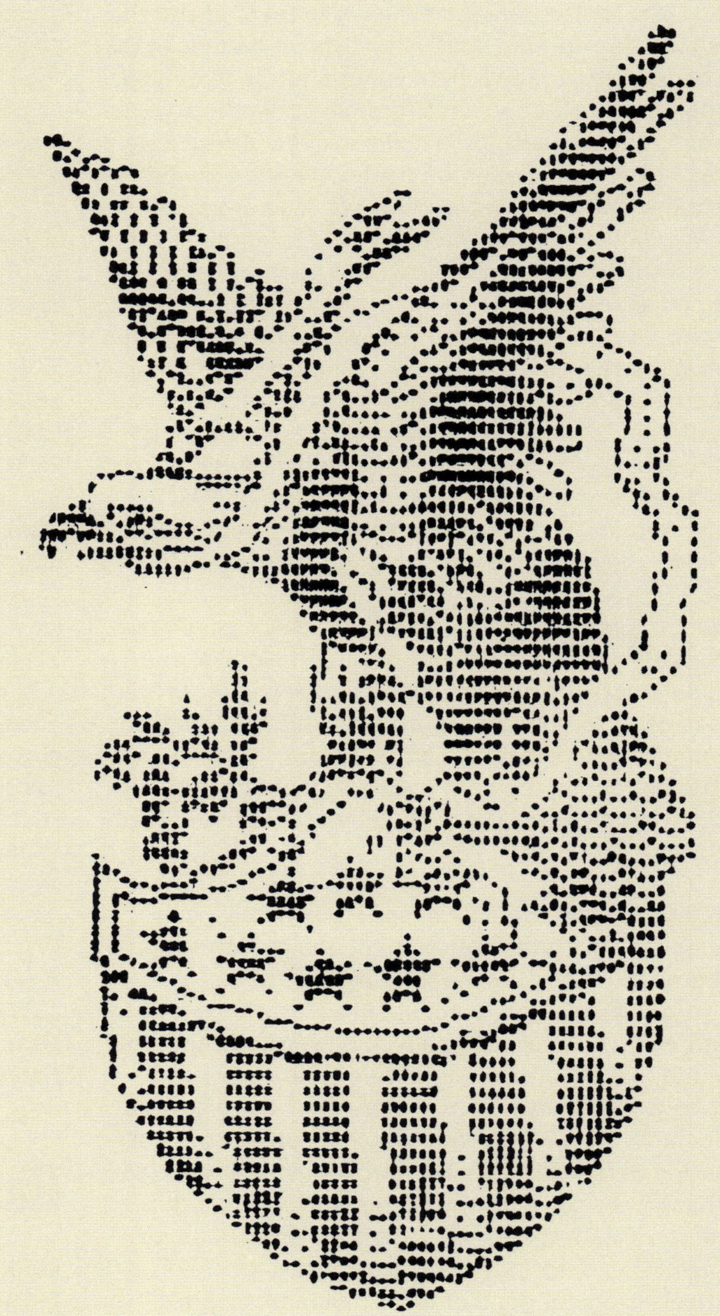

"Peace Through Victory," originated for RTTY by Don WA6PIR, art by XYL Maxine.

RTTY ART MADE EASY

broadcast [17], pirate radio [18], two-way radio [20], packet radio network [26], microbroadcast [27], software defined radio [28]

DESCRIPTION Radioteletype (RTTY), also known as radio automatic teletype by the US Navy and single-channel radio teletype by the Army Signal Corps, is an early networking protocol insofar as it establishes a set of rules via a five-bit character code (sometimes called Baudot code) for the radio transmission of data between teletype machines. RTTY was invented in roughly the early 1920s by John B. Brady while he was an employee of the Morkrum Company (makers of printing telegraph machines); in 1921, Brady filed a patent for a "radio telegraph system" on behalf of Morkrum. Instead of Morse code, Brady's system transmitted the aforementioned five-bit character code "upon the operation of a letter key board at the transmitter," at which point the code was then "received simultaneously in printed type at the receiver." His system soon became known as "radioteletype" because of its reliance on teletype machines. Because of the teletype keyboard, RTTY was considerably easier to use than those telegraphic systems that used Morse code; in Brady's words, it could be controlled by "an operator unskilled in the art of radio telegraphy . . . and by one without knowledge of codes." While Brady reports in his patent filing that his system had already been successfully tested between the US Naval Aircraft Radio Laboratory at Anacostia, D.C., and the US Naval Radio Research Laboratory in Washington, D.C., the US Navy also successfully tested the system in 1922 between an airplane and a ground station. When one presses a teletype key, the system converts the key into a five-bit character, which in turn is converted to a serial format (represented by one of two states—usually off or on, 0 or 1) and transmitted. When it arrives at the receiving teleprinter, the serial format is converted back into a five-bit character and passed on to the printer. A specialized RTTY protocol called amateur teletype over radio (AMTOR) was developed in 1978; it uses a seven-bit character code and incorporates error detection and error correction.

Since World War II, RTTY has become most commonly associated with amateur radio [16] operators who learned how to make use of commercial-grade teletype equipment that was being disposed of. The earliest record of a two-way amateur RTTY transmission is from May 1946 between a user in Brooklyn, New York, and another in Long Island, New York. RTTY is now used to also refer to any system of teleprinters (or computers using software that emulates teleprinters) over radio.

EXPERIMENTS Thanks to radio amateurs experimenting with RTTY, many instances of RTTY art have been produced since at least the early 1970s but likely earlier. Unlike ASCII code, which uses upper and lowercase letters, numbers, and certain punctuation characters, RTTY uses only uppercase letters along with numbers and punctuation characters. RTTY can be used to create an image or translate a preexisting image into characters, and thus can also be considered a form of concrete or visual poetry. It was originally transmitted on long paper tapes and, given the relatively slow transmission speed, could take up to an hour to transmit an entire image. The last known average speed of an RTTY transmission is 45.45 baud, or sixty words per minute.

SOURCES John J. Brady, "Radio Telegraph System," US Patent 1,485,212 (December 28, 1921); "Typing in Airplane Received by Radio," *The New York Times* (August 10, 1922); Jim Haynes, "The Start of Amateur RTTY," Southwest Museum of Engineering, Communications and Computation website; Ed Muns, "Getting Started in RTTY," *On All Bands* blog (March 11, 2021); John Evans Williams, "The Story of Amateur Radio Teletype," *QST* 32:10 (October 1948); Don Royer, "RTTY Art Made Easy," *73 Magazine* (January 1972)

Amateur Television [16.3]

COUNTRY OF ORIGIN USA
CREATOR(S) Charles Francis Jenkins
EARLIEST KNOWN USE 1923

BASIC INFRASTRUCTURE/MATERIALS Transmitter (including power supply, oscillator, amplifier, and modulator), receiver, antennas, television set or monitor, computer (optional), microphone (optional), earphones (optional)

RELATED Amateur radio [16], radiotelegraphy [16.1], radioteletype [16.2], amateur television [16.3], Hellschreiber [16.4], earth-moon-earth communication [16.5], slow-scan television [23], amateur radio satellite [16.6], amateur packet radio [16.7], pirate television [25], microbroadcast [27], television broadcast network [36], cable television [48]

DESCRIPTION Amateur television (ATV), also known as HAMTV or fast-scan TV, uses amateur wide radio bands in the ultrahigh frequency range (or higher) to send and receive live-action images. Given the bandwidth required for ATV, it is almost always only used for local communications. While the word "television" was coined in 1900 and a successful demonstration of

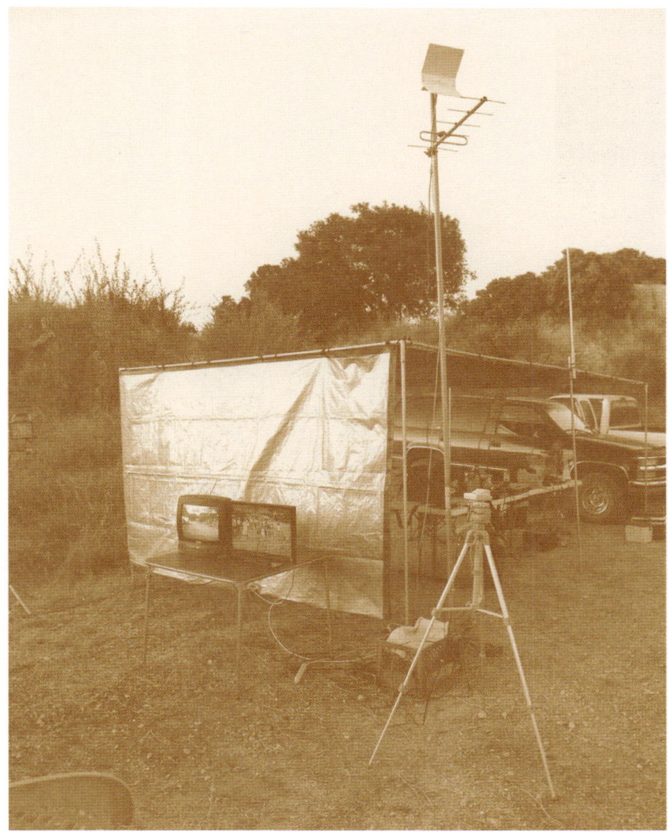

An aid station at the Angeles Crest 100 Mile Trail Race in California that used ATV to monitor runners approaching and leaving the station at the 2020 Angeles Crest.

the transmission of images took place in 1909 using wires, it was not until 1923 that American engineer Charles Francis Jenkins was able to successfully demonstrate the transmission of images wirelessly using several Nipkow disks (mechanical and rotating image scanning devices), lens prisms, a radiophone transmitter, camera, lens, and photoelectric cells. Hugo Gernsback reported on his witnessing of this demonstration in *Radio News*: "I have been able to see my hand projected by radio and being received by radio." He ends by speculating about what eventually evolves into broadcast television [47]: "The day will come when you will be able to sit at home and witness a baseball game as it is being played five thousand miles away or you will be able to sit at home and not only listen to, but actually see an opera as it is being sung and acted." Shortly after, Scottish engineer John Logie Baird also used Nipkow disks for what is considered the first public demonstration at the Royal Institution in London, in January 1926. From this point on, the technology for ATV transmissions changed rapidly, but it remained quite costly until the advent of electronic television in the mid to late 1930s, at which point amateur television gradually became more accessible and more popular. By 1951, *Television and Radio Manual* featured an article with the headline "Radio 'Ham' Builds TV Station: California amateur sends voice and picture over transmitter made from $500 worth of war-surplus parts." Contemporarily, amateur TV images are often exchanged using digital video broadcast standards along with video streaming technologies. ATV is also used to help coordinate events such as parades and races; for example, it was used in 2020 at the Angeles Crest 100 Mile Trail Race in California as a way to help track runners needing medical attention. It's also used for information sharing during natural disasters, as well as for transmitting images of remote control aircraft and/or drones.

SOURCES Philip W. Sewell, *Radio in the Age of Television* (Rutgers University Press, 2014); Doron Galili, *Seeing by Electricity: The Emergence of Television, 1878–1939* (Duke University Press, 2020); Hugo Gernsback, "Radio Vision," *Radio News* (December 1923); Donald G. Godfrew, *C. Francis Jenkins: Pioneer of Film and Television* (University of Illinois Press, 2014); *Television and Radio Manual: Everybody's Television and Radio Manual* (Popular Science Publishing, 1951)

Hellschreiber [16.4]

COUNTRY OF ORIGIN Germany
CREATOR(S) Rudolph Hell
EARLIEST KNOWN USE 1929

BASIC INFRASTRUCTURE/MATERIALS Transmitter (including power supply, oscillator, amplifier, and modulator), receiver, antennas, teleprinters (optional), keyboard (optional), punch card reader (optional), paper tape printer (optional), computer (optional), software (optional)

RELATED Amateur radio [16], radiotelegraphy [16.1], radioteletype [16.2], amateur television [16.3], earth-moon-earth communication [16.5], slow-scan television [23], amateur radio satellite [16.6], amateur packet radio [16.7], radiofax [19], electrical printing telegraph [33.1], image telegraph [33.2], pantelegraph [33.4], telautograph [36], software defined radio [28]

DESCRIPTION Hellschreiber (or "Hell Writer" in English)—also known as Feldhellschreiber, Typenbildfeldfernschreiber, or Hell-Schreiber—is a simple, inexpensive, error-free method of directly printing text transmission sent wirelessly or, more rarely, over telegraph/telephone wires. In basic terms, it sends pictures of letters rather than the more popular method of sending codes. Patented in 1929 by German electrical engineer Rudolf Hell,

[16.4] A later model of the Hellschreiber, the Hell Feldfernschreiber, from 1938.

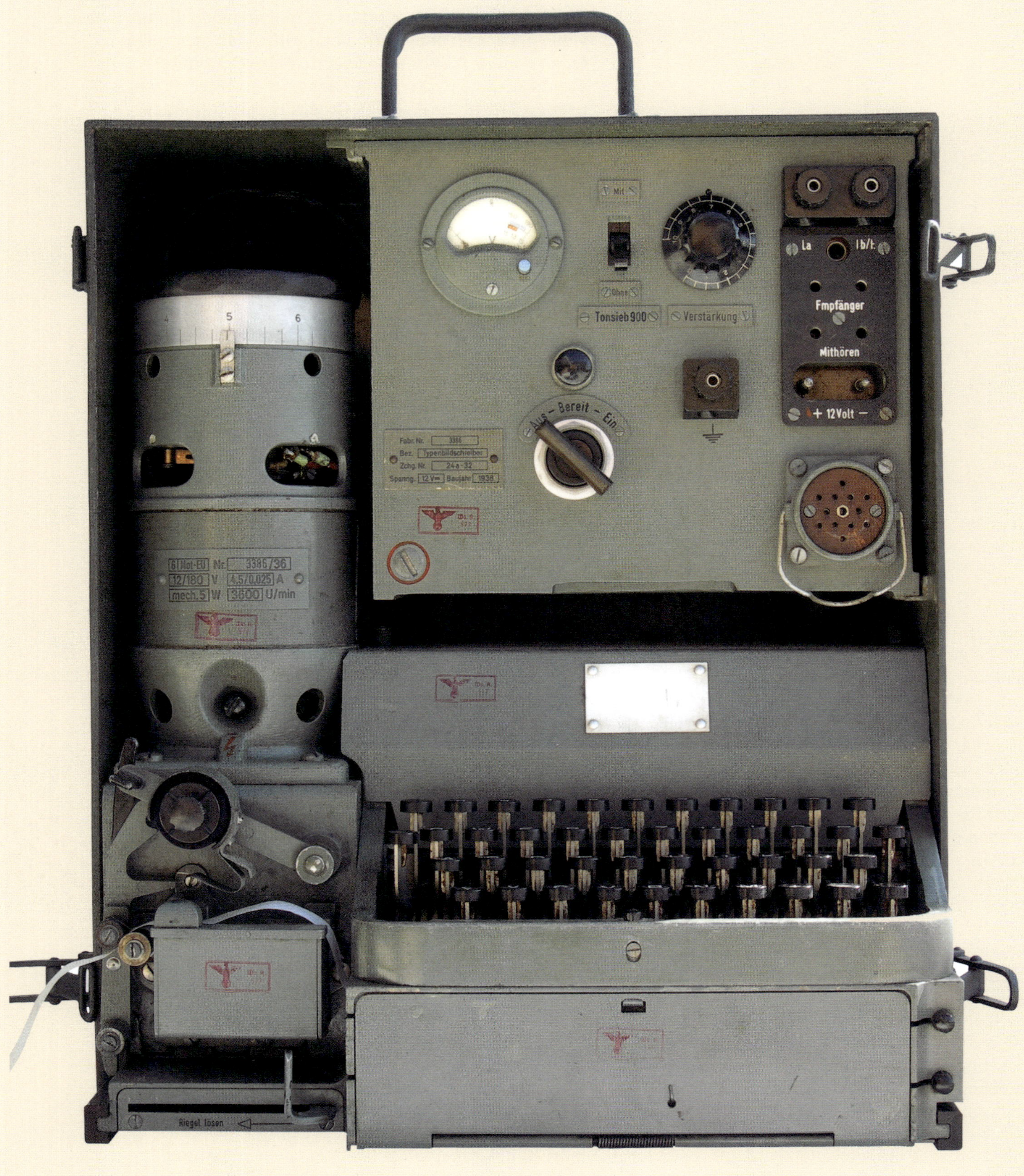

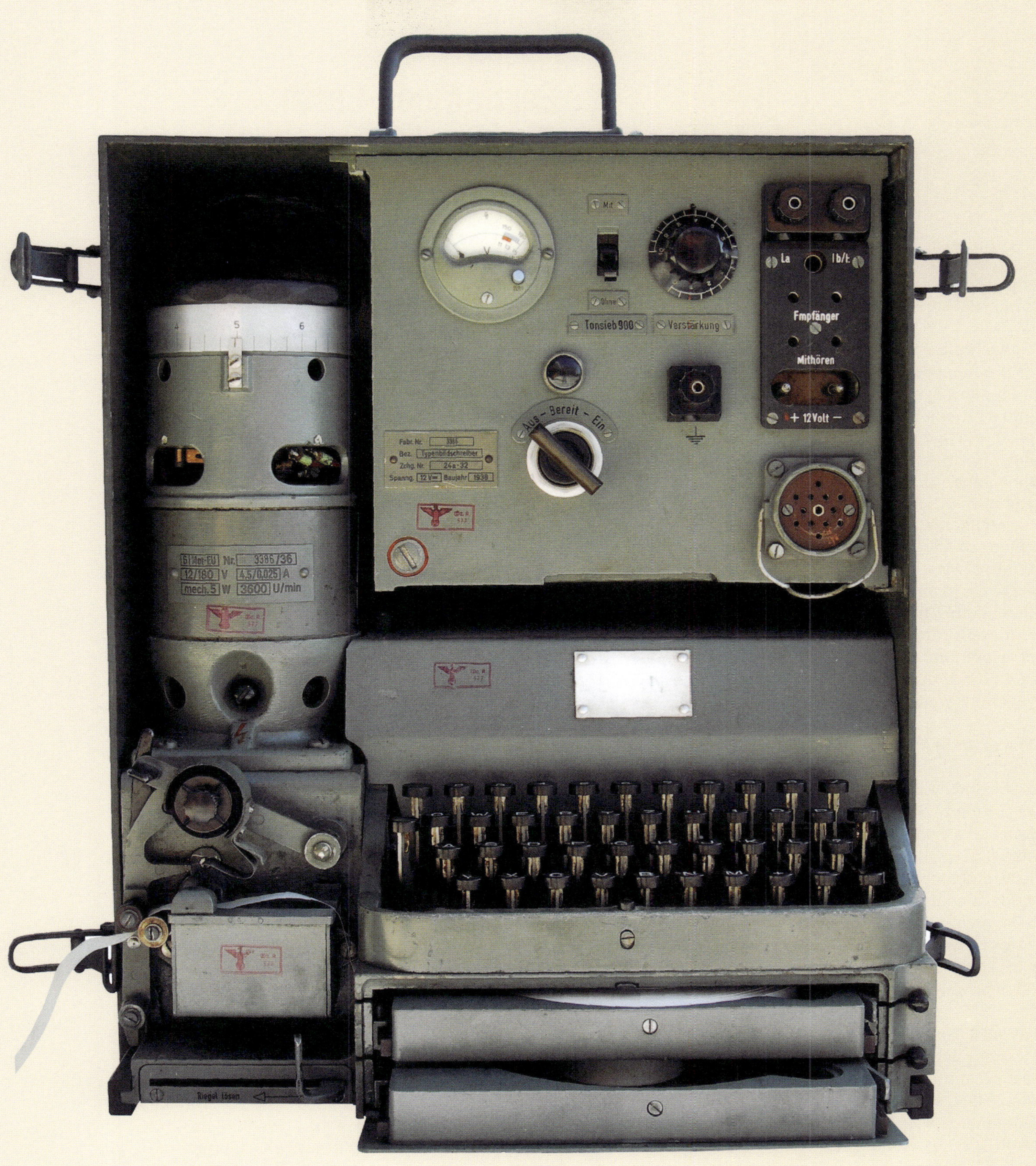
Tonsieb 900
Verstärkung
Aus – Bereit – Ein
La
Ib/t
Empfänger
Mithören
+ 12 Volt −
Fabr. Nr.
Bez.
Zchg. Nr. 24a-32
Spanng. 12 V Baujahr 1938
Nr. 3386/36
12/180 V
4,5/0,025 A
mech. 5 W
3600 U/min
Riegel lösen

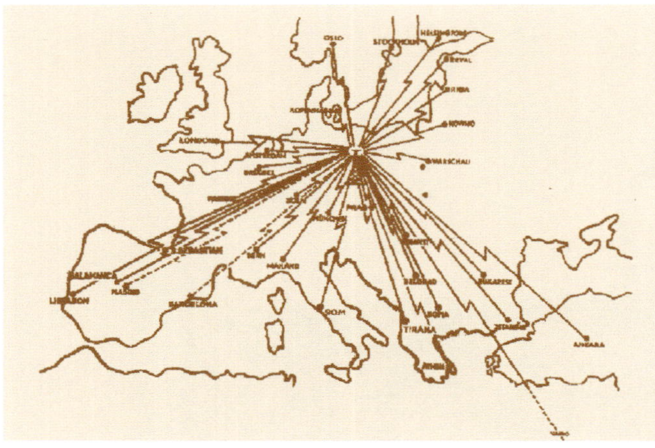

A map from 1939 of the network of Hellschreiber machines used by the German news agency Deutsches Nachrichtenbüro.

Hellschreiber was designed to be, in Hell's words, "a practical device for the reception of messages from news agencies." Each character is scanned into a bitmap format consisting of a seven-by-seven pixel grid. Each pixel corresponds to a radio wave pulsing either on or off. The receiving Hellschreiber then simply prints the pulsations, which appear, once again, as bitmapped characters. It is worth noting the Hellschreiber printer has a thread that is wound twice around the spindle, resulting in text that is printed twice—such that if one part of the text is not readable in one line, it is readable in the other line. Remarkably, unlike radioteletype [16.2] and radiofax [19], Hellschreiber involves no encoding or decoding, no synchronization of transmission and reception, and no photochemical processes, and it uses only a very narrow bandwidth for transmission.

The readability and lack of errors in Hellschreiber text transmissions soon made the device popular for diplomatic and military communications in countries such as Germany, Sweden, Spain, Italy, and Finland, as well as for police communications. Hellschreiber was also first used by amateur radio [16] enthusiasts in 1959, and today, with the use of computers and sound cards, it continues to be a popular form of transmission; amateur radio enthusiasts have even created new variants of Hellschreiber including PSK (phase shift key) Hell, FM (frequency modulation) Hell, C/MT (concurrent multi-tone) Hell, and others.

SOURCES Rudolf Hell, "Vorrichtung zur elektrischer Übertragung von Schriftzeichen" [Device for the electrical transmission of text characters], Patent #54,0849, German Patent and Trademark Office; F. Dörenberg, "Hellschreiber," Nonstopsystems.com (April 3, 2022); Rudolf Hell, "Die Entwicklung des Hell-Schreibers," *Hell: Technische Mitteilungen*, Gerätentwicklungen aus den Jahren 1929–1939 (May 1940); "A History of Hellschreiber," QSL.net (June 27, 2008); Al Williams, "Messages from Hell," Hackaday.com (December 30, 2015)

Earth-Moon-Earth Communication [16.5]

COUNTRY OF ORIGIN USA
CREATOR(S) James H. Trexler
EARLIEST KNOWN USE 1954

BASIC INFRASTRUCTURE/MATERIALS Transmitter (including power supply, oscillator, amplifier, and modulator), receiver, antennas, computer (optional), microphone (optional), earphones (optional)

RELATED Amateur radio [16], radiotelegraphy [16.1], radioteletype [16.2], amateur television [16.3], Hellschreiber [16.4], slow-scan television [23], amateur radio satellite [16.6], amateur packet radio [16.7], meteor burst communication [22]

DESCRIPTION Earth-Moon-Earth communication (or EME, also known as "moon bounce") is the use of the moon as a "natural satellite" off of which radio waves can be bounced. Radio amateurs use the surface of the moon to send and receive messages between stations on earth. The technique was first developed not by amateurs but by James H. Trexler, an engineer at the US Naval Research Laboratory, as part of the Department of the Navy's Communication Moon Relay project (later known as Passive Moon Relay), which was dedicated to relaying electronic intelligence. In 1954, Trexler successfully sent and received a voice transmission via the moon. As David K. Van Keuren details, "The completed system used eighty-four-foot-diameter . . . steerable parabolic antennas and 100-kilowatt transmitters installed at Annapolis, Maryland, and Opana, Oahu, with receivers at Cheltenham, Maryland, and Wahiawa, Oahu. The system operated at frequencies around 400 megahertz, it could accommodate up to sixteen teleprinter channels operating at the rate of sixty words per minute, and it was capable of processing teletype and photographic facsimiles." Reportedly, the system was moderately successful, as it was used in 1964 to intercept a signal from a radio dish in the southeastern Soviet Union. Even though creating a radio station capable of sending messages to the moon and back can cost up to $3,000, since the 1960s EME has been used almost entirely by amateurs all over the world because it poses such formidable (and therefore enjoyable) challenges as figuring out how to bounce signals off of an irregular lunar surface that is also about 250,000 miles away from the Earth.

EXPERIMENTS In November 2007, Scottish artist Katie Paterson used EME to send her own version of John Cage's *4'33"* of silence from Nagano, Japan, to and from the moon. In 2009, Dutch media artist Daniela de Paulis proposed using EME for a live image performance; the result of her proposal was an image transmission to/from the moon with radio amateurs Jan van Muijlwijk and Daniel Gautchi. From 2010 and on, de Paulis has used her pioneering technology (which she calls "visual moonbounce") for live performances, such as *OPTICKS*. De Paulis writes that she considers *OPTICKS* "a contemporary form of Happening and Mail Art, for which people from all over the world exchange images via the Moon during a playful and experimental event streamed live on the web from the cabin of the Dwingeloo radio telescope. For viewers in the live OPTICKS performance, the technology becomes a tool for experiencing virtual space travel."

SOURCES H. Ward Silver, *The ARRL Antenna Book: For Radio Communications*, 22nd edition (ARRL, 2011); David K. Van Keuren, "Moon in Their Eyes: Moon Communication Relay at the Naval Research Laboratory, 1951–1962," in *Beyond the Ionosphere: Fifty Years of Satellite Communication* (NASA, 1997); Jeffrey T. Richelson, ed., "Soldiers, Spies and the Moon: Secret US and Soviet Plans from the 1950s and 1960s," National Security Archive (July 20, 2014); Daniel Oberhaus, "How the Moon Was Turned Into a Cold War Spy Weapon," *Vice* (June 12, 2017); Allen Katz and Marc Franco, "Targeting the Moon," *IEEE Microwave Magazine* (June 2011); Katie Paterson, "Earth-Moon-Earth, *4'33"*," https://katie-paterson.org/artwork/earth-moon-earth/; Daniela de Paulus, "*OPTICKS* and Visual Moonbounce in Live Performance," *LEONARDO* 49:5 (2016)

Amateur Radio Satellite [16.6]

COUNTRY OF ORIGIN USA
CREATOR(S) TRW Radio Club of Redondo Beach, CA
EARLIEST KNOWN USE 1961

BASIC INFRASTRUCTURE/MATERIALS Transmitter (including power supply, oscillator, amplifier, and modulator), receiver, antennas, diplexer, rocket, satellite tracking app (optional), telemetry beacon (optional), transponder (optional), digital repeater (optional)

RELATED Amateur radio [16], radiotelegraphy [16.1], radioteletype [16.2], amateur television [16.3], Hellschreiber [16.4], earth-moon-earth communication [16.5], slow-scan television [23], amateur packet radio [16.7], packet radio network [26], commercial communications satellite [32]

DESCRIPTION A satellite is an entity launched into the Earth's atmosphere for the purposes of orbiting around the Earth and receiving/transmitting information between other satellites or to/from the Earth using radio waves. Over a period of roughly four years, a group of amateur radio [16] enthusiasts built the first amateur radio satellite, OSCAR 1, from $61 worth of scavenged parts. It was launched into low Earth orbit from the Vandenberg Air Force Base (near Santa Barbara, California) in December 1961. The first OSCAR (or "orbiting satellite carrying amateur radio") weighed only ten pounds and carried a simple beacon that transmitted "HI" in Morse code for twenty-two days. While this launch was followed by the launch of numerous OSCAR satellites, OSCAR 3 was the first amateur radio satellite to carry a transponder, one that acted as a repeater—receiving signals and retransmitting them at a different frequency. This development allowed amateurs to communicate over long distances such that about one thousand amateurs in countries around the world were able to operate through it. Today, amateur radio satellite communication remains popular precisely because of its affordability and the way in which the communication reach of these satellites is much larger than anything provided by an earthbound repeater. The transponder capabilities of these satellites also make possible communication not only via Morse code (or continuous wave) but also voice communication via single sideband and frequency modulation and sometimes even image transmission via slow-scan television [23]. The Radio Amateur Satellite Corporation (or AMSAT) maintains a website showing the current status of all amateur satellites based on user reporting. As of 2021, more than twenty-one amateur radio satellites are in orbit.

EXPERIMENTS The Satellite Art Project began in 2010 as a collaboration between more than seventy individuals at Tama Art University, the University of Tokyo, and the Japan Aerospace Exploration Agency; it was led by artist and professor Akihiro Kubota. The project launched three ultra-small satellites that each weighed between one and two kilograms: ARTSAT1: Project in 2010, ARTSAT1: INVADER in 2014, and ARTSAT2: DESPATCH in December of the same year. ARTSAT2 represents the pinnacle of the project, as the satellite itself was conceived as a work of deep space sculpture in the form of a 3D-printed metal

[16.5] Postcard confirming receipt of Katie Paterson's transmission of *4'33"* from the moon back to Nagano, Japan, with a photo of the antenna used for the transmission.

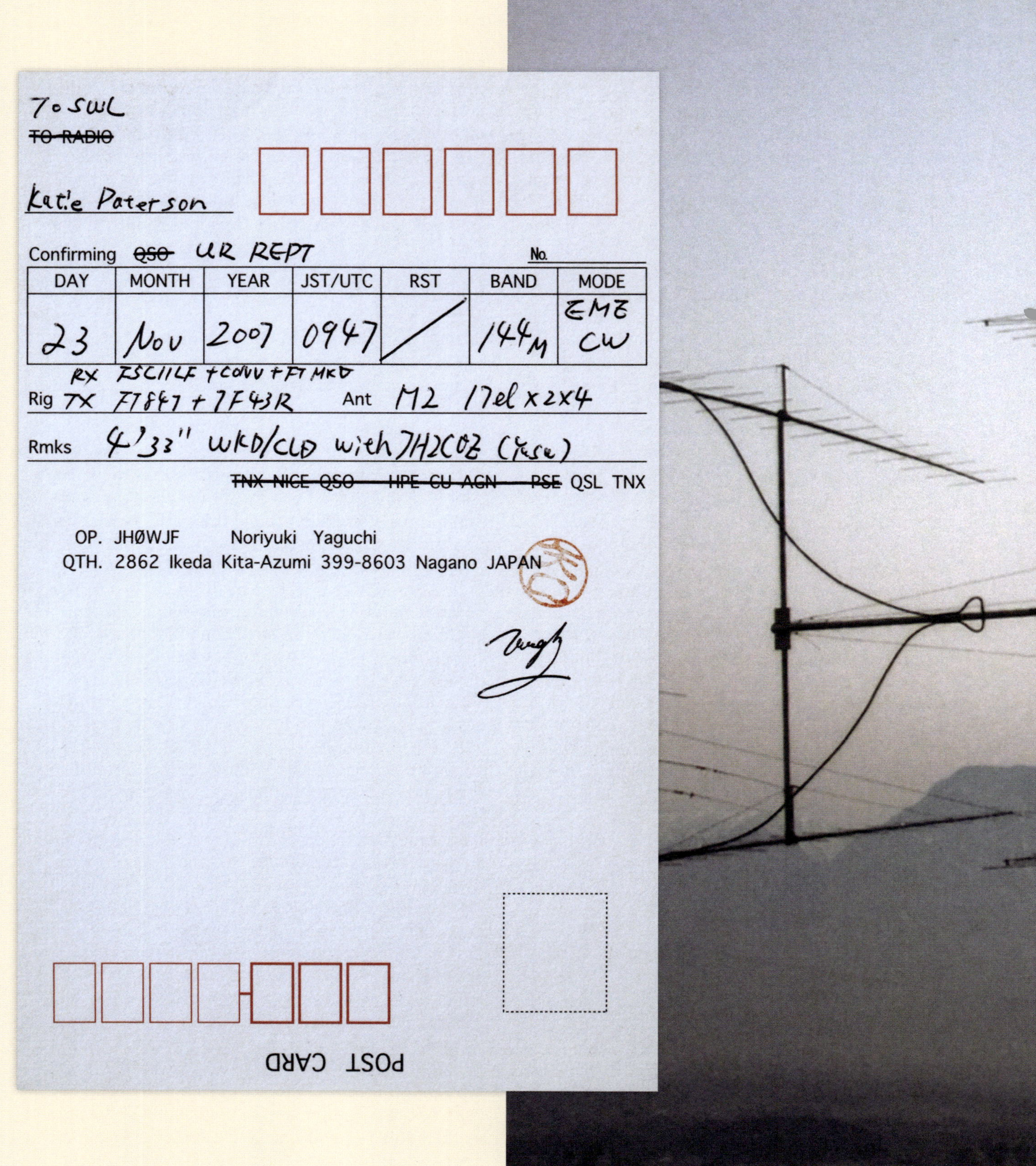

TO SWL
~~TO RADIO~~

Katie Paterson

Confirming ~~QSO~~ UR REPT No.

DAY	MONTH	YEAR	JST/UTC	RST	BAND	MODE
23	Nov	2007	0947	/	144M	EME CW

Rig RX TS-711LF + CONV + FT MKD
TX FT847 + 7F43R Ant M2 17el x2x4

Rmks 4'33" WKD/CLD with JH2COZ (Yasu)

~~TNX NICE QSO HPE CU AGN PSE~~ QSL TNX

OP. JHØWJF Noriyuki Yaguchi
QTH. 2862 Ikeda Kita-Azumi 399-8603 Nagano JAPAN

POST CARD

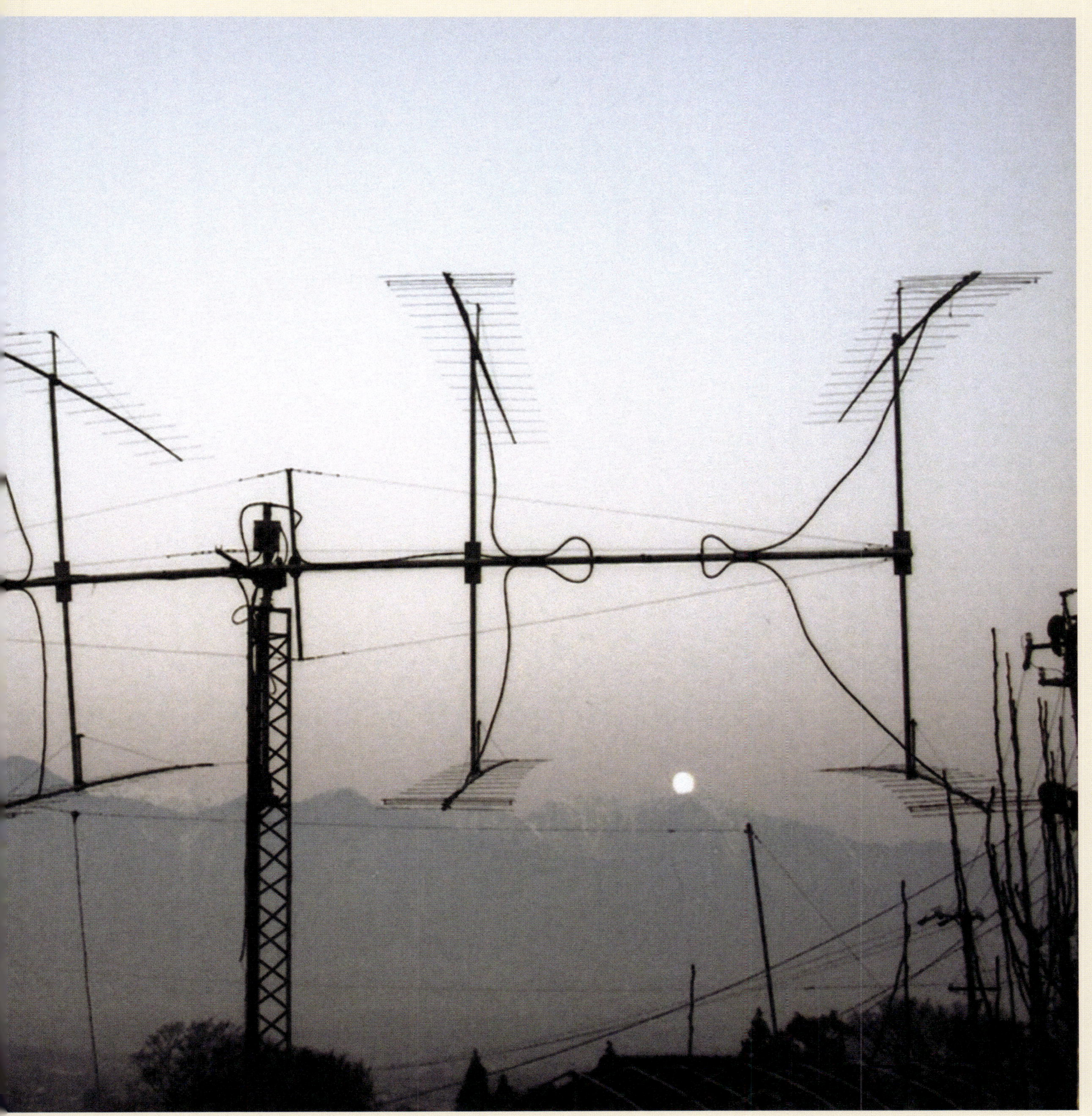

spiral. It also carried a small computer designed to both withstand the extreme environment of deep space and generate poetry, which it then transmitted back to earth via a Morse code beacon. A few years later, another experiment involving amateur radio satellite emerged as "spacekeet," which translates roughly to "space shack" in English. From May to June 2017, Dutch artist Roel Roscam Abbing built the drop-in satellite ground station with assistance from local amateur radio hobbyists; the station then hosted workshops, lectures, and satellite observations—aspiring to be a place "where citizen science can happen, where everybody is an amateur and collective learning is the goal." Finally, since early 2021, a transnational feminist collective called Radio Amatrices (including artists Audrey Briot, Afroditi Psarra, Adriana Knouf, and Sasha Engelmann) have been working on a proposal for FemSat, a feminist amateur radio satellite that, in their words, "would enable new forms of decentralised collaboration, data encoding / codings and cyborg-queer-xeno encounters through the radio spectrum."

SOURCES Donald H. Martin, *Communication Satellites, 1958–1995* (Aerospace Corporation, 1996); Keith Baker, "A Brief History of AMSAT," Radio Amateur Satellite Corporation website; Martin Davidoff, *The Satellite Experimenter's Handbook* (American Radio Relay League, 1984); Steve Ford, "An Amateur Satellite Primer," *QST* (April 2000); "AMSAT Live OSCAR Satellite Status Page," AMSAT website; "ARTSAT: Art Satellite Project," Japan Aerospace Exploration Agency's Institute of Space and Astronautical Science website; Radio Amatrices, "FemSat: Propositions for Feminism in Radiophonic Space," *MONDAY* 6 (Winter 2022); Roel Roscam Abbing, "spacekeet," Roel Roscam Abbing website

Amateur Packet Radio [16.7]

COUNTRY OF ORIGIN Canada
CREATOR(S) Robert Rouleau, Bram Frank, Norm Pearl, Jacques Orsali
EARLIEST KNOWN USE 1978

BASIC INFRASTRUCTURE/MATERIALS Transceiver, computer, terminal node connector or packet modem and software

RELATED Amateur radio [16], radiotelegraphy [16.1], radioteletype [16.2], amateur television [16.3], Hellschreiber [16.4], earth-moon-earth communication [16.5], slow-scan television [23], amateur radio satellite [16.6], packet radio network [26]

DESCRIPTION Amateur packet radio is an inexpensive and error-free mode of transmitting so-called datagrams or packets of data using a transceiver, a Terminal node connector (TNC), and a computer. The TNC includes a modem, a computer processor, and the means by which to convert communications between a computer and the amateur packet radio protocol AX-25. The TNC is responsible for automatically dividing messages into packets, converting the packets into audio tones, and, while receiving and decoding the tones back into packets, automatically decoding the message, checking for errors, and displaying the message. Up to 256 characters may be included in each packet, making it possible to send more than three lines of text in a few seconds. Further, since packets are digitally encoded, they can transmit text, voice, and image; specifically, packet radio can be used for interactive group chats and the transmission and storage of messages using packet BBSs, and it can even be used with slow-scan television [23]. While packet radio was first used in the early 1970s to connect to the ARPANET (the Advanced Research Projects Agency Network, an early wide-area network developed by the United States Department of Defense), a feasible way for amateur radio [16] enthusiasts to use packet radio didn't emerge until the first amateur radio packets were sent in May 1978 by a group of ham radio enthusiasts in Montreal, Canada. Throughout 1980–81, this group collaborated with amateurs in Tucson, Arizona, and together they developed a protocol for amateur packet radio transmission called AX-25 (or Amateur X-25). The last known average speed of an amateur packet radio transmission is 1,200 bits per second.

EXPERIMENTS In 2022, the Wellesley Amateur Radio Society in Massachusetts created a packet radio network (called the WARS LoRa Birdhouse Project) composed of solar-powered nodes in the form of birdhouses. Because of its low power and solar design, each node or house is inexpensive and autonomous as each requires only a transceiver and microcontroller. Users can access the network with a computer and USB cable and interact with it using a serial terminal program. Packets are transmitted from one birdhouse to the next to reach their final destination.

SOURCES Robert T. Rouleau, "The Packet Radio Revolution," *73 Magazine* (December 1978); Robert Rouleau and Ian Hodgson, *Packet Radio* (Tab Books, 1981); Clifford A. Lynch, *Packet Radio Networks* (Pergamon Press, 1987); John Goerzen, "Packet Radio," Complete.org (2022); Greg Jones, "Introduction to Packet Radio," *Bug Out Bag Builder* Blog (September 4, 2020); Roel Roscam Abbing, *Oh Really? Messing Around with Packet Radio* (self-published, November 2015)

[16.6] Design of ARTSAT2: DESPATCH by Akihiro Kubota (2014).

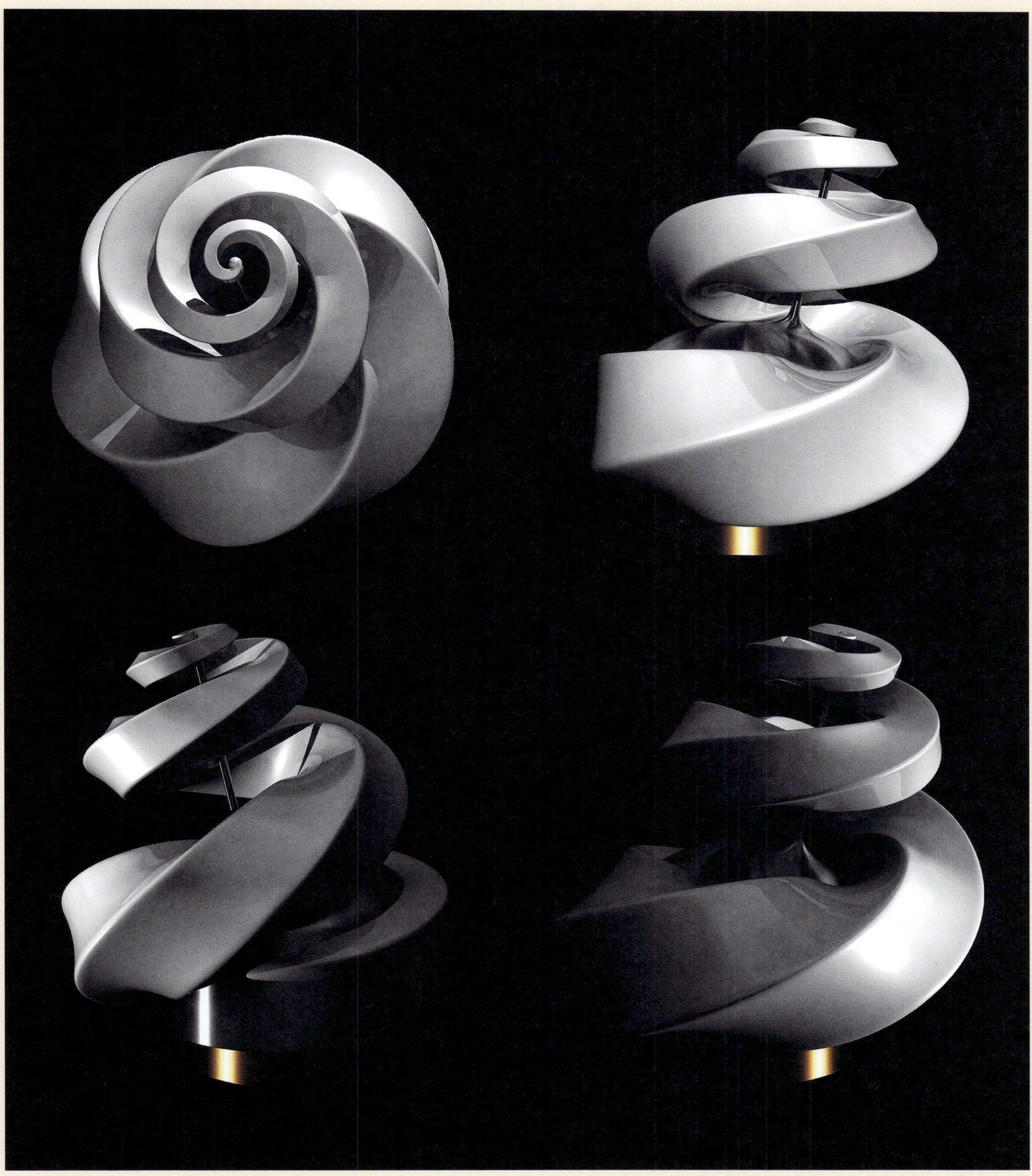

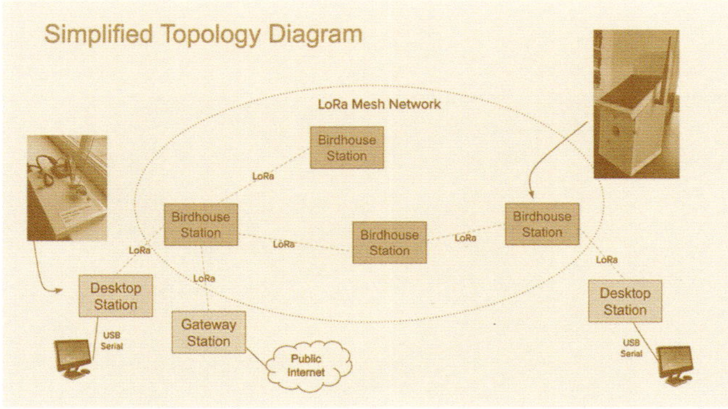

Architecture of the WARS LoRa Birdhouse Project (2022).

Radio Broadcast [17]

COUNTRY OF ORIGIN USA
CREATOR(S) Reginald Aubrey Fessenden
EARLIEST KNOWN USE 1906

BASIC INFRASTRUCTURE/MATERIALS Transmitter (including power supply, oscillator, amplifier, and modulator), receiver, antennas and/or transmitting tower, cables, computer (optional), microphone (optional), earphones (optional), audio processor (optional), tuner (optional), multiplexer (optional)

RELATED Amateur radio [16], microbroadcast [27], communications satellite [32], television broadcast network [36], teletext [51], wired radio [35], world brain [59]

DESCRIPTION Radio broadcast refers to the one-to-many wireless transmission of sound via a radio station for educational, commercial, or nonprofit/state-sponsored purposes and whose geographical reach ranges from the local to the international, depending on the power of the transmitter. While the general techniques for this type of broadcasting are essentially the same as for amateur radio [16], setting up a broadcast radio station involves a fairly lengthy and potentially expensive process of licensing, obtaining equipment, and building the station itself. Broadcasting modes generally include transmission using amplitude modulation (AM), frequency modulation (FM), and digital radio. Listeners receive the broadcast by way of a radio receiver, now simply called a radio.

The term "broadcast" originally referred to the broad scattering of seeds; by 1916, American general David Sarnoff was envisioning individuals receiving wireless signals in their homes via a "radio music box," and by 1922 the term "broadcast" had moved far enough away from its agricultural meaning that it appeared in the London *Times* in reference to the "distribution of information by wireless telephony." However, while the point at which a narrowcast becomes a broadcast is now largely determined by national and international regulatory bodies, the development of wireless radio in the early twentieth century predates this distinction, which makes the first radio broadcast a matter of debate. Although the consensus is that Canadian-born inventor Reginald Fessenden was the first to clearly broadcast "phonographic talking records and music" from Brant Rock, Massachusetts, to listeners as far away as Norfolk, Virginia, it is worth mentioning that, after performing deep and thorough archival research, James O'Neal reported in 2006 that there is strangely no physical evidence for Fessenden's broadcast, and that it is more likely that American electronics pioneer Lee De Forest transmitted the first broadcast in 1907. However, here the historical record is once again unclear.

Radio broadcasting played an enormously important role around the world in the early twentieth century in building and promoting national identities; for example, the world's most powerful radio transmitter was installed in Moscow by 1922 and, as Doron Galili puts it, "The transmission of Moscow's messages to every remote village in the union played a crucial role in informing the modernization of everyday life." By 1935, most American homes likely had at least one radio receiver in their home, and the radio remained the country's most popular home electronic apparatus until around 1955, when it was supplanted by television. Since the advent of affordable satellite radio in 1998, it has seen a surge in popularity in the twenty-first century; as of 2019, there were 34.9 million total SiriusXM satellite radio subscribers.

EXPERIMENTS While women have been an essential part of radio since its beginnings, as both creators and listeners, scant documentation remains of women who experimented with radio broadcasting or who created radio art prior to the 1980s. (Even in Heidi Grundmann's 1995 "Past and Present of Radio Art," no more than two women are mentioned in passing throughout the entire piece.) It's worth noting that the number of women celebrated as radio artists increased throughout the 2000s, but not dramatically so. For example, in the collection *Transmission Arts*, roughly seven women were among the thirty-nine artists whose work was highlighted in the section dedicated to broadcast radio and television, including avant-garde composer Judy Dunaway, who creates electronic, multimedia, and sound installation works for balloons; experimental poet Friederike Mayröcker, who, with Ernst Jandl, created works such as *Fünf Mann Menschen*, a series of poetic sonic experiences composed for Southwest German Radio; and visual

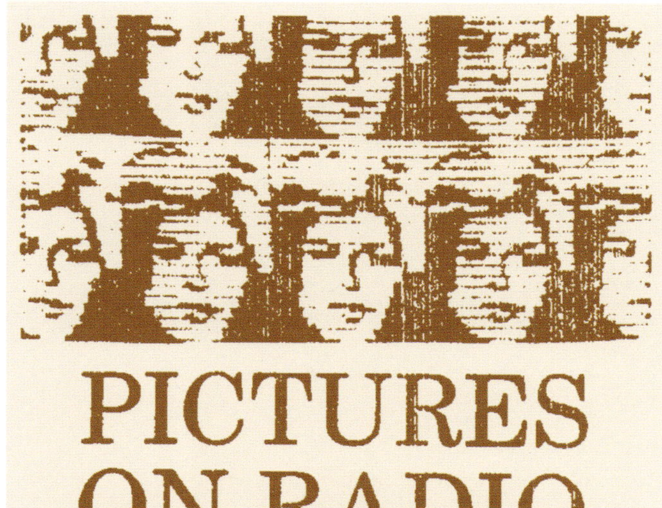

Teletext-style image of actor Cheryl Ladd broadcast over Bristol-based radio show *Datarama* in 1983.

artist Alison Knowles, who created pieces such as *Bohnen Sequenzen* (or *Bean Sequences*, from 1982), which explored, in Galen Joseph-Hunter's words, "the effects of resonant sounds produced by beans and hard surfaces" for the German radio station West Deutscher Rundfunk.

Other than the obvious explanation that women have simply been ignored, one possible reason for this imbalance is that it likely requires resources and privilege to access, let alone experiment with, the massive infrastructure and apparatus surrounding broadcast radio. Another possible reason is that Italian Futurists such as F.T. Marinetti and Fortunato Depero—who have become known as among the first to undertake radio art experiments—were unabashedly misogynistic, as when Marinetti, in his 1909 *Futurist Manifesto*, openly declared his "contempt for woman." Marinetti and Depero broadcast their sound poetry (nonrepresentational, experimental poetry that explored the sounds and sculptural qualities of letters, phonemes, and words) over Italian radio in 1933; however, there is no known recording of this broadcast. That same year, Marinetti and Pino Masnata described "radio art" in their *Manifesto futurista della radio*, declaring it constituted "freedom from all point of contact with literary and artistic tradition" and a "pure organism of radio sensations." The following year, Depero published *Liriche Radiofoniche*—lyrics he described as "radiophonic" because, according to Giovanna Ginex's translation, many of them "were created specifically for the radio and because the others also contain the necessary elements that radio broadcasting requires." Marinetti also created numerous scores for radio that he called Sintesi, some of which simply consisted of a list of timed events: "ten seconds of dishwashing, one second of rustling, eight seconds of dishwashing."

By contrast, in the US, the interest in radio broadcasting as an art form was more oriented toward producing radio dramas that were as realistic and believable as possible; for example, Orson Welles's 1938 radio drama *The War of the Worlds* caused at least one million people (according to Friedrich Kittler's recounting, which has since been challenged) to react hysterically to the dramatization of a Martian invasion. However, by the 1950s, artists such as the American composer John Cage moved away, once again, from the conventional form of the radio broadcast and instead experimented, for example, with open-ended, live radio remixes using multiple radios and performers, as in Cage's piece *Imaginary Landscape No. 4* (1951). At the same time, artists around the world noticeably returned to the principles of the Italian Futurists and attempted to disrupt and even subvert broadcast radio's function as a one-to-many medium. As just one example among many, artist Max Neuhaus managed to access the National Public Radio network to create *Radio Net* in 1977. Neuhaus describes the complex setup for *Radio Net* as follows:

> In those days radio programs on NPR were distributed by what they called a Round Robin—telephone lines connecting all two hundred stations into a large loop stretching across the country. Any station in the system could broadcast a program on all the others by opening the loop and feeding the program around it. I saw that it was possible to make the loop itself into a sound-transformation circuit and tried a few things with it in several preliminary studies in 1974. For the broadcast I decided to configure it into five loops, one for each call-in city, all entering and leaving the NPR studios in Washington. Instead of being open loops as usual during a broadcast, though, I wanted to close them and insert a frequency shifter in each so that the sounds would circulate; it created a sound-transformation "box" that was literally fifteen hundred miles wide by three thousand miles long with five ins and five outs emerging in Washington. In all the previous works I had left the nature of the sounds phoned in for each caller to decide. Here I wanted to provide an indication to try and move them past the "Listen, it's my voice on the radio" stage and towards listening to one another. The question was what kind of indication—how does one indicate something to perhaps half a million people with their diverse backgrounds, intentions, and ways of interpreting? I decided to ask them simply

> to whistle . . . During the broadcast, the sounds phoned into each city passed through its self-mixer and started looping. With each cross-country pass, each sound made another layer, overlapping itself at different pitches until it gradually died away. It was quite a beautiful Sunday afternoon— two hours over which ten thousand people found their way into the work and made sounds.

With Neuhaus's and others' growing interest in networks, it is probably not a coincidence that within just a few years of his *Radio Net*, radio stations, particularly in Europe, had started experimenting with radio broadcasts of computer software. Three eight-bit computers were released in 1977 (the Apple II, the Commodore PET, and the TRS-80) and all three came with cassette players whose tapes could be used for data storage. Five years later, in 1982, the first software broadcasts over radio took place. Transmitted via a series of audio tones (a practice often called "buzzing"), the tones could be recorded by listeners on cassette and converted by their computers into 0s and 1s, turning the practice into a pre-internet form of widespread file sharing. In anticipating the problem of incompatible file formats between different brands of computer, engineers at the Dutch broadcasting organization Nederlandse Omroep Stichting (NOS) created a format called BASICODE that made it possible to transmit software over the radio that would be readable by a wide range of normally incompatible computer platforms. As Ralf Homann describes it, "A 0-bit was represented by a single period of a 1200 Hz tone and a 1-bit was represented by two periods of a 2400 Hz tone. The whole program was broadcast as one large block with five seconds 2400 Hz leaders and trailers." The first BASICODE broadcast on NOS took place in 1982 and was soon followed by broadcasts in the UK, West Germany, and East Germany.

Other broadcasts of computer software took place for some years in Finland, Yugoslavia, Italy, and other countries. One noteworthy experiment emerged from the Bristol, UK, radio station Radio West. In 1983, producer Joe Tozer managed to broadcast on the show *Datarama* a forty-by-eighty-pixel black-and-white image of actor Cheryl Ladd that had appeared in the *London Evening Standard* in 1975. The image was readable on BBC Micro and ZX81 computers. By the late 1980s, the practice of software broadcasting came to an end as 8-bit computing was replaced by 16-bit computing, which involved more storage than a cassette tape could handle. However, the practice has been recently revived. A program produced and presented by Dr. Kim Andrew Elliott called *Shortwave Radiogram* currently transmits digital text and images on an analog shortwave broadcast transmitter.

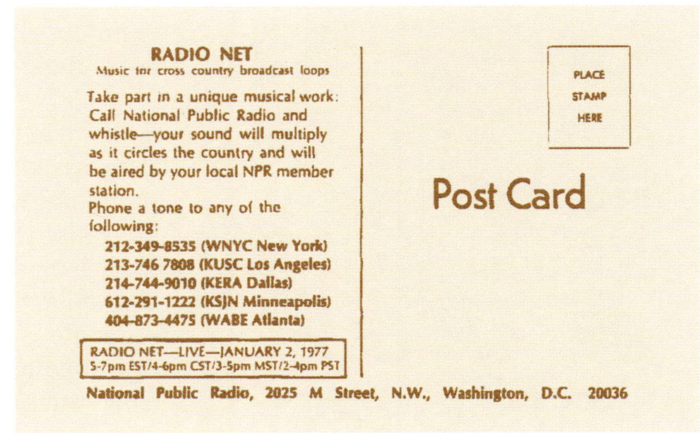
RADIO NET
Music for cross country broadcast loops
Take part in a unique musical work: Call National Public Radio and whistle—your sound will multiply as it circles the country and will be aired by your local NPR member station.
Phone a tone to any of the following:
212-349-8535 (WNYC New York)
213-746 7808 (KUSC Los Angeles)
214-744-9010 (KERA Dallas)
612-291-1222 (KSJN Minneapolis)
404-873-4475 (WABE Atlanta)
RADIO NET—LIVE—JANUARY 2, 1977
5-7pm EST/4-6pm CST/3-5pm MST/2-4pm PST
PLACE STAMP HERE
Post Card
National Public Radio, 2025 M Street, N.W., Washington, D.C. 20036

A postcard for Max Neuhaus's *Radio Net* (1977) inviting individuals to call a National Public Radio station and whistle a tone.

SOURCES Susan J. Douglas, *Inventing American Broadcasting, 1899–1922* (Johns Hopkins University Press, 1989); Louise Benjamin, "In Search of the Sarnoff 'Radio Music Box' Memo: Nally's Reply," *Journal of Radio Studies* 9:1 (2002); Reginald A. Fessenden, "Recent Progress in Wireless Telephony," *Scientific American* 96:3 (January 19, 1907); Doron Galili, *Seeing by Electricity: The Emergence of Television, 1878–1939* (Duke University Press, 2020); James O'Neal, "Fessenden: World's First Broadcaster?" *Radioworld* (October 25, 2006); Sirius XM Holdings Inc., "2019 Form 10-k Annual Report" (2019); John Dunning, *Tune in Yesterday: The Ultimate Encyclopedia of Old-Time Radio, 1925–1976* (Prentice-Hall, 1976); Heidi Grundmann, "Past and Present of Radio Art," in *Re-Inventing Radio: Aspects of Radio as Art*, Heidi Grundmann, Elisabeth Zimmerman, et al., eds. (Revolver, 2008); Galen Joseph-Hunter, *Transmission Arts: Artists & Airwaves* (PAJ Publications, 2011); F.T. Marinetti, "The Futurist Manifesto," in *Futurism: An Anthology*, Lawrence Rainey et al., eds. (Yale University Press, 2009); F.T. Marinetti and Pino Masnata, "The Radio," in *Futurism: An Anthology*; Fortunato Depero, *Liriche radiofoniche* (Morreale, 1934); Giovanna Ginex, "Not Just Campari! Depero and Advertising," *Italian Modern Art* 1 (January 2019); Friedrich Kittler, "The Last Radio Broadcast," in *Re-Inventing Radio: Aspects of Radio as Art*, Heidi Grundmann, Elisabeth Zimmerman, et al., eds. (Revolver, 2008); John Cage, *Silence* (Wesleyan University Press, 1967); Max Neuhaus, "The Broadcast Works: Radio Net," Kunstradio.at; Kaushik Patowary, "People Once Downloaded Games from the Internet," Amusing Planet website (April 6, 2019); Jörgen Skågeby, "The Media Archaeology of File Sharing: Broadcasting Computer Code to Swedish Homes,"

Popular Communication 13:1 (2015); Ralf Homann, "Digital Folklore," Wavefarm website (2008)

Pirate Radio [18]

COUNTRY OF ORIGIN USA
CREATOR(S) Unknown
EARLIEST KNOWN USE 1907

BASIC INFRASTRUCTURE/MATERIALSW Transmitter (including power supply, oscillator, amplifier, and modulator), receiver, antennas, telegraph key (optional), teleprinters (optional), computer (optional), microphone (optional), earphones (optional)

RELATED Amateur radio [16], radiotelegraphy [16.1], radio broadcast [17], pirate television [25], microbroadcast [27]

DESCRIPTION Pirate radio—also known as bootleg radio, clandestine radio, and free radio—refers to the unlicensed use of radio waves to send and/or receive communications. In this way, pirate radio has existed as long as radio licensing has existed; licenses for radiotelegraphy [16.1] were first required in the UK once the Wireless Telegraphy Act became law in 1904, whereas, in the wake of the sinking of the *Titanic* and the need for reliable and interference-free communication, licenses were first required in the US with the Radio Act of 1912. While there are records of individuals deliberately interfering with wireless radio transmissions even before 1904, the first (English language) record of an illegal radio transmission is from a 1907 issue of *Electrical World*: "A youth living near by [Washington], the son of a policeman, has set up a station of his own, and takes delight in interpolating messages during official [US Navy] exchanges." The first English language record of the term "pirate" to describe one who makes such unauthorized radio transmissions appeared in a 1913 article titled "Life as a Wireless Telegraphist." After this point, however, the past and present of pirate radio is extremely heterogeneous; practices, interventions, and philosophies vary widely on a country-by-country basis. Certainly one can see the political and philosophical reasons for radio piracy as ranging from, in the words of Ron Sakolsky and Stephen Dunifer, "plundering and hijacking the airwaves from their rightful state and corporate owners" to "state-free rebels using culture jamming tactics to challenge the power of the media monopoly and the authority granted by government's normalizing regulations." For example, one of the first instances of "sea-based piracy" in radio was Radio Mecur (founded by Peer Jansen and Ib Fogh), which attempted to undermine the monopoly of the Danish National Broadcasting Corporation by broadcasting pop music from ships off the coast of the Netherlands in 1958. However, many other pirate broadcasts around the world have been more invested in supporting small communities, protecting dying languages, and advocating for more overt political action. For example, the predominantly Inuit community near Pond Inlet in Nunavut used scavenged government equipment in 1964 to set up a two-way radio [20] system for "sending messages between communities, broadcasting news, important publication information, and music." Another example is *The Voice of Peace*, which broadcast music and messages of peace across Palestine and Israel from 1973 to 1993. Finally, journalists and activists in Burkina Faso set up *Radio Resistance* in September 2015 as a way to urge the public to resist a military coup that had just taken place; as a result of its broadcasts, members of the public flooded the streets and installed barricades, eventually leading to the surrender of members of the presidential guard.

EXPERIMENTS In 2005, neuroTransmitter (Angel Negarez and Valerie Tevere) created an installation called *12 Miles Out*. As Galen Joseph-Hunter, Penny Duff, and Maria Papadomanolaki describe it in *Transmission Arts*, *12 Miles Out* "merges analog radio transmission with line-drawing constructed from wire and nails. This work continues the artists' exploration of offshore pirate radio practices prevalent in the 1960s and 1970s, specifically referencing Radio Caroline, which, in defiance of international broadcasting regulations, transmitted off the European cost from 1964 through the late 1960s. Installed in a gallery setting, the drawing-as-antenna represents a blueprint of one in a fleet of Radio Caroline mobile pirate radio ships. The wire drawing is attached to a radio transmitter and serves as an antenna for the transmission of *12 Miles Out*. The transmitted sound collage, which is created from archived broadcasts of Radio Caroline and ambient sound recordings of a voyage taken out to sea, is received by radios within the exhibition space."

SOURCES "Wireless and Lawless," *Electrical World* (May 25, 1907); Sungook Hong, "Syntony and Credibility: John Ambrose Fleming, Guglielmo Marconi, and the Maskelyne Affair," in *Scientific Credibility and Technical Standards* (Springer, 2012); "Life as a Wireless Telegraphist," *The Marconigraph* (February 1913); Ron Sakolsky and Stephen Dunifer, eds., *Seizing the Airwaves: A Free Radio Handbook* (AK Press, 1998); Peter Nkanga, "Resistance Over the Airwaves: Pirate Station's Vital Role During the Bukina Faso Coup," Committee to Protect Journalists (October 6, 2015); Galen Joseph-Hunter, Penny Duff, and Maria Papadomanolaki, *Transmission Arts: Artists & Airwaves* (PAJ Publications, 2011)

Radiofax [19]

COUNTRY OF ORIGIN Great Britain
CREATOR(S) Hans Knudsen
EARLIEST KNOWN USE 1908

BASIC INFRASTRUCTURE/MATERIALS Transmitter (including power supply, oscillator, amplifier, and modulator), receiver, antennas, electric motor, paper (tape or chemically treated or printing paper), scanner, recorder, keyboard (optional), headphones (optional), or computer, software, sound card

RELATED Amateur radio [16], radiotelegraphy [16.1], amateur television [16.3], Hellschreiber [16.4], slow-scan television [23], electrical printing telegraph [33.1], image telegraph [33.2], pantelegraph [33.4], telautograph [36]

DESCRIPTION Radiofax is short for "radio facsimile" and it is sometimes referred to as "radiophotogram," "message facsimile," "document facsimile," "fax," "weather fax," "WEFAX" or "HF fax." Radiofax refers to a cluster of analog and digital technologies and techniques for transmitting monochrome images by scanning them line by line and encoding the scan into a signal that is transmitted via high-frequency (HF) radio waves. Even though radiofax is a predecessor to slow-scan television [23] and even though both use different transmission modes, the differences between radiofax, SSTV and also WEFAX appear less significant now that operators can use all modes with a digital computer; in fact, radiofax and WEFAX are often now referred to as forms of SSTV. Further complicating matters is the fact that throughout the long history of radiofax and its related forms, it has sometimes been developed and used by corporations and other times by amateur radio [16] operators.

The first wireless image transmission likely happened in London in 1908 as an experiment conducted by Dutch inventor Hans Knudsen. Knudsen used radiotelegraphy [16.1] as part of a complex process of first photographing an image of King Edward VII, dividing it up into small pixels, and then transferring it to a collodion plate upon which the shape appeared embossed; he then used a suspended needle to travel across the raised surface of the image and transmit each rise and drop of the needle as electrical pulses; the pulses were then received wirelessly and a suspended stylus was used to re-create the image. Not surprisingly,

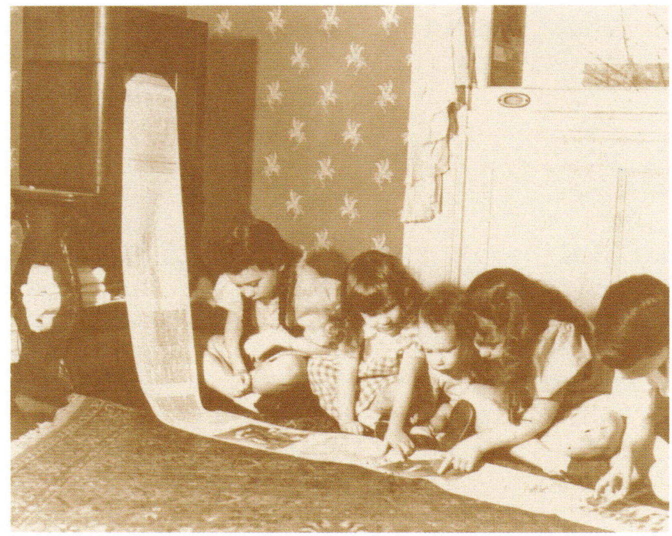

A photograph, likely from 1938, of children reading a radiofaxed Missouri paper transmitted from radio station WOR in New York.

the image quality of this early experiment was quite poor. While the next, more successful wireless transmission of an image took place in 1920 as Danish watchmaker Thorvald Andersen transmitted photographs from Copenhagen to London, Édouard Belin's 1921 adaptation of his Bélinographe to wirelessly transmit a picture of US President Harding from Annapolis, Maryland, to Paris, France, over the span of twenty minutes most closely resembles many radiofax devices invented thereafter. Bélin's wireless transmission process was also similar to that of early mechanical television. It featured an image wrapped around a cylinder that, as the cylinder rotated, was scanned by a pulsating light beam which then reflected back onto a photoelectric cell. As Dirk Bladt writes, "By revolving the cylinder and moving the light beam and photocell from one end of the cylinder to the other, the entire image is scanned in helical form, a line per revolution at a time. After being reflected off the picture, the pulsating beam's intensity is a measure of the brightness of the area under focus. The light beam is pulsating at a constant frequency by turning a perforated disc in front of the light source. This pulse is necessary in order to get a constant frequency to transmit. Otherwise the signal frequency generated by uniformly coloured areas in the image would be too low to transmit and/or pass the amplifiers. The pulse frequency is between 1000 and 1500 Hz." At the receiving end, the image was reproduced by a stylus that moved over a sheet of thermal or electrochemically treated paper that was also wrapped around a revolving cylinder.

The Bélinographe was followed by many other devices with varying techniques such as the photoradiogram

[19] A sheet of thermal paper from Lucy Helton's *WEFAX: Life Began on Earth in Hydrothermal Vents*.

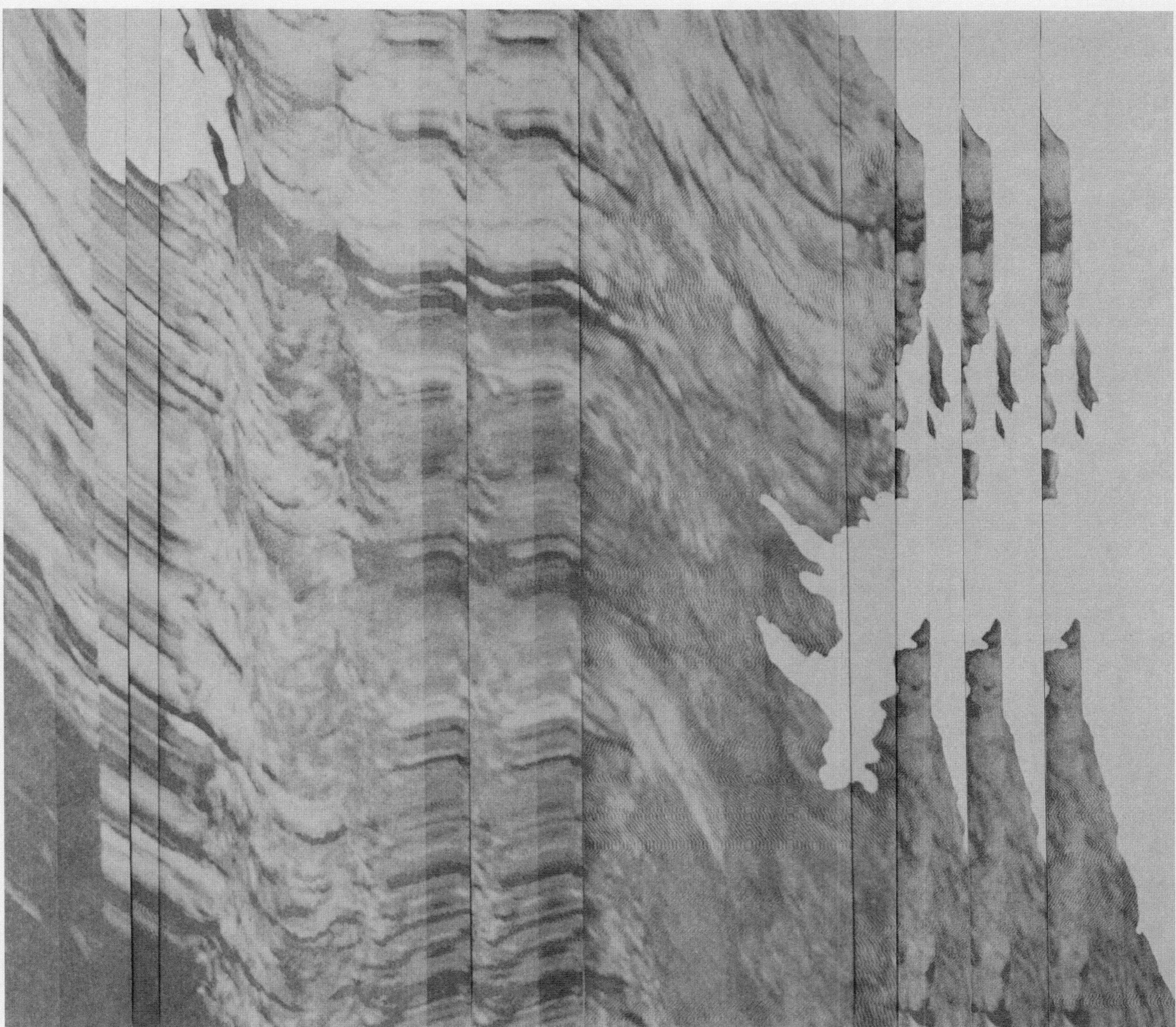

in 1924 by Richard Ranger, the Fultograph by Otho Fuller in 1927, the Cooley-Ray Foto System by Austin G. Cooley in 1928, the Hellschreiber by Rudolf Hell in 1929, and the Finch Facsimile system by William G.H. Finch in 1933. Finch was particularly successful for a time at bringing radiofax into individual homes, as he pioneered a system whereby local AM radio stations transmitted newspapers overnight to subscribers' Finch receivers. The first experimental transmissions took place in 1933 on W10XDF and by 1937 the US 's FCC authorized KSTP in St. Paul, Minnesota, to broadcast radiofaxed newspapers. By the end of 1940, most newspapers had ceased radiofax delivery, as FM radio and television eclipsed the public's interest in facsimile. Furthermore, because of the complexity and cost involved, after this point most systems were then released by corporations such as RCA, AT&T, and Nippon Electric. Notably, from the late 1940s until the mid-1950s, Western Union in the US experimented with a Telecar service. Initially launched in Baltimore, Maryland, the Telecar service was "a modern message center on wheels which receives telegrams on radio and facsimile while cruising in outlying areas . . . virtually a telegram center on the doorstep."

Radiofax is still used by *Kyodo News* to broadcast news in Japanese and English. It is also used for the transmission of maps and weather information to ships at sea using "a dedicated radiofax receiver or a single sideband shortwave receiver connected to an external facsimile recorder or PC equipped with a radiofax interface and application software." The latter is now mostly referred to as weather fax, or WEFAX. The last-known average speed of a radiofax transmission is 120 lines per minute.

EXPERIMENTS In 2020–2021, UK artist Lucy Helton created *WEFAX: Life Began on Earth in Hydrothermal Vents* using weather fax. Helton writes that meditating on the future of the environment led her "to a discovery and fascination with basic image transmission using low energy radio waves," which scientists speculate will also be how humans will communicate once living on other planets is a possibility. For this particular work, Helton used high-resolution images from videos of deep-sea scientific expeditions as the basis for her transmissions. She then "connected computer to radio, used specialized software to convert image to sound, and transmitted across [her] apartment to a nautical Radiofax . . . like most analogue fax machines, the information is printed out onto a roll of thermal paper. This work is made with a new thermal technology, which instead of having a chemical reaction to heat, it has a physical one, causing the top blue layer to collapse into the bottom black layer, making the paper more archival and recyclable. To reference the fixed eye of traversing satellites which use line scanning to build images of the Earth, [Helton] layered the paper using it as a mapping tool, and replaced film in large-format camera with an early scanning technology to make the preliminary high-resolution, non-static images."

SOURCES Jonathan Coopersmith, *Faxed: The Rise and Fall of the Fax Machine* (Johns Hopkins University Press, 2015); "Knudsen's Process of Transmitting Pictures by Wireless Telegraphy," *Scientific American* 98:23 (June 6, 1908); Wilfred S. Ogden, "How the World's First Wireless News-Picture Was Flashed Across the Atlantic Ocean," *Popular Science Monthly* 99:6 (December 1921); Dirk Bladt, "belinographe: BEL-1; Amateur," Radiomuseum website (July 4, 2015); "'Ray-foto' Tests Please Inventor," *The New York Times* (February 1, 1928); John Schneider, "The Newspaper of the Air: Early Experiments with Radio Facsimile," *Monitoring Times Magazine* (February 2011); Transfilm, "Telegram for America," Internet Archive (1956); Fred Shunaman, "Telecar Speeds Telegrams," *Radio-Electronics* (July 1951); G. H. Ridings, "A Facsimile Transceiver for Pickup and Delivery of Telegrams," *Western Union Technical Review* 3:1 (January 1949); "Radiofax Hardware Companies," National Weather Service website, National Oceanic and Atmospheric Administration; Lucy Helton, "Life on Earth Began in Hydrothermal Vents," Artist statement (2021)

Two-Way Radio [20]

COUNTRY OF ORIGIN USA
CREATOR(S) Radio Engineering Laboratories, Inc.
EARLIEST KNOWN USE 1930

BASIC INFRASTRUCTURE/MATERIALS Transmitter (including power supply, oscillator, amplifier, and modulator), receiver, antennas, microphone, speaker, earphones (optional)

RELATED Photophone [12], amateur radio [16], pager [21], cellular network [49], telephone [34], fence phones [41]

DESCRIPTION Two-way radio refers to radio devices called "transceivers" that both send and receive; many two-way radios are limited in that only one user may transmit at a time, known as "half-duplex." The most well-known examples of two-way radio are portable or handheld transceivers such as walkie-talkies, mobile transceivers such as Citizens Band Radio and fire/police radio, and nonportable systems such as intercoms in schools and hotels. Further complicating matters, regulatory bodies such as the ITU and the FCC in the US classify two-way radios by a wide range of personal or amateur and commercial services, ranging from Citizens Band to Multi-Use Radio Service, Family Radio Service, Low Power Radio Service, and General Mobile Radio Service.

It is not clear when the first two-way radio transmission took place. Some sources assert the earliest instance took place in Melbourne, Australia, in 1922 at the instigation of Constable Frederick William Downie, who installed a modified Marconi receiver in a patrol car; however, the system appeared to be set up for one-way transmissions of mostly Morse code. Bell Laboratories also claims they sent the first two-way mobile radio transmission in 1924; however, while the device was a radio mounted in the backseat of a car, it did not receive and transmit voice. Thus, the first voice-based two-way radio transmission most likely took place in 1930 in Long Island City, New York, as part of an initiative by amateur radio [16] enthusiasts who were part of Radio Engineering Laboratories and who created a line of portable transmitters and receivers for other amateurs. The equipment was reportedly used to communicate to and from cars and airplanes, and eventually led to its adoption in police cars in Bayonne, New Jersey, in 1933.

One of the best-known types of two-way radio is the walkie-talkie, also known as a handheld transceiver—a type of portable two-way radio that was also an important precursor to the cell phone. While Polish engineer Henryk Magnuski was the first to file a patent for a handheld transceiver, in 1936, Canadian engineer Donald Hings likely built the first portable two-way radio in 1937 for the mining company CM&S to help workers communicate with each other across remote areas. Originally called a "handie-talkie," this handheld two-way radio system was called the SCR-536 and was built by Galvin Manufacturing, which later became Motorola. Weighing roughly five pounds, it was further developed by the Canadian and US military and was in use by 1942 as a means for more reliable communication between infantry and commanders than the communications provided by the wired field telephone [34]. However, given the SCR-536's limited range and general fragility, Galvin released the SCR-300; even though this thirty-five-pound device was not handheld and instead had to be transported via backpack, the SCR-300 gave birth to the term "walkie-talkie," as the US Army described it in its technical manual as "primarily intended as a walkie-talkie for foot combat troops." Walkie-talkies continue to be popular as toys as well as communication devices for those involved in the military or with

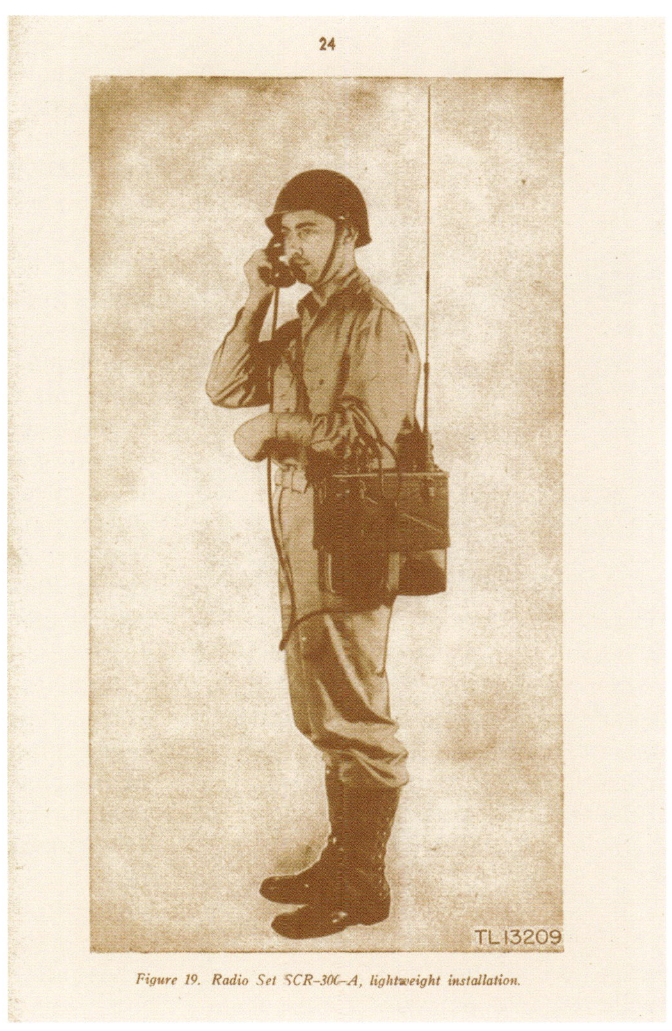

Image of a military-issue Walkie-Talkie, the SCR-300-A, from a 1945 US War Department technical manual.

aviation, marine, or general outdoor pursuits, largely because they are inexpensive and do not depend on cell towers to function.

EXPERIMENTS In the US, a type of two-way radio called the Citizens Band radio service was established by the FCC in 1945 for personal communication. As Cheryl Higashida points out, "typical CB users were stranded drivers calling for help, farmers radioing their barn from the field, housewives reaching husbands on the road, and above all, freewheeling truckers." However, in the mid-1960s, African American civil rights organizers Nettie and Isaiah Sellers built and maintained CB radios for the Student Nonviolent Coordinating Committee and also helped deploy CB radio across Mississippi, Alabama, and Louisiana "for communications, self-defense, and mutual aid in voter registration drives and community organizing." Building on Simone Browne's term "dark sousveillance" to refer to "Black epistemologies of antisurveillance, countersurveillance, and other

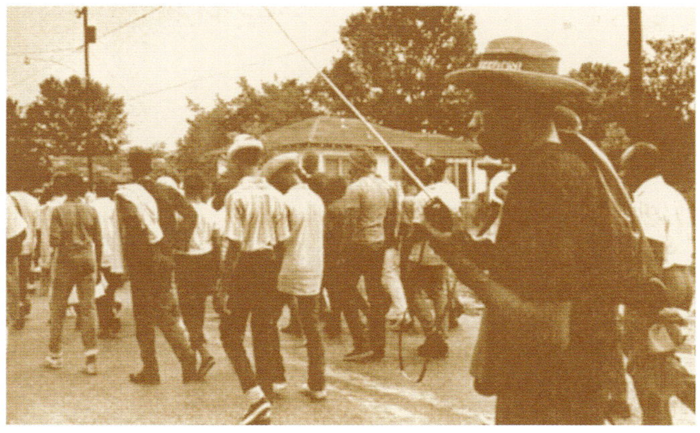

Photograph of members of the African American self-defense group the Deacons for Defense and Justice using CB radios, featured in a September 1965 issue of *Ebony*.

freedom acts," Higashida writes that "These dark sousveillant CB networks were essential to Black self-defense in organized and guerrilla struggles against white terrorism." These CB radio networks also became essential to "Black, Chicanx, Filipinx, and Native people fighting differentially shared conditions of racial capitalist and colonial violence in Los Angeles, Oakland, and Delano, California, and Minneapolis."

SOURCES Robert Haldane, *The People's Force: A History of the Victoria Police* (Melbourne University Press, 1995); "Land-Mobile Radio Milestones," *Proceedings of the Radio Club of America* 61:2 (November 1987); Richard Brewster, "The SCR-536 Handie-Talkie Was the Modern Walkie-Talkie's Finicky Ancestor," *IEEE Spectrum* (September 25, 2020); TM-11-242 War Department Technical Manual, Radio Set SCR-300-A, War Department (February 1945); Harry Mark Petrakis, *The Founder's Touch: The Life of Paul Galvin of Motorola* (McGraw-Hill, 1965); Cheryl Higashida, "Citizens Band: Surveillance, Dark Sousveillance, and Social Movements," *American Quarterly* 74:2 (June 2022)

Pager [21]

COUNTRY OF ORIGIN USA
CREATOR(S) Al Gross
EARLIEST KNOWN USE 1949

BASIC INFRASTRUCTURE/MATERIALS Receiver, transmitter (optional and includes power supply, oscillator, amplifier, and modulator), batteries, antennas, speaker, microphone (optional), earphones (optional), microprocessor (optional), screen (optional), LEDs (optional)

RELATED Amateur radio [16], two-way radio [20], microbroadcast [27], cellular network [49], telephone [34], fence phones [41]

DESCRIPTION A pager—also known as a radiopager, beeper, or bleeper—is a small, portable radio receiver that has been assigned a personal code or phone number that can be dialed or quoted when anyone wants to contact the owner of the pager. Depending on the type of pager, they may be connected to a wired telephone [34] network, an RF link transmitter system, communications satellites [32], or the internet. Before internet access became ubiquitous, the general process for paging was as follows: once the caller contacts the central switchboard or transmitter with the appropriate number or code for the pager, the same number/code is encoded in radio waves and broadcast at the same time as thousands of others to all pagers connected to the transmitter; each pager then automatically filters out any messages that do not include its unique number or code. When a message is transmitted with the correct number or code, the pager buzzes or beeps and may even display an alphanumeric message. For decades, pagers were one-way communication devices without display screens; eventually they became two-way devices that could send, reply to, and acknowledge messages using an internal transmitter. Pagers can also belong to a limited range network (usually consisting of just one lower power transmitter) or a wide range network (usually consisting of hundreds of high power transmitters).

In 1949, while living in Cleveland, Canadian-born engineer Al Gross created the first pocket-sized wireless transmitter, for which he submitted several patents in 1950. The invention, originally intended to be used by doctors, was adopted by New York's Jewish Hospital the same year. By January 1951, the magazine *Radio Electronics* announced Gross's units, then called "Aircall Radiopaging": "The doctor carries a small radio receiver . . . This receiver repeats once every minute a series of numbers, one of which may be his special code call. If, during the minute, the subscriber hears his number, he calls his office from the nearest phone." In 1955, Motorola announced what would become the most dominant model of pager: the handie-talkie radio pocket pager. Bell Systems introduced the first wide area paging system with their Bellboy pager in 1962, and then other countries in Europe and Japan soon introduced their own paging services. That said, Motorola initiated most of the major innovations in paging, including the introduction of a pen-sized pager called the Sensar in 1982 and the introduction of a two-way pager that could be connected to a personal computer in 1995. Cell phones largely replaced pagers in the early 2000s, after which

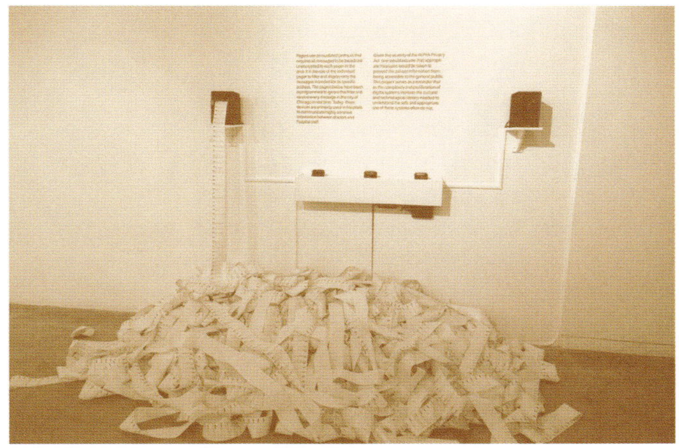

Brannon Dorsey's installation *Holy Pager* in the Sullivan Galleries, Chicago, 2016.

most national paging networks were phased out. However, because pagers can be faster, more reliable (especially in remote areas), and simpler to use than cell phones and their location cannot be tracked, they are still used by fire, ambulance, and medical workers. Most garage door openers are also one-way pagers. More recently, restaurants have also returned to using pagers as a way to contact diners, within usually a two-kilometer radius of the desktop transmitter. They are also popular among radio amateurs: there is, for example, DAPNET in Germany, a decentralized amateur paging network only requiring a low cost, lower power transmitter as an add-on to a Raspberry Pi microcontroller. Finally, paging has also transformed into a one-way broadcasting technique that is built into cellular networks [49]. For example, many national public alerting systems, such as the Wireless Emergency Alerts sent out by the Federal Emergency Management Agency in the USA, are essentially pages sent out over cellular networks on a broadcast control channel to individual cell phones.

EXPERIMENTS In 2016, artist and researcher Brannon Dorsey exhibited the installation *Holy Pager* that featured a system, Holypager, which intercepts all paging messages sent using the protocol POCSAG and forwards the anonymized messages to one of three pagers on display. As Dorsey describes it, "each message is also printed on a contiguous role of receipt paper amassing a large pile of captured pages for gallery goers to peruse." Importantly, the project tries to make visible the fact that "pagers use an outdated protocol that requires all messages to be broadcast unencrypted to each pager in the area. It is the role of the individual pager to filter and diplay only the messages intended for its specific address." However, the pagers used in this installation have been "reprogrammed to ignore this filter and receive every message in the city in real-time."

SOURCES Al Gross, Papers 1909–2000, Ms2001-001, Virginia Tech Special Collections and University Archives; "Doctor Always on Call with Radiopaging Unit," *Radio Electronics* (January 1951); "Al Gross," Lemelson-MIT Program website; Jeremy A. Greene, *The Doctor Who Wasn't There: Technology, History, and the Limits of Telehealth* (University of Chicago Press, 2022); R.J. Holbeche, ed., *Land Mobile Radio Systems* (Peter Peregrinus, 1985); "Technical Realization of Cell Broadcast Service", Specification #23.041, 3rd Generation Partnership Project (3GPP) website; Brannon Dorsey, *Holy Pager* (2016)

Meteor Burst Communication [22]

COUNTRY OF ORIGIN Canada
CREATOR(S) P.A. Forsyth
EARLIEST KNOWN USE 1954

BASIC INFRASTRUCTURE/MATERIALS Transmitter (including power supply, oscillator, amplifier, and modulator), receiver, antennas, microphone, speaker, earphones (optional), computer (optional)

RELATED Amateur radio [16], earth-moon-earth communication [16.5], project west ford [24]

DESCRIPTION Meteor burst communications (MBC), originally known as ionospheric scatter and also later known as meteor scatter communications, is a technique used by the military, scientists, and amateurs to send radio communications from one point to another using the trails of ions left in the wake of a meteor's entry into the upper atmosphere; as a meteor burns up, it leaves a trail of ions in its wake that can be used to reflect radio waves. A transmission may only pass from the transmitter to the receiver when the meteor trail is visible to both antennas. It is worth noting that meteor bursts happen sporadically throughout the year, usually peaking during the summer months. The distance over which a radio communication can travel is also highly variable (depending on the altitude of the ionization, the location of the meteor, etc.), with the maximum distance of a single hop being a little over 2,000 kilometers.

The possibility of communication via meteor burst was first postulated in the 1930s, particularly by T. L. Eckersley in 1932. However, it wasn't until after World War II that MBC was successfully demonstrated. In 1954, while working on the Canadian Defence Research Board's Radio Physics Laboratory, P.A.

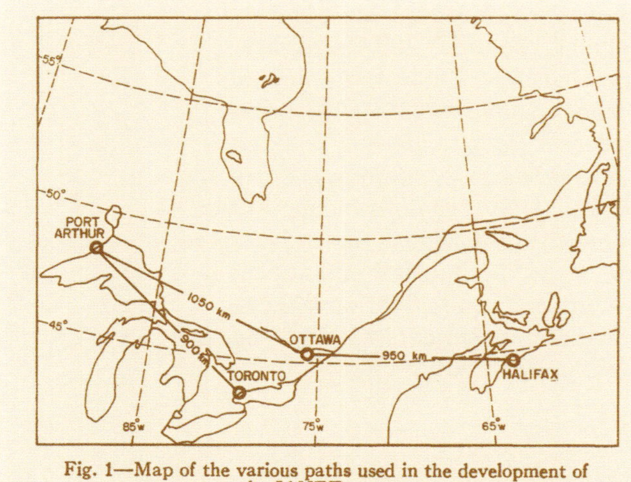

A map of the MBC stations linking Halifax, Ottawa, Port Arthur, and Toronto that were part of the JANET system developed by the Canadian Defence Research Board's Radio Physics Laboratory in 1954.

Forsyth and others successfully demonstrated that it was possible to communicate via VHF signals reflected from meteor trails. The project was named JANET. Another major deployment of MBC took place in Europe in 1965 under the name COMET (communication by meteor trails)—a system operated by NATO that aimed to provide communication between the Netherlands, France, Italy, West Germany, the United Kingdom, and Norway. The system reportedly only transmitted teletype messages and was quickly phased out with the arrival of satellite technology in the late 1960s. However, once it became clear that satellites were both expensive and unreliable at high altitudes, interest in MBC was renewed with the creation of, for example, SNOTEL (automated snow telemetry sites set up at remote sites across the USA) in the 1960s, and the Alaska MBC System, which collected data for the National Weather Service.

While MBC may not be ideal given the limited distance of a transmission, the time one must spend finding a usable meteor trail, limitations on sending large amounts of data, and the difficulty of transmitting voice, MBC is popular among amateurs for the transmission of Morse code. MBC is also notably inexpensive, reliable, requires low power, and is not controlled by third parties (as is the case with satellites). No doubt its popularity during the Cold War was also due to the fact that MBC systems should be able to survive a nuclear war. The last-known average speed of an MBC transmission ranges from seventy-five to a hundred words per minute if unencrypted and fifteen words per minute encrypted.

EXPERIMENTS Arguably all MBC communication is experimental as it never became an established method of communication outside of scientific, military, and amateur radio [16] circles. However, there have been numerous experiments with visualizing and sonifying meteor bursts themselves—surely a type of nonverbal communication. Citizen scientist and artist Thomas Ashcraft built an observatory in 1992 that also acts as a laboratory and a studio. In 2021, Ashcraft released *Meteor Fireballs in Light and Sound*—a video that captures paths taken by four fireballs that appeared across the sky in New Mexico, USA, along with the subsequent meteor streaking; the video also includes the sound Ashcraft captured from the radio reflection off the streaking.

SOURCES T. L. Eckersley, "Studies in Radio Transmission," *Journal of IEEE* 71 (September 1932); P.A. Forsyth, E.E. Vogan, D.R. Hansen, and C.O. Hines, "The Principles of JANET: A Meteor-Burst Communication System," *Proceedings of the IRE* (1957); Jay A. Weitzen, "Meteor Scatter Communication: A New Understanding," in Donald D. Schilling, ed., *Meteor Burst Communications: Theory and Practice* (John Wiley & Sons, 1993); "Meteor Burst Communications: An Ignored Phenomenon?," *Cryptologic Quarterly* 9:3 (Fall 1990); Thomas Ashcraft, *Meteor Fireballs in Light and Sound*, NASA Astronomy Picture of the Day (March 15, 2021)

Slow-Scan Television [23]

COUNTRY OF ORIGIN USA
CREATOR(S) Copthorn Macdonald
EARLIEST KNOWN USE 1957

BASIC INFRASTRUCTURE/MATERIALS Stand-alone SSTV transceiver; or receiver, transmitter, antenna, modem, cathode-ray tube, scanner, or analog camera; or digital camera, computer, software, sound card

RELATED Amateur radio [16], radiotelegraphy [16.1], radioteletype [16.2], amateur television [16.3], Hellschreiber [16.4], earth-moon-earth communication [16.5], amateur radio satellite [16.6], amateur packet radio [16.7], pirate television [25], radiofax [19], television broadcast network [36], electrical printing telegraph [33.1], image telegraph [33.2], pantelegraph [33.4], telautograph [36], cable television [48], communications satellite [32], microwave radio-relay [31]

DESCRIPTION In 1957, Copthorn Macdonald began work on a method called slow-scan television (because of the length of time it took to transmit an image) in an

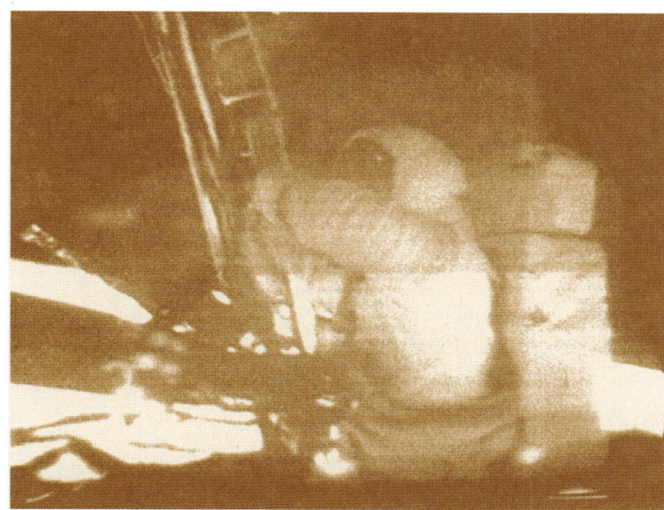

Slow-scan television image of Neil Armstrong on the moon during the Apollo 11 landing on July 21, 1969.

attempt to reduce the bandwidth of a static picture transmission so that it could be sent over a voice communication system, first on the AM band, and then later on the FM band. In 1960, the US's FCC granted Macdonald the first temporary authorization to send and receive images over radio, and in 1968 slow-scan television (or SSTV) was officially authorized. Thereafter, especially because of the affordability of this mode of picture transmission, SSTV gained popularity amongst amateur radio [16] enthusiasts around the world. It also became more well-known because it was used to transmit images of, for example, the far side of the moon during the Apollo 11 mission in July 1969. By 1977, Dave Ingram wrote, "The seventies is the era of sight and SSTV is the window of view." While contemporary SSTV systems are digital and thus involve the use of a computer, a sound card, and specialized software, SSTV was originally analog. It worked by slowly scanning each line of an image from left to right to convert color and brightness into audio tones; if it was a color image, then each of the red, green, and blue color components were sent separately as audio tones. The tones were fed into a transceiver and sent to other amateur radio users, who would then translate them back into an image with the same process. SSTV may also be used in conjunction with the telephone [34], communications satellites [32], microwave radio-relay [31], and cable television [48].

EXPERIMENTS From 1978 through the early 1980s, the US Library of Congress experimented with using SSTV to facilitate interlibrary loan, as well as communications between federal libraries; experiments with distance education using SSTV were also taken up in US public schools in the early 1980s. From the late 1970s through the early 1990s (particularly as stand-alone SSTV systems became available), artists around the world enthusiastically took up SSTV experiments. Many of the SSTV experiments took place between Open Space (Victoria, Canada), Western Front (Vancouver, Canada), School of the Art Institute (Chicago, USA), Digital Art Exchange and Carnegie Mellon University (Pittsburgh, USA), and São Paulo (Brazil).

For example, as early as 1978, Bill Bartlett along with other members of the Direct Media Association and artists associated with the nonprofit Open Space began hosting experimental SSTV demonstrations and workshops in Victoria, Canada. The groups organized point-to-point and multi-point sessions with artists, students, and educators from Victoria (Canada), Vancouver (Canada), Toronto (Canada), New York City (USA), Memphis (USA), and San Francisco (USA). In 1979, participant Peggy Cady described the process as follows: "Using a Robot Research Model 530 PhoneLine Transceiver, on loan from Estron Industries, Open Space Artists and Direct Media Association artists transmitted and received live video time-lapse images over telephone lines. This video telephone set-up takes an image generated by a standard video camera and, with an 8.5 second digital scan converts the frozen image into coordinate audio-FM tones. This 'audio image' is then transmitted over the telephone network to a similar receiver. The receiving video monitor reconstructs the audio information into a visual display. The resulting program is an 8.5 second time-lapse series, a montage of frozen images produced at 8.5 second intervals, creating an effect like animation or pixillation" [*sic*]. Also in 1979, Bartlett gained access to NASA's ATS-1 satellite, which, in conjunction with SSTV, made possible events such as *Pacific Rim*, which took place in the

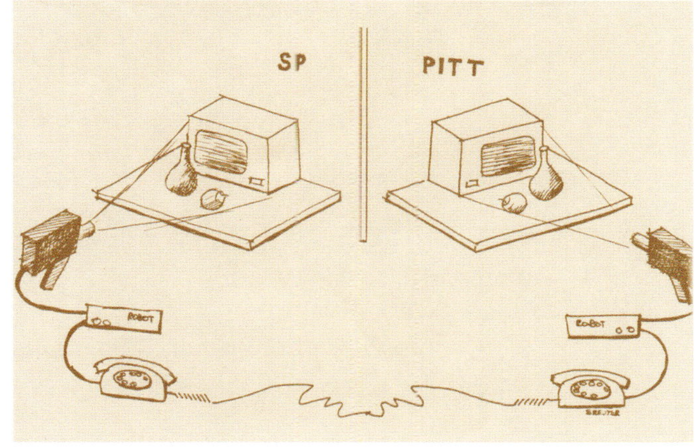

Illustration of SSTV event Still Life Alive, which also included *Intercities São Paulo / Pittsburgh* from 1988, organized by the Digital Art Exchange (headed by artist Bruce Breland).

[23] Diagram by Bill Bartlett from 1978–1979 detailing the arrangement of SSTV equipment (including a scanning monitor, live monitor, Robot 530 transceiver, audio tape recorder, telephone, camera, power outlets, and cables), alongside a photograph of Bartlett and others gathered around an SSTV setup for a public event in Open Space (Victoria, Canada), likely from 1979.

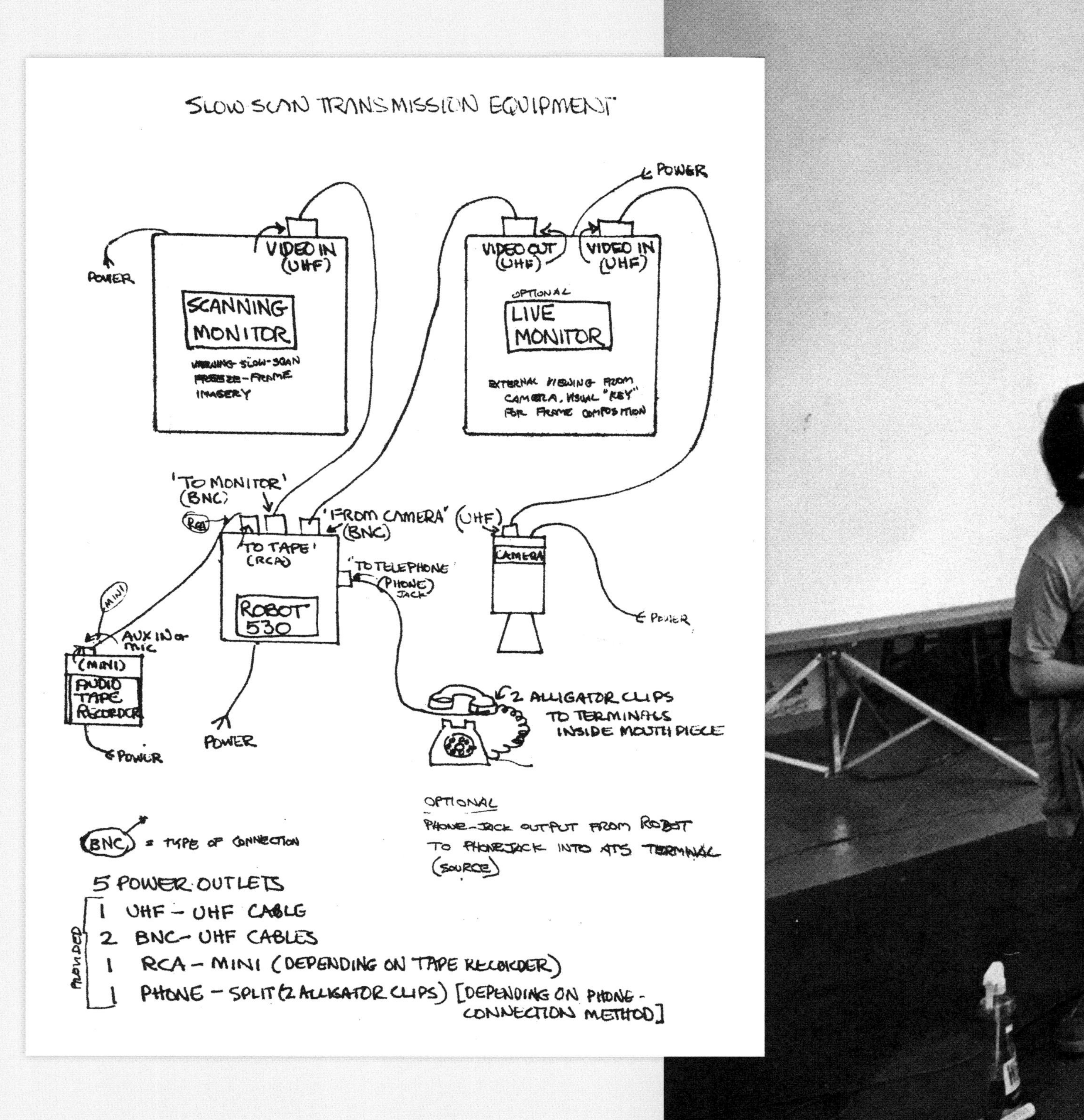

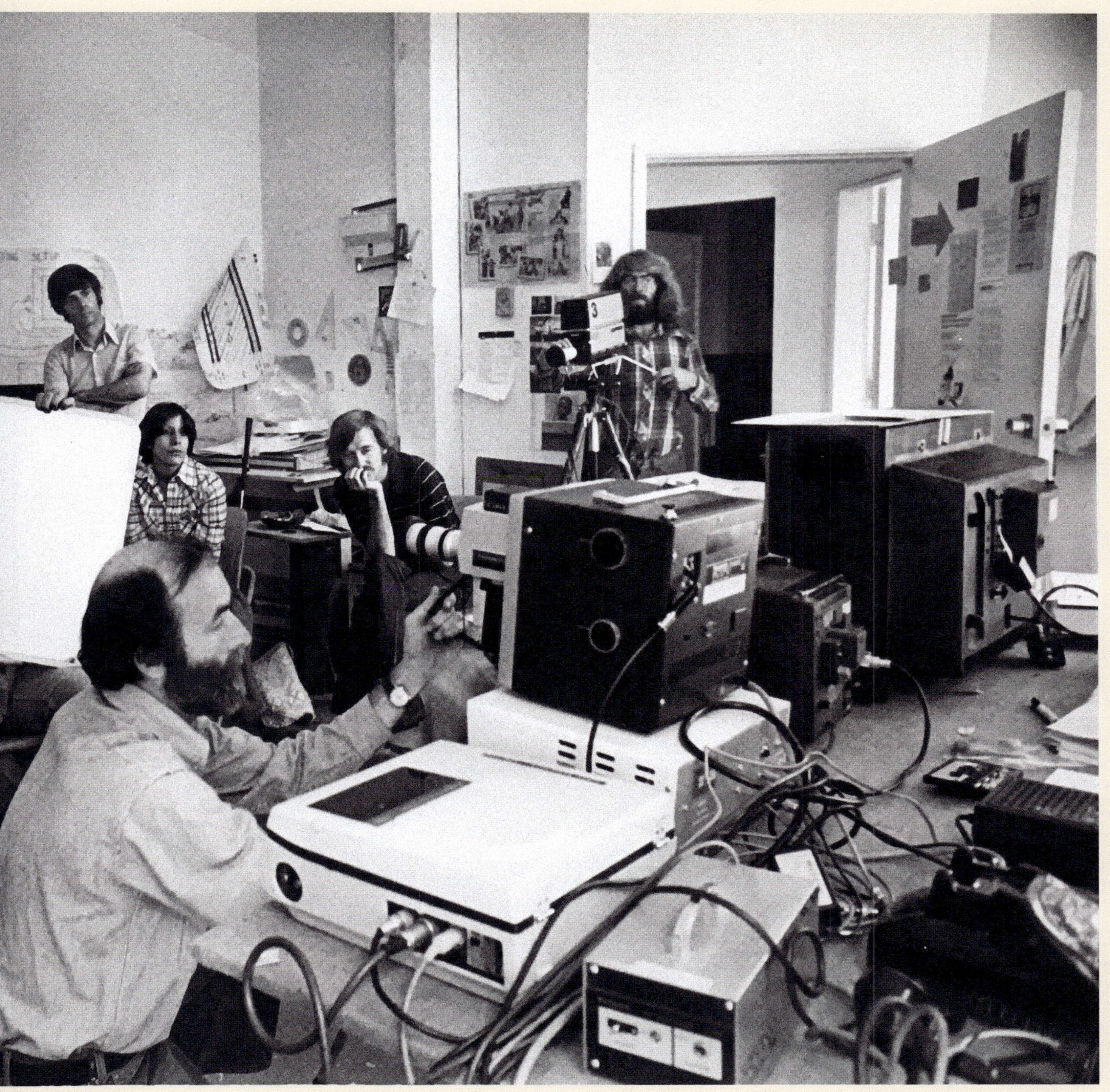

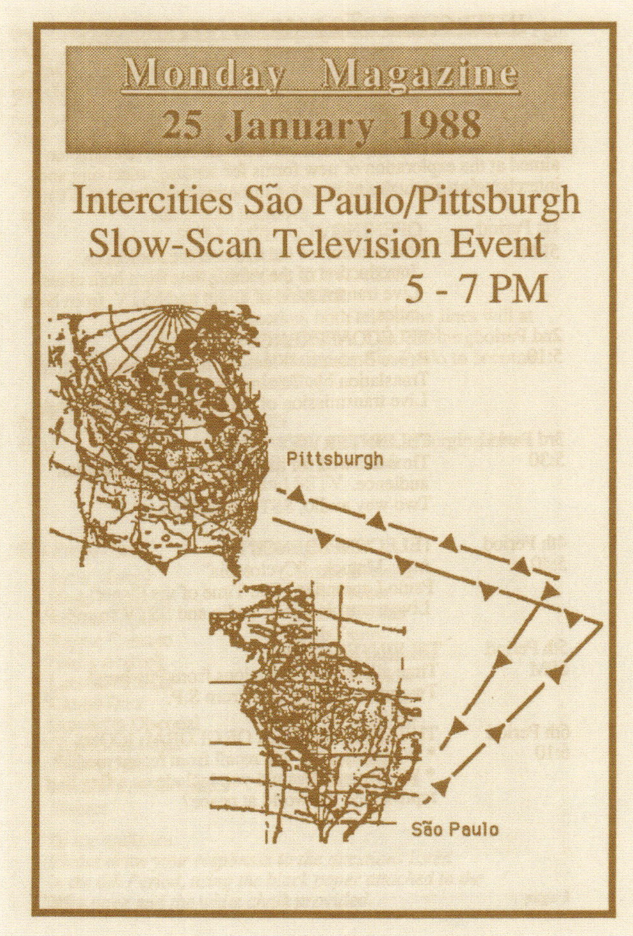

Promotional poster for an SSTV event in 1988, *Intercities São Paulo / Pittsburgh*, organized by the Digital Art Exchange.

Vancouver Art Gallery and involved other groups from the Cook Islands and New Zealand. He also organized the better-known *Artists' Use of Telecommunications* event, which took place in the San Francisco Museum of Modern Art in 1980 and included a three-hour, five-node SSTV exchange between writers, art curators, and artists from the US, Japan, Canada, and Austria.

The aforementioned Digital Art Exchange, or DAX, was formed in 1979 by Bruce Breland at Carnegie Mellon University in Pittsburgh (USA). They too experimented with giving artists access to SSTV along with telefacsimile [37], amateur radio [16], and computer networks such as IPSA, Bitnet, and EARN. As for their SSTV experiments, DAX hosted a two-hour long event in January 1988 called *Intercities São Paulo / Pittsburgh* whereby images were exchanged between these two cities over SSTV and audio was exchanged over telephone. As they wrote in their promotional materials, "the event will have an experimental and self-reflexive character, aiming the exploration of new forms for sending, receiving, and interchanging aesthetic information through telecommunications." One part of the project was called Still Life Alive, in which objects were set in front of television sets which formed the background, while the foreground consisted of the addition or subtraction of different objects.

In 1990, Brazilian-American artist Eduardo Kac organized a live event called *Interfaces* between artists at the School of the Art Institute in Chicago and DAX in Pittsburgh. The event consisted of what Kac called a "visual dialogue" between the two groups. Writes Kac: "Participants in Chicago were not aware of the exact images that would be transmitted by the Pittsburgh group and vice-versa. This unpredictable situation added an element of surprise to the process. As images overlapped on the screen, parts of a face (from Pittsburgh, for example) were slowly scanned over another face (previously sent by the Chicago group). Successive faces were created in the virtual space of the screen as the performance progressed. It took approximately eight seconds to form each image on the screen. Participants in one location transmitted an image as soon as they received an image from the other location. While a face was scanned, eyes of a woman, for example, overlapped with nose and mouth of a man. Parts of one face composed dynamically an 'in-between face' with parts of another face."

SOURCES Don C. Miller and Ralph Taggart, *Slow-Scan Television Handbook* (73 Inc., 1972); Dave Ingram, *The Complete Handbook of Slow-Scan TV* (Tab, 1977); Ward Silver, *Ham Radio for Dummies*, 4th Edition (John Wiley & Sons, 2021); Ramon L. Glidden, "Getting Started with Slow-Scan Television," *QST* (September 1990); Saul Herner, "Evaluation of the Use of Slow Scan Television and Telefacsimile," Report for the Federal Library Committee (Herner & Co., December 1980); Kathleen Kelleher, "Tele-Education: Teaching Over the Telephone with Slow Scan Video," *The Business Communication Magazine* 2:8 (1983); Bill Bartlett, "Telecommunication Field-Trial Projects," Open Space archives (undated); Peggy Cady, "Talk-Back," Open Space archives (1979); Carl Eugene Loeffler and Roy Ascott, "Chronology and Working Survey of Select Telecommunications Activity," *Leonardo* 24:2 (1991); *Intercities São Paulo / Pittsburgh*, DAX archives, Carnegie Mellon University (1988); Eduardo Kac, *Interfaces* website

Project West Ford [24]

COUNTRY OF ORIGIN USA
CREATOR(S) Walter E. Morrow and MIT Lincoln Laboratory
EARLIEST KNOWN USE 1963

BASIC INFRASTRUCTURE/MATERIALS Transmitter (including energy and/or power suppliers, oscillator, amplifier, and modulator), receiver, antennas, 480 million copper needles, naphthalene gel, communications satellite

RELATED Radio broadcast [17], meteor burst communication [22], communications satellite [32]

DESCRIPTION Project West Ford, originally named Project Needles, was an intentionally short-lived Cold War–era experiment first imagined by electrical engineer Walter Morrow and then undertaken by MIT's Lincoln Laboratory in the early 1960s. In the wake of concerns about natural phenomena or, more pressingly at the time, nuclear war disrupting high-frequency radio communications, Morrow and members of the Lincoln Laboratory proposed the creation of an "artificial ionosphere" that would consist of a donut-shaped belt of 480 million copper fibers or needles, each of which was .7 inches, or 0.0018 centimeters, long, orbiting the earth at an altitude of about 2,237 miles, or 3,600 kilometers. The first experiment was launched by piggybacking on another rocket launch, but the fibers failed to deploy. The second launch, which took place on May 8, 1963, was successful and took roughly forty days for the fibers to form a belt around the Earth. Reportedly, the belt enabled radio communication between California and Massachusetts at data rates of up to 20,000 bits per second. Gradually, however, the fibers fell out of orbit and by 1966 the belt was no longer functional. Although the copper needles "were embedded in a naphthalene gel designed to evaporate quickly once it reached the vacuum of space," this design "allowed metal-on-metal contact, which, in a vacuum, can weld fragments into larger clumps." The result is that Project West Ford's needle clusters still orbit the earth as space debris.

SOURCES William W. Ward and Franklin W. Floyd, "Thirty Years of Space Communications Research and Development at Lincoln Laboratory," *Beyond the Ionosphere: Fifty Years of Satellite Communication* (NASA, 1997); C. F. J. Overhage and W. H. Radford, "The Lincoln Laboratory West Ford Program: An Historical Perspective," *Project West Ford Issue, Proceedings of the IEEE* 52:5 (May 1964); Martin P. Brown, "Project West Ford," *Compendium of Communication and Broadcast Satellites, 1958 to 1980* (Institute of Electric and Electronic Engineers, 1981); Joe Hansen, "The Forgotten Cold War Plan That Put a Ring of Copper Around the Earth," *Wired* (August 13, 2013)

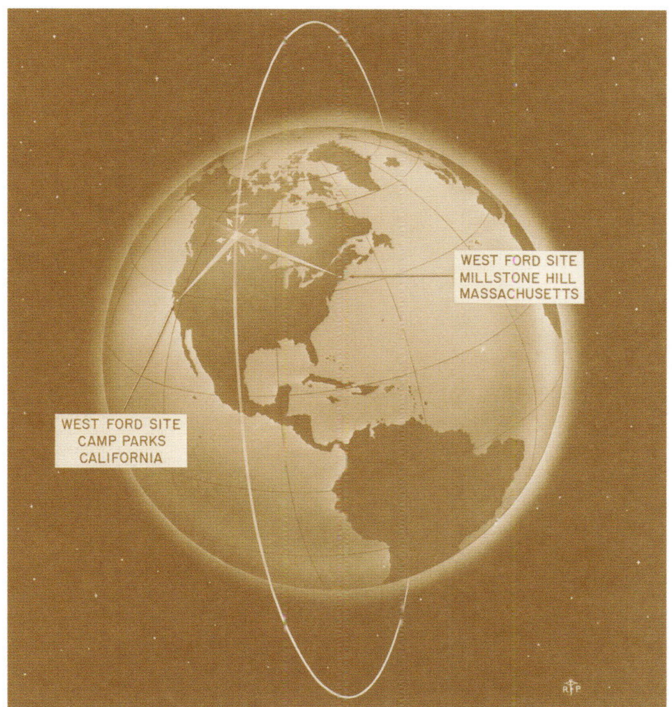

An illustration of the Project West Ford belt of copper needles orbiting earth.

Pirate Television [25]

COUNTRY OF ORIGIN USA
CREATOR(S) Skip Blumberg, Nancy Cain, David Cort, Bart Friedman, Davidson Gigliotti, Chuck Kennedy, Mary Curtis Ratcliff, Parry Teasdale, Carol Vontobel, Ann Woodward
EARLIEST KNOWN USE 1971

BASIC INFRASTRUCTURE/MATERIALS Transmitter (including power supply, oscillator, amplifier, and modulator), receiver, antennas and/or transmitting tower, cables, computer (optional), microphone (optional), earphones (optional), audio processor (optional), tuner (optional), multiplexer (optional)

RELATED Amateur television [16.3], slow-scan television [23], pirate radio [18]

DESCRIPTION Pirate television, also known as bootleg TV and guerilla TV, is similar to pirate radio [18] insofar as it also refers to the unlicensed use of

[25] Videofreex in 1977 with their multichannel video playback system.

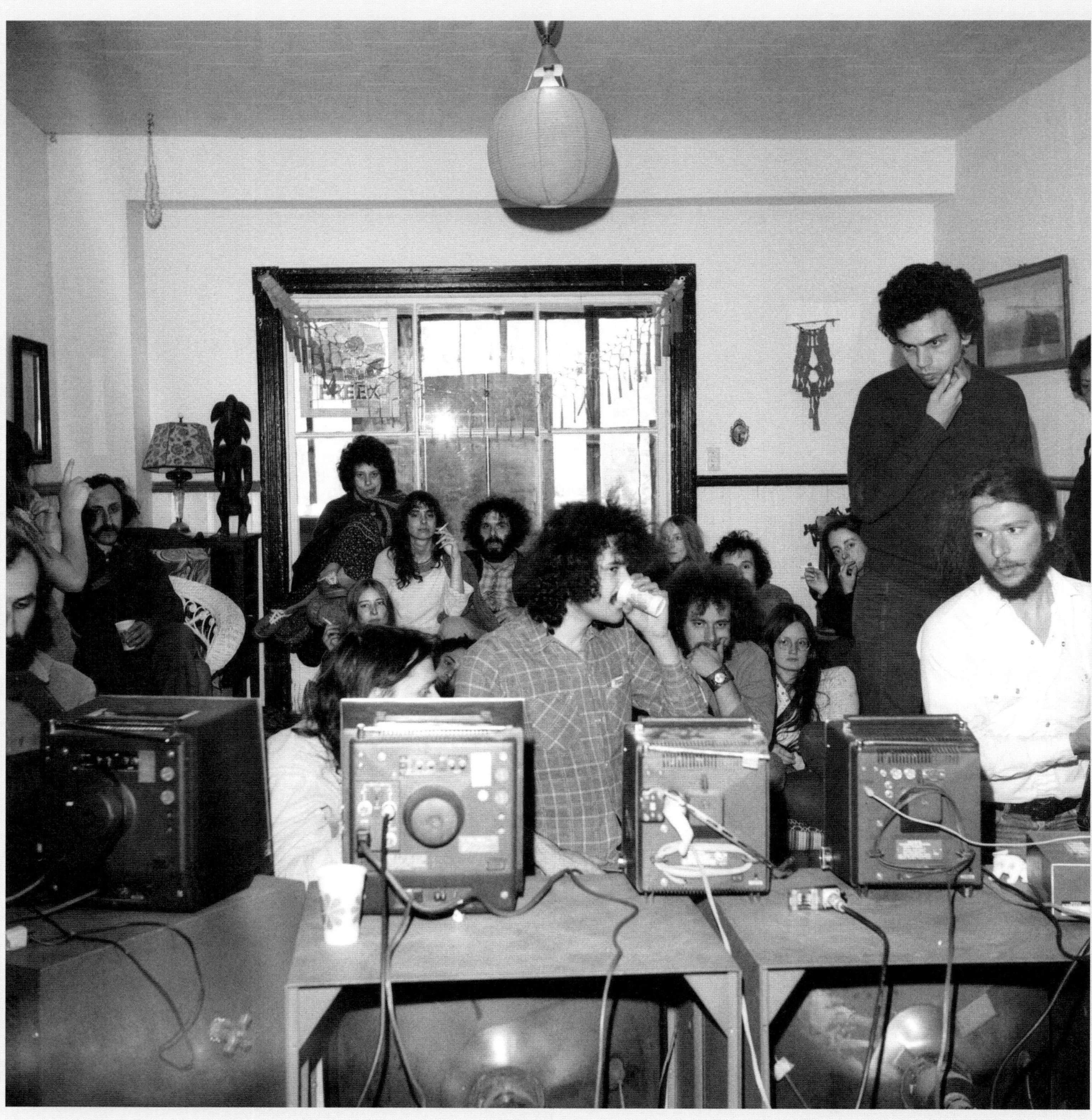

radio waves to send and/or receive communications; the only difference between the two is that pirate TV involves the transmission of video material—an important difference given the way television eclipsed radio during the 1950s as one of the most important twentieth-century media. The main author of *Steal This Book* (1971), Abbie Hoffman, wrote that "Guerrilla TV is the vanguard of the communications revolution, rather than the avant-garde cellophane light shows and the weekend conferences. One pirate picture on the sets in Amerika's living rooms is worth a thousand wasted words." Throughout the 1970s and 1980s (before satellite feeds and digital television made most pirate TV technology obsolete), two techniques were commonly used for analog pirate TV broadcasts: individuals would use a weak relay transmitter to detect when a more powerful transmitter ceases broadcasting, usually for the night, and they would then relay the same signal to broadcast pirate content; or, individuals would simply build powerful VHF/UHF transmitters and use the same process as one would for a legal, local television broadcast. Even with the difficulties imposed by digital television on pirate televising, pirate TV stations continue to operate around the world, including the sophisticated network beoutQ in Saudi Arabia, which involves satellites and set-top boxes and illegally broadcasts content from Qatar's BeIN network. Poor relations between the Qatari and Saudi regimes, as well as disputes over beoutQ's possible Saudi state sponsorship, show that the politics of pirate broadcasting continue to be felt in complex ways.

EXPERIMENTS The first pirate television station in the US was Lanesville TV which lasted from 1971 to 1977. This station was created by members of the video collective Videofreex after several years of successful experimentation with the newly released Sony Portapak—a portable, battery-powered videotape analog recording system that was seen as, in Whitney Kimball's words, "a tool for complete social upheaval." In exchange for Videofreex founding member Parry Teasdale writing a section of *Steal This Book* on pirate TV broadcasting, Abbie Hoffman paid for parts that the collective could use to build a television transmitter, which Videofreex soon installed along with an antenna and an amplifier at Maple Tree Farm in Lanesville, New York. The group covered local events and broadcast mockumentaries, call-in shows, and variety performances. Lanesville TV is now seen as the forerunner to public access TV. Telestreet was another important pirate TV experiment, based in Italy and lasting from 2002 until 2010. Telestreet stations transmitted local content as well as sporting events to neighborhoods at ultrahigh frequencies. As Alessandra Renzi describes it, "Each of Telestreet's do-it-yourself (DIY) transmission systems was connected to an online peer-to-peer (P2P) network, sharing a web archive from which they could download autonomously produced broadcasting material, considerably reducing production and distribution costs." Early on, striking workers at a Fiat manufacturing plant in Termini Imerese used a Telestreet node to "bear witness to their labor struggle."

SOURCES Abbie Hoffman, *Steal This Book* (Pirate Editions, 1971); Alex Ritman, "Could This Be the World's Biggest State-Sponsored Piracy Operation?," *The Hollywood Reporter* (June 20, 2019); Whitney Kimball, "Their Own Private Lanesville: The Videofreex on a Decade of Pirate TV in the Catskills," ARTFCITY website (October 8, 2014); Liz Flynt, "Media Ecology and Cultural Climate Change: A Weekend with the Videofreex," *Afterimage* 47:1 (March 2020); Alessandra Renzi, *Hacked Transmissions: Technology and Connective Activism in Italy* (University of Minnesota Press, 2020)

Packet Radio Network [26]

COUNTRY OF ORIGIN USA
CREATOR(S) Norman Abramson and Franklin Kuo
EARLIEST KNOWN USE 1971

BASIC INFRASTRUCTURE/MATERIALS Transceivers, antennas, computers, terminal connection unit (TCU), repeaters, communications satellite (optional)

RELATED Amateur packet radio [16.7], Wi-Fi [29], communications satellite [32], postal system [44], cellular network [49]

DESCRIPTION Packet radio networks, occasionally referred to as "radionets" in the late 1970s, use packet switching techniques for large-scale wireless computer networking. American Paul Baran first proposed "distributed adaptive message block switching" while he was at the RAND Corporation in the early 1960s and soon thereafter the concept of switching was further developed (and named "packet switching") by British engineer Donald Davies at the National Physical Laboratory (UK) in 1965. Both individuals sought to develop a network topology that, unlike the centralized topology of a telephone network [34] (whose dependence on a central switchboard makes it vulnerable to failure in the event of a catastrophic event), would continue to function even in the wake of, say, a nuclear blast. As Baran put it, "The distributed network is a network without any hierarchical structure; thus, there is no single point of vulnerability to bring down much of the network." It was also becoming evident that a more efficient use of bandwidth than that offered by telephone networks had to be developed to accommodate growing demand partly driven by the rapid development of more inexpensive digital electronics. With packet switching, many users can share the same transmission line that previously could only be used by one user at a time. Packet radio switching involves the transmission of data that is broken up into "packets" and distributed wirelessly across multiple paths in the network; similar to a bundle of letters that includes information about each sender and the receiver and that ought to be delivered in the most efficient manner, each packet contains an address indicating its destination, often an address indicating the source of the transmission, and a message that is usually broken up across multiple packets and re-assembled at its destination. In order to facilitate efficient delivery, the packets may also be stored and then forwarded to the recipient.

The first packet-switched radio network, ALOHAnet, was created by a team of researchers led by Norman Abramson and Franklin Kuo at the University of Hawaii between 1968 and 1976. They provided the first public demonstration of a packet radio network in June 1971. ALOHAnet used two different ultra-high frequencies for transmissions between terminal control units stationed around campus (later the units were located across the wider Honolulu area and eventually across other Hawaiian islands) and the base station, an HP 2115A computer dubbed Menehune, located at the University Computing Center. Its method of packet switching was called the "ALOHA multiple access protocol"; this method was eventually utilized as an essential element of "internetting" or communicating with other networks. In an effort to build on ALOHAnet's innovations, the Defense Advanced Research Projects Agency (DARPA) created another packet radio network in 1973 called PRNET with nodes around the San Francisco Bay Area as a way to try to connect to ARPANET and SATNET (a satellite packet radio network) using mobile vehicles equipped to be packet radio nodes. The PRNET's mobile vehicle was known as the Packet Radio Van, and in 1977 it participated in the first three-way transmission between the van, SATNET, and the ARPANET; as the transmission connected three different networks, it is often considered the beginning of what's now known as the internet. ALOHA protocol lives on, as it was incorporated into Ethernet cable networks, Wi-Fi [29], and mobile phone networks. Packet radio also continues via the General Packet Radio Service, which is a data transmission standard for 2G and 3G cellular communications [49]. The technique of packet switching itself was also

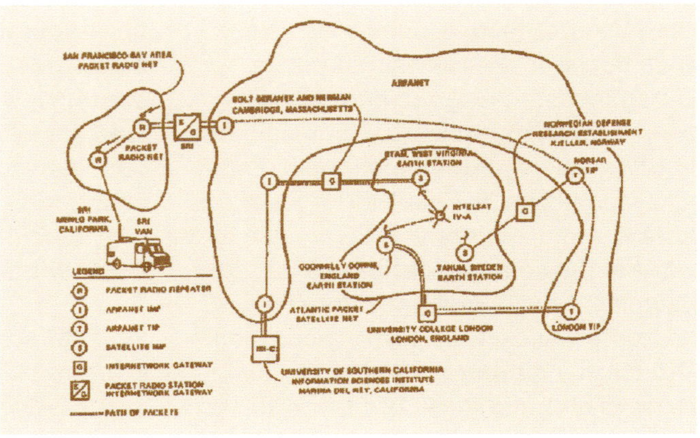

First demonstration of an internet that linked ARPANET and SATNET with PRNET via its packet radio van, from November 22, 1977.

incorporated into the design of ARPANET, widely considered the precursor to the contemporary internet; it too is now the basis of most digital transmissions.

SOURCES Janet Abbate, *Inventing the Internet* (MIT Press, 2000); Paul Baran, "The Beginnings of Packet Switching: Some Underlying Concepts," *IEEE Communications Magazine* (July 2002); Paul Baran, "On Distributed Communications: I. Introduction to Distributed Communications Networks" (RAND Corporation, 1964); Norman Abramson, "The ALOHA System: Another Alternative for Computer Communications," Fall Joint Computer Conference (1970); Tyler Morgenstern, *Colonial Recursion and Decolonial Maneuver in the Cybernetic Diaspora* (dissertation, University of California Santa Barbara, 2021); Franklin F. Kuo, "Computer Networks: The ALOHA System," Report of the Office of Naval Research (1981); John Schoch and Larry Stewart, "Internetwork Experiments with the Bay Area Packet Radio Network," *Internet Experiment Note* 78 (February 1979); Don Nielson, "The SRI Van and Computer Internetworking," *CORE* 3:1 (February 2002)

Microbroadcast [27]

COUNTRY OF ORIGIN Italy
CREATOR(S) Franco "Bifo" Berardi, Maurizio Torrealta, Filippo Scòzzari, Paolo Ricci, Carlo Rovelli
EARLIEST KNOWN USE 1976

BASIC INFRASTRUCTURE/MATERIALS Transistor, registers, capacitors, coil, trimmer capacitor, copper-clad circuit board, audio cable, 9-volt battery

RELATED Amateur radio [16], radio broadcast [17], pirate radio [18]

DESCRIPTION "Microbroadcasting" is often used interchangeably with "micro-radio," "miniFM," and sometimes even "free radio." These are a collection of practices involving the use of a low power transmitter (either as an aesthetic or political choice or out of necessity) over a limited distance, thereby reaching a limited number of people. It is also considered a type of "community media" because of its local and noncommercial nature. Given the low power (as much as one hundred watts but often as low as one watt) and short distances involved (as much as five kilometers from the transmitter but often much shorter), microbroadcasting can be both unlicensed and legal. However, regulations determining the legal status and power of microbroadcasting vary significantly over time and from country to country; if national regulatory bodies prohibit individuals from transmitting to their local community, microbroadcasting may turn into pirate radio [18]. The fact that regulatory bodies determine whether and how microbroadcasting is legal means it is not a practice determined strictly by the constraints of one's transmitter. Rather, as Tetsuo Kogawa writes in "A Micro Radio Manifesto," "micro means diverse, multiple, and polymorphous. If micro does not mean small in physical size, then even physically bigger radio station [*sic*] could become micro. Micro radio is an alternative to mass medium and global communications that could cover the globe with the qualitatively same and patterned information." Microbroadcasting is therefore primarily a social practice aimed at providing diverse points of view while also resisting the commodification of these points of view. Its origins can thus be traced to specific forms of media activism rather than, for example, early wireless radio experiments with low power over short distances.

Based in Bologna, Italy, and lasting from 1976 to 1981, the unlicensed radio station Radio Alice was likely the first instance of microbroadcasting as defined above. Even though it was referred to as "free radio" and predated the emergence of the term "microbroadcasting," its main founders (students/activists Franco "Bifo" Berardi, Maurizio Torrealta, Filippo Scòzzari, Paolo Ricci, and Carlo Rovelli) essentially envisioned Radio Alice as a conscious micro-radio experiment that sought to distribute control of the airwaves across many small transmitters as a way to flatten hierarchies between sender and receiver, embrace localism, and use art to unsettle if not unseat capitalism. As the editors of the Toronto-based magazine *The Red Menace* described it in 1978, "Radio Alice broadcast news of the events as they occurred, often by airing telephone calls from militants who described events, called for assistance in a given sector, and reported police movements. The station was twice raided and

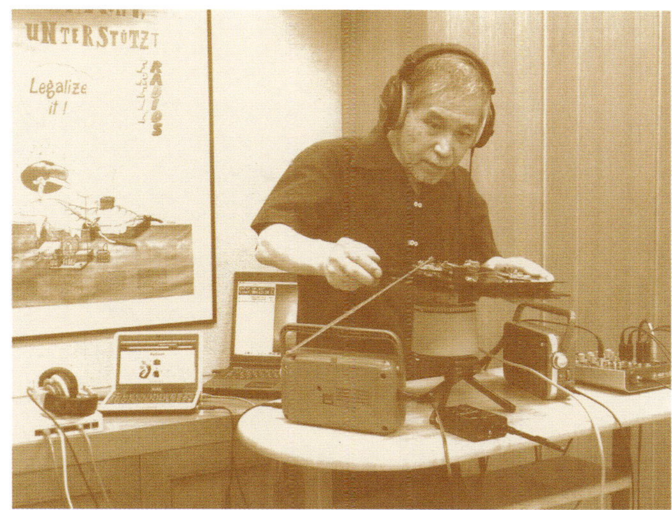

Microbroadcasting pioneer Tetsuo Kogawa running a workshop on MiniFM.

closed down by police, but resumed broadcasting by switching locations and resorting to a transmitter powered by a car battery."

A few years later, in the early 1980s, Tetsuo Kogawa introduced free radio to Japan, calling it "miniFM" as he led the way to hand-building tiny FM transmitters that used less than a hundred milliwatts and only had a half-mile radius. The term "micro-radio" or "micro-broadcasting" then emerged in the US in 1983 in the wake of the police beating of African American Dewayne Readus in a public housing development in Springfield, Illinois. Readus, a blind man who changed his name to Mbanna Kantako, first created the Tenants Rights Association (TRA) and, to make sure the TRA could reach as many residents of the development as possible, he created radio station WTRA using a one-watt transmitter and broadcast from his living room. In 1988, WTRA became Zoom Black Magic Liberation Radio, then Black Liberation Radio, followed by Human Rights Radio.

EXPERIMENTS Microbroadcasting is practically experimental by definition. In addition to any of the examples described above, "Talking Homes" was another microbroadcasting experiment that lasted throughout 2006 and 2007. Led by Jon Brumit, "Talking Homes" repurposed "Talking House" FM radio transmitters commonly used by real estate agents to give potential home buyers audio tours of properties. With transmitter sites in San Francisco and Oakland (USA), one part of the project was called Bringing Down the Neighborhood and was, in Brumit's words, "designed as a 'reality-based radio game' in which listeners listened in and voted for their favorite 'WORST NEIGHBOR' story or alternately tried to guess the exact location of the transmitters. There were 'worst

Grand piano and software-defined radio
for Floris Vanhoof's *Antenna* (2022).

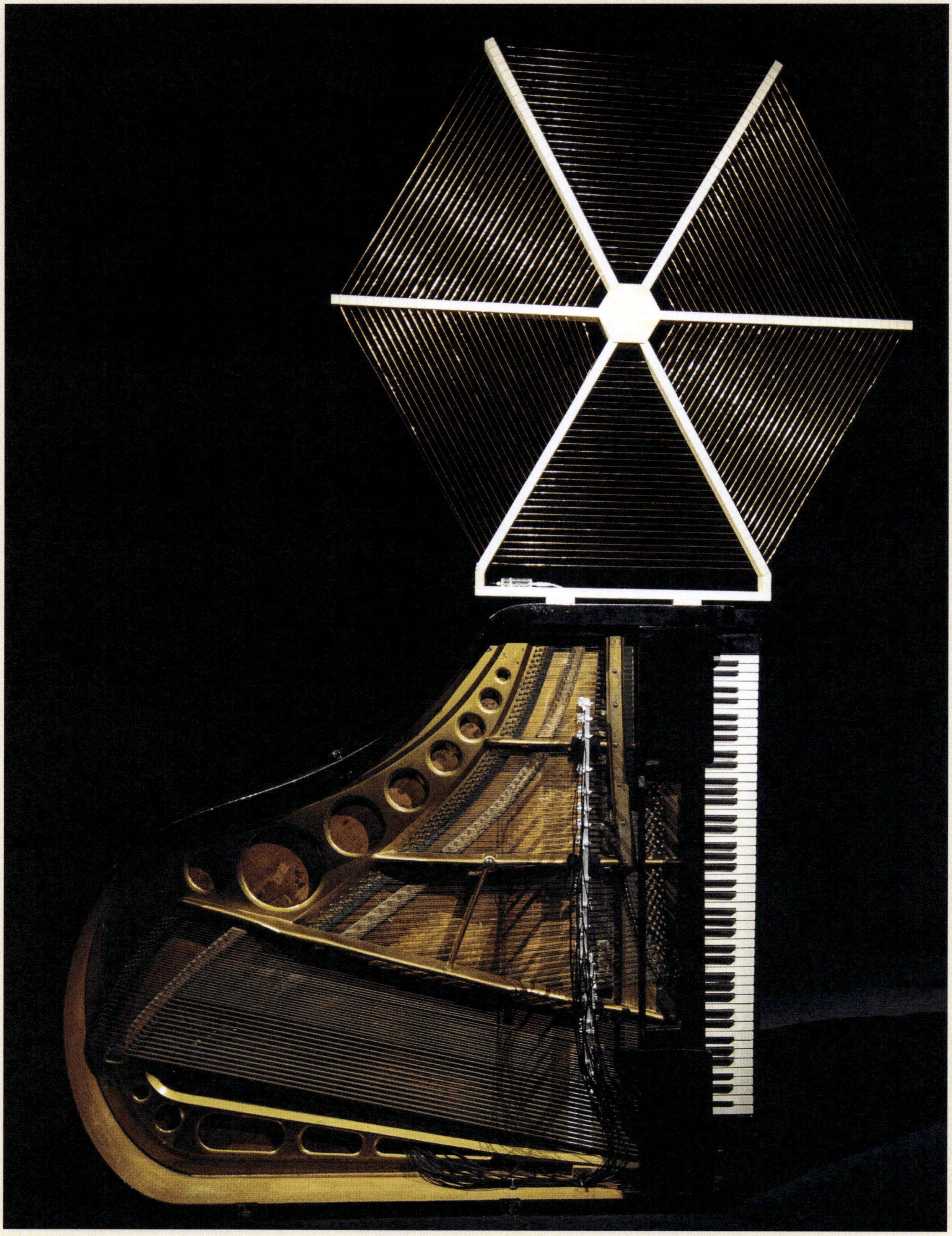

neighbor' stories . . . broadcast continually from 14 different transmitters in select SF and East Bay sites." Another iteration of the project was called Lake Merritt Loops and used the same radio transmitters to present "a kind-of 'drive through audio collage' around Oakland's Lake Merritt in which quiet field recordings were looped and broadcast continually for two months such that visitors and listeners could hear—via their car or portable radios—the quiet sounds of early morning wildlife and foliage throughout each day regardless of the usual noise pollution."

A few years later, in 2014, Ebru Kurbak and Irene Posch created *The Knitted Radio* as part of a residency at Eyebeam Art + Technology Center (New York City). The project emerged in the wake of the 2013 protests in Gezi Park in Istanbul, Turkey. As the creators put it, "Clouds of gas that [surrounded] people [inspired] the idea of creating invisible clouds made of radio waves for enabling communication among people in physical proximity." *The Knitted Radio* is a sweater that is also an FM transmitter, the design for which is drawn from Tetsuo Kogawa's instructional "How to Build the Simplest FM Transmitter." The sweater makes it possible for the wearer to transmit any audio information within a six-meter radius. As the creators put it, "The sweater holds all electronic components necessary for an FM radio transmitter circuit, except a transistor, a battery, and an audio source. Those additional elements are designed to be attached to the sweater easily. Once those parts are detached from the sweater, it ceases to be a transmitter and becomes electronically undetectable."

SOURCES David J. Hess and Robert Gottlieb, *Localist Movements in a Global Economy: Sustainability, Justice, and Urban Development in the United States* (MIT Press, 2009); Tetsuo Kogawa, "A Micro Radio Manifesto," Polymorphous Space website (2002, 2006); Marco Briziarelli, "Tripping Down the (Media) Rabbit Hole: Radio Alice and the Insurgent Socialization of Airwaves," *Journal of Radio & Audio Media* 23:2 (October 2016); "Radio Alice: Radio in Action in Italy," *The Red Menace* 2:2 (Spring 1978); Tetsuo Kogawa, "Toward Polymorphous Radio," Daina Augaitis and Dan Lander, eds., *Radio Rethink: Art Sound and Transmission* (Walter Phillips Gallery, 1994); Lawrence Soley, *Free Radio: Electronic Civil Disobedience* (Routledge, 2018); Steven O. Shields and Robert Ogles, "Black Liberation Radio: A Case Study of Free Radio Micro-broadcasting," *Howard Journal of Communications* 5 (1995); Andy Opel, *Micro Radio and the FCC: Media Activism and the Struggle over Broadcast Policy* (Praeger, 2004); Christina Dunbar-Hester, "Spectral Utopias: Community Radio in the United States, 1970 to Present," *Historia Actual Online* 54:1 (2021); Christina Dunbar-Hester, *Low Power to the People: Pirates, Protest and Politics in FM Radio Activism* (MIT Press, 2014); Jon Brumit, "Talking Homes" website; Ebru Kurbak, *The Knitted Radio*, website

Software Defined Radio [28]

COUNTRY OF ORIGIN Germany
CREATOR(S) Peter Hoeher and Helmuth Lang
EARLIEST KNOWN USE 1988

BASIC INFRASTRUCTURE/MATERIALS Software-defined radio transceiver, computer, sound card or analog-to-digital converter, software, antenna

RELATED Amateur radio [16], radiotelegraphy [16.1], radioteletype [16.2], amateur television [16.3], Hellschreiber [16.4], earth-moon-earth communication [16.5], slow-scan television [23], amateur radio satellite [16.6], amateur packet radio [16.7], radio broadcast [17], pirate radio [18], two-way radio [20], packet radio network [26], microbroadcast [27]

DESCRIPTION Software-defined radio (SDR) is a form of wireless radio communication whereby analog components of radio communication are replaced with computer software. While SDR is expensive and complex to develop, it is relatively inexpensive to purchase and the fact that it can be used for almost any kind of wireless radio communication without the need for individual parts makes it especially appealing to amateur radio [16] enthusiasts. Depending on the unit, SDR may also (but doesn't necessarily) require an internet connection. While Ulrich L. Rohde developed an SDR prototype in 1982 and then a team at E-Systems Inc. in 1984 developed an SDR receiver, it seems likely that the first SDR transceiver was created by Peter Hoeher and Helmuth Lang in Germany in 1988. The term "software-defined radio" was coined in 1995 by Stephen Blust, who was working at Bell South Wireless. While SDR has been used widely by the military, industry, and amateurs, it is increasingly used for wildlife tracking, radio astronomy, and medical imaging.

EXPERIMENTS As the cost of SDR lowers and it becomes easier to implement, artists are also using SDR more frequently. For example, in 2018, Nicholas A. Knouf's *they transmitted continuously / but our times rarely aligned / and their signals dissipated in the æther* (exhibited at Wichita State University) collected the sounds of weather satellites, CubeSats, and amateur radio repeaters using SDR and played them back within the gallery space. Writes Knouf, "This 20-channel sound installation represents the results of collecting hundreds of transmissions from satellites orbiting the earth. Using custom antennas that I built from scratch, I tracked the orbits and frequencies of satellites using specialized software. This software then allows me to collect the radio frequency signals and translate them into sound." Another example is Floris Vanhoof's installation in Leuven, Belgium, from 2022 called *Antenna*. The installation consists of a grand piano on its side with a hexagonal antenna affixed to it; the piano produces sound from radio waves picked up by an SDR and translated into vibrations, which in turn produce sounds from the piano's strings.

SOURCES John Bard and Vincent J. Kovarik Jr., *Software Defined Radio: The Software Communications Architecture* (John Wiley & Sons, 2007); Peter Hoeher and Helmuth Lang, "Coded-8PSK Modem for Fixed and Mobile Satellite Services Based on DSP," *Proceedings of the First International Workshop on Digital Signal Processing Techniques Applied to Space Communications* (November 1988); Joseph Mitola, Preston Marshall, Kwang-Cheng Chen, Markus Mueck, and Zoran Zvonar, "Software Defined Radio — 20 Years Later: Part 1," *IEEE Communications Magazine* (September 2015); Kurt VonEhr, Seth Hilaski, Bruce Dunne, and Jeffrey Ward, "Software Defined Radio for Direction-finding in UAV Wildlife Tracking," *IEEE International Conference on Electro Information Technology* (2016); R. Keller, B. Staufenbiel and N. Aderhold, "Software Defined Radio in Radio Astronomy; Limits and Application in RFI Detection," URSI Atlantic Radio Science Conference (2015); K.S. Bialkowski, J. Marimuthu, and A. M. Abbosh, "Biomedical Imaging System Using Software Defined Radio," IEEE International Symposium on Antennas and Propagation & USNC/URSI National Radio Science Meeting (2015); Brett Balogh, *Entropy Pool*, website (2016); Nicholas A. Knouf, *they transmitted continuously / but our times rarely aligned / and their signals dissipated in the æther*, Zeitkunst.org (2018); Floris Vanhoof, *Antenna* website (2022)

Wi-Fi [29]

COUNTRY OF ORIGIN N/A
CREATOR(S) Vic Hayes and the Institute of Electrical and Electronics Engineers
EARLIEST KNOWN USE 1997

BASIC INFRASTRUCTURE/MATERIALS Transmitter (including power supply, oscillator, amplifier, and modulator), receiver, antennas, router (optional), Wi-Fi® certified device(s), digital device (optional)

RELATED Amateur packet radio [16.7], packet radio network [26], Bluetooth [30], sneakernet [45], cellular network [49]

DESCRIPTION Wi-Fi (which is not an abbreviation but was dubbed as such by the marketing firm Interbrand) is the term used by the Institute of Electrical and Electronics Engineers to refer to devices that adhere to the standard known as IEEE 802.11. This is a standard for implementing wireless local area networks on unlicensed common channels 2.4 or 5.8 gigahertz with a transmission range of up to around one hundred meters; connected devices also only need to use about 250 milliwatts or less. Even though Wi-Fi uses modulated and demodulated radio waves as a means of sending and receiving packets of data like other wireless radio networks (such as Packet Radio Networks [26]), Wi-Fi is unique insofar as it is both a standard and a trademark. The first version of Wi-Fi was released by the IEEE in 1997, led by committee founder and Chair Vic Hayes, on behalf of what was then known as the Wireless Ethernet Compatibility Alliance, which later became the Wi-Fi Alliance. The original name of the alliance points to the fact that Wi-Fi was originally intended to be a method for wirelessly extending Ethernet.

IEEE 802.11 defines three main modes for establishing a Wi-Fi network: infrastructure mode, ad hoc mode, and mesh mode. Today, the most common mode is infrastructure mode, which consists of a central access point (AP) or a piece of hardware (usually a router) that Wi-Fi devices connect to; the AP then connects the devices to a wired or wireless network. Infrastructure mode is usually used to access the internet. Ad hoc mode is more properly an "other network," insofar as it is used to connect devices directly to each other, wirelessly, in a peer-to-peer mode. Ad hoc mode does not require a central access point to connect to a fixed or permanent network; however, not all devices can be connected to each other in this mode. By contrast, mesh mode, often referred to as "mesh Wi-Fi," allows every device or node to act as a router and retransmit packets for any device.

EXPERIMENTS Netless is a series of projects by Danja Vasiliev starting in 2012 that involve the creation of ad hoc Wi-Fi networks attached to trams, buses, taxis and pedestrians. Inspired by Sneakernets [45], Netless uses small, low-power wireless digital transponders as nodes; every time a node comes near another, a wireless link is created using Wi-Fi and data stored on each is swapped. As Vasiliev puts it, Netless

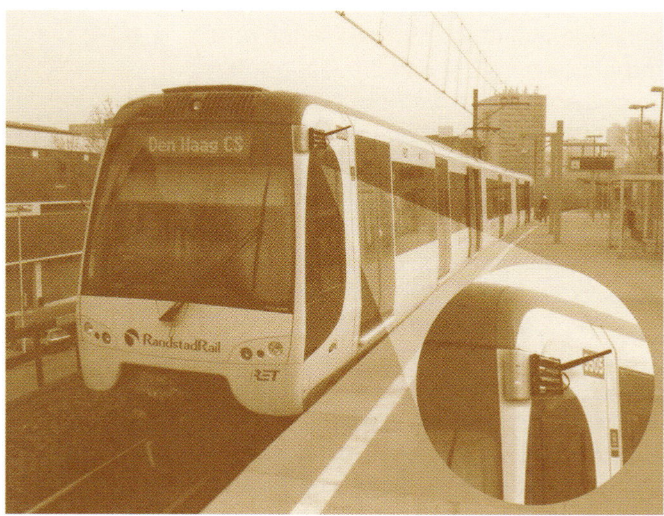

Image of Netless affixed to a tram in The Hague, Netherlands.

is "an exploit of the existing IEEE 802.11bg (2.4GHz WiFi) stack . . . By repurposing 802.11 Beacon Frame specification (beacon frames are long range broadcast radio /frames/ or in Ethernet lingo broadcast network /packets/) Netless could send out radio signals that any WiFi capable device was capable of receiving and interpreting." Netless is, then, "an automated version of SneakerNet, a WiFi-enabled SneakerNet." It thus exists outside of government and corporate controlled/monitored telecommunications infrastructure. Also created in 2012, Dan Phiffer's Occupy.here emerged at the same time as the Occupy Wall Street protests, and likewise used Wi-Fi's ad hoc mode. The network consisted of a wireless router near Zuccotti Park in New York City that anyone with a smartphone or laptop within range could access through a portal website, which opened up onto what the creator described as a bulletin-board system style message board, on which users could share messages and files without fear of surveillance.

SOURCES Cory Doctorow, "WiFi Isn't Short for Wireless Fidelity," *Boing Boing* (November 8, 2005); Julian Thomas, Rowan Wilken, Ellie Rennie, *Wi-Fi* (Polity, 2021); Houda Labiod, Hossam Afifi, and Constantino De Santis, *Wi-Fi, Bluetooth, ZigBee, and WiMax* (Springer 2007); Wolter Lemstra and Vic Hayes, "Unlicensed Innovation: The Case of Wi-Fi," *Competition and Regulation in Network Industries* 9:2 (June 2008); Nathan J. Muller, *Wireless A to Z* (McGraw-Hill, 2002); Danja Vasiliev, Netless website; Dan Phiffer, "Occupy.here: A Tiny, Self-Contained Darknet," *Rhizome* blog (October 1, 2013)

Bluetooth [30]

COUNTRY OF ORIGIN N/A
CREATOR(S) Ericsson, Intel, Nokia, Toshiba, IBM
EARLIEST KNOWN USE 1997

BASIC INFRASTRUCTURE/MATERIALS Bluetooth® transceiver microchip, digital device

RELATED Amateur packet radio [16.7], packet radio network [26], Wi-Fi [29], sneakernet [45], cellular network [49]

DESCRIPTION Bluetooth, for a brief period also known as RadioWire, is a standard managed by the Bluetooth Special Interest Group for a type of wireless personal area network (PAN). Bluetooth networks are implemented using UHF radio waves, ranging from 2.402 gigahertz to 2.48 gigahertz with a transmission range from 0.5 meters up to 240 meters and power consumption from 0.5 miliwatts to 100 miliwatts. Similar to Wi-Fi [29], Bluetooth is unique among wireless radio networks insofar as it is both a standard and a registered trademark. In 1994, two employees of the Swedish telecommunications company Ericsson, Nils Rydbeck and Jaap Haartsen, began developing a short-range radio link that would connect cell phones with electronic devices and allow them to share voice and data. By 1997 the team at Ericsson had developed what became known as Bluetooth, named after Danish King Harald Bluetooth, who was, in the words of Jim Kardach from Intel, "famous for uniting Scandinavia just as we intended to unite the PC and cellular industries with a short-range wireless link." In 1998, Ericsson brought in Intel, Nokia, Toshiba, and IBM as project partners to create the Bluetooth Special Interest Group that would oversee and manage the industry-wide standard.

There are two types of Bluetooth: Bluetooth Classic Radio (or Bluetooth Basic Rate/Enhanced Data Rate) which is a type of low power radio that transmits across eighty channels in the 2.4-gigahertz frequency band, predominantly for direct communication between one digital device and another; and Bluetooth Low Energy radio, which is a type of very low power radio that transmits across forty channels also in the 2.4-gigahertz frequency band, for anything ranging from direct communication between one digital device and another, the creation of mesh networks for multiple devices, as well as device positioning indoors. Communication between Bluetooth devices involves the creation of ad hoc networks known as "piconets" whereby one device (usually the device with the least computation power) becomes the primary device and

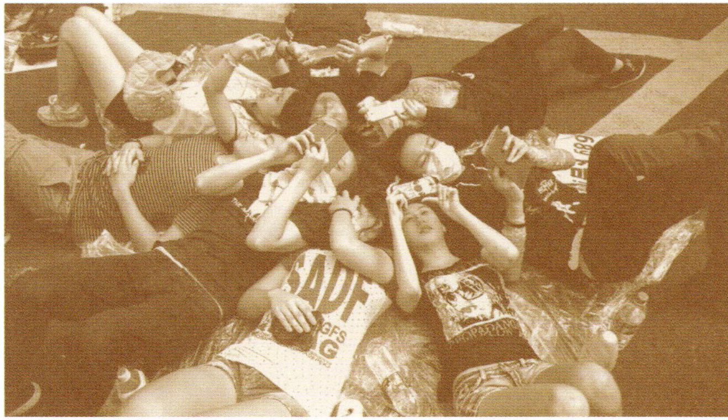

Hong Kong protestors in 2014, many of whom used the Bluetooth-enabled app Firechat.

the others become secondary devices; as Khaled Salah Mohamed writes, "Piconets are established dynamically and automatically as Bluetooth devices enter and leave radio proximity." Since Bluetooth uses channels in the unlicensed frequency band, to avoid interference it uses a transmission technique that is a type of frequency-hopping spread spectrum called "adaptive frequency hopping" whereby packets of data are transmitted by hopping between channels, while avoiding channels that are already busy.

EXPERIMENTS Mexico-based artist, programmer, and researcher Eugenio Tisselli began work on a program called MIDIPoet in 1999. As he writes, "MIDIPoet is a software tool that allows real-time manipulation of text and images on the computer screen. It is made up of two programs: composer and performer with which it is possible, respectively, to compose and interpret pieces of text and/or manipulable images. These pieces may or may not respond to external impulses, such as MIDI messages or the computer keyboard, and generate visual manifestations that involve the manipulation of the different attributes of the text (content, font, position, size, etc.), the image (content, position, etc.) and other elements and visual effects." In 2008, he further experimented with MIDIPoet by using a smartphone as a MIDI controller that connected his phone to his computer using Bluetooth. In short, Tisselli's phone would generate a Bluetooth signal when a key was pressed on the phone keypad; these signals were then received on his computer as ASCII code, converted into MIDI messages, and then sent to MIDIPoet.

Another important experiment with Bluetooth was the release of the mobile app Firechat in 2014 by the company Open Garden; Firechat made it possible to create ad hoc mesh networks using only Bluetooth, Wi-Fi, or Apple's Multipeer. Given its ability to function outside the internet, it became a crucial communications tool in protests around the world.

When it was first released, Open Garden's vice president of sales and marketing, Christophe Daligault, declared that "in a year or two from now, I think people won't even remember that you had to be on Wi-Fi or get a cell signal to be able to communicate." The app was discontinued in 2018.

SOURCES Thomas G. Robertazzi, *Introduction to Computer Networking* (Springer, 2017); Khaled Salah Mohamed, *Bluetooth 5.0 Modem Design for IoT Devices* (Springer, 2022); "The Story Behind How Bluetooth® Technology Got Its Name," Bluetooth website; Dave Hollander, "2 Ways Bluetooth Technology Makes Wireless Connections Reliable," *Bluetooth Blog* (October 27, 2020); Jason Marcel, "How Bluetooth Technology Uses Adaptive Frequency Hopping to Overcome Packet Interference," *Bluetooth Blog* (November 15, 2020); Jaap Haartsen, "How We Made Bluetooth," *Nature Electronics* 1 (December 13, 2018); Jeanette Irekvist, "Bluetooth: Born in Our Backyard, Raised by the World," *The Ericsson Blog* (September 26, 2022); Eugenio Tisselli, "MIDIPoet," motorhueso website; Eugenio Tisselli, "How to Control MIDIPoet with a Mobile Phone," motorhueso website; "FireChat 'Off-the-Grid' Messaging App: What You Need to Know," *CBC News* (October 3, 2014); Peter Shadbolt, "FireChat in Hong Kong: How an App Tapped Its Way into the Protests," *CNN* (October 16, 2014); Mark Milian, "Russians Are Organizing Against Putin Using FireChat Messaging App," *Bloomberg* (December 30, 2014); Elise Hu, "How Hong Kong Protesters Are Connecting, Without Cell or Wi-Fi Networks," *National Public Radio*, All Tech Considered (September 29, 2014)

MICROWAVE NETWORKS

Microwaves, sometimes considered a type of radio wave, are electromagnetic waves that are slightly shorter than radio waves and slightly longer than infrared waves. Because of their inability to diffract around the curvature of the earth, the transmission of microwaves to a receiver almost always has to take place over a line-of-sight path via relay towers that are often eighty to one hundred feet tall. Microwave links have the ability to handle a very high volume of audio, video, or data with minimal delay (known as "low latency"). The average distance between towers is twenty-five to forty miles; however, if they are high enough, the towers can be spaced as much as ninety miles apart. Microwave-based networks are often used in conjunction with other major networks such as the public switched telephone network [34], radio broadcast network [46], and broadcast television [47].

Microwave Radio-Relay [31]

COUNTRY OF ORIGIN UK and France
CREATOR(S) Standard Cables and Telephones Ltd. and Les Laboratoires de Matériel Téléphonique
EARLIEST KNOWN USE 1931

BASIC INFRASTRUCTURE/MATERIALS Power source, transmitter, receiver, transmission lines (usually coaxial cables but could be power lines, telephone wires, or speaker cables), antennas (usually a parabolic dish or horn antenna)

RELATED Radio broadcast [17], communications satellite [32], radio broadcast network [46], television broadcast network [36], cellular network [49], telephone [34], telautograph [36], cable television [48]

DESCRIPTION The first known long-distance microwave relay link was demonstrated in 1931; spanning the English Channel, the relay was a single two-way telephone [34] channel, and was the result of a collaboration between engineers working for the British Standard Cables and Telephones Ltd. and the French Les Laboratoires de Matériel Téléphonique. In 1951, AT&T launched what was then advertised as the longest transcontinental microwave relay in the world; composed of a chain of 107 microwave towers spaced about thirty miles apart, the relay spanned the distance between New York City and San Francisco (roughly 3,000 miles) and, significantly, had six channels that could be used to transmit either telephone or television signals. Shortly after, microwave links became essential to telecommunications in the US , Canada, Europe, and Japan—at least until they were largely supplanted by satellite and fiber optic in the 1980s and 1990s. However, given their low latency, microwave links are still used to help carry cellular

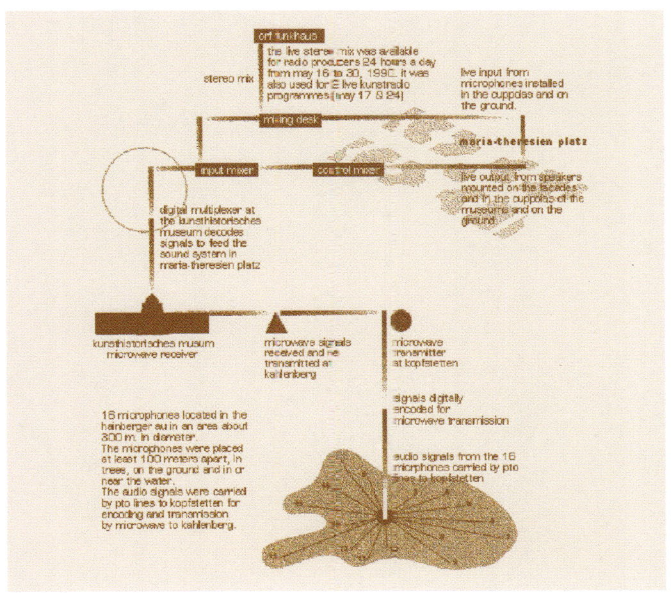

Signal path diagram for Bill Fontana's *Landscape Soundings* (1990).

network [49] calls between cell towers, to provide nearly instantaneous communication between local company or government facilities, for satellite communications, and even for deep space radio communications. The transmission speed for a microwave radio relay ranges between three hundred megabits per second and one gigabit per second.

EXPERIMENTS In 1990, American-born artist Bill Fontana used digital microwave transmission as part of a complex sound sculpture called *Landscape Soundings* that was transmitted over KunstRadio in Vienna over a two-week period. Fontana placed sixteen microphones in a marshy area thirty kilometers away from the Vienna city center and, with significant assistance from the ÖPT (Austrian Post and Telecommunications), had phone lines installed to

[31] Brochure for AT&T's first transcontinental microwave radio route (1951) and map depicting its long routes across the USA (March 1960).

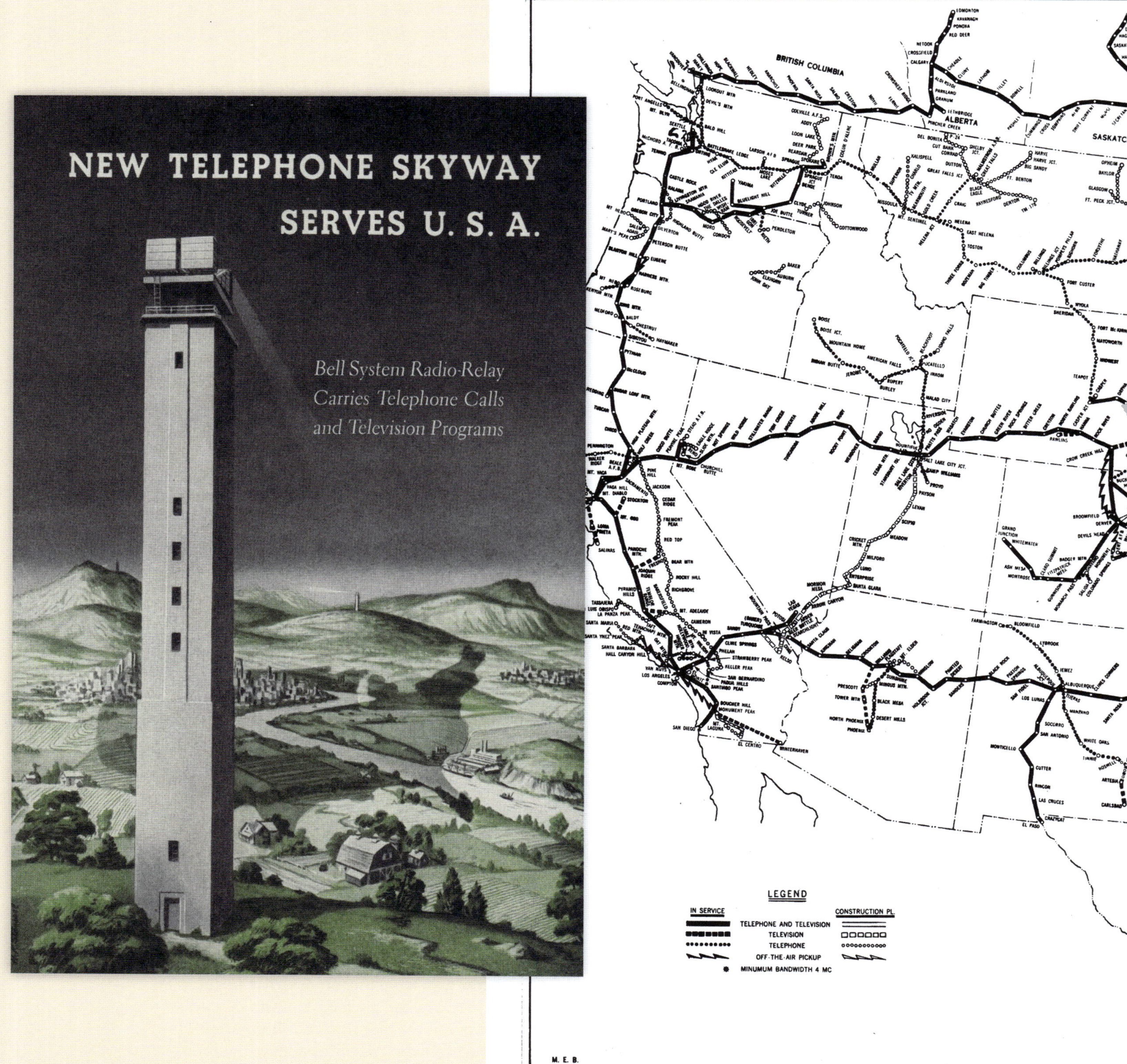

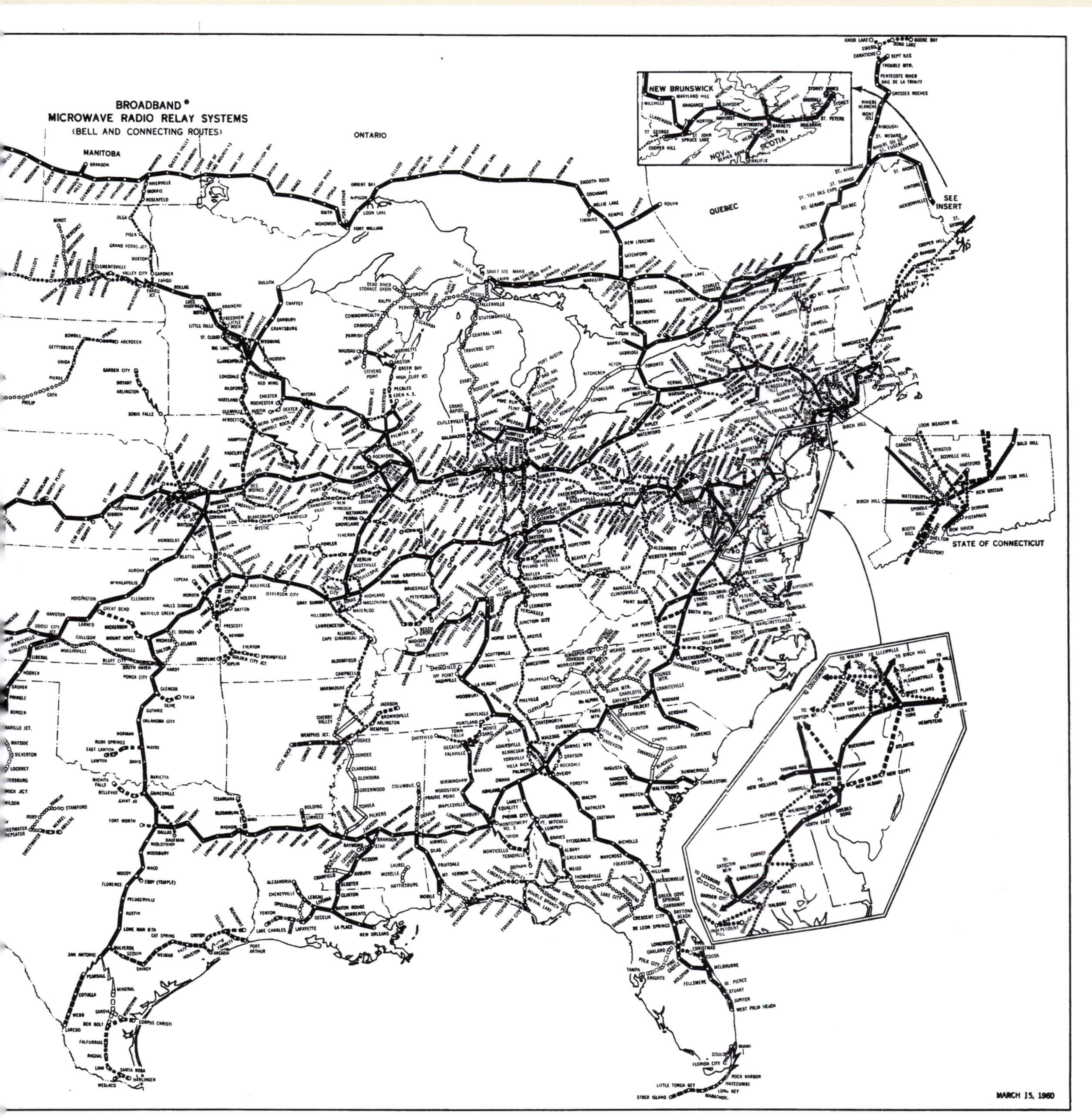
BROADBAND*
MICROWAVE RADIO RELAY SYSTEMS
(BELL AND CONNECTING ROUTES)
MANITOBA
ONTARIO
QUEBEC
NEW BRUNSWICK
SEE INSERT
STATE OF CONNECTICUT
MARCH 15, 1960

connect each microphone from the marsh to a nearby village; according to Heidi Grundmann, he then had the lines

> connected . . . to the new ORF digital microwave transmission equipment which had been mounted on a farm for line-of-sight transmissions in Vienna. The sounds from the microphones in the marshes were converted to video signals for digital transmission to the ÖPT Kahlenberg relay station, a high point in the Vienna Woods, and from there to the roof of the Kunsthistorische Museum Vienna, and finally down to an improvised studio in the museum. The digital signals were changed back into analogue and distributed live to seventy loudspeakers installed in shrubbery and topiary of the Maria Theresia Platz, a formal garden . . . and on the facades of the Kunsthistorische Museum and Naturhistorische Museum which flank the park. The work had another dimension: the signals from sixteen lines went into a stereo mixer and at the same time were transmitted in stereo to the radio station. Bill Fontana controlled the live mix both in the park and for the radio at all times—either personally or by means of a "score" based on his intimate knowledge of what sounds would arrive when and on which lines.

SOURCES John Bray, *The Communications Miracle* (Plenum Press, 1995); "The Microwave Radio and Coaxial Cable Networks of the Bell System," long-lines.net; E.E. Free, "Searchlight Radio with the New 7 Inch Waves," *Radio News* 8:2 (1931); "Microwaves Span the English Channel," *Short Wave Craft* 6:5 (1935); Heidi Grundmann, "The Geometry of Silence," in *Radio Rethink*, Daina Augaitis and Dan Lander, eds. (Banff Centre for the Arts, 1994)

Communications Satellite [32]

COUNTRY OF ORIGIN USA
CREATOR(S) US Army / Advanced Research Projects Agency
EARLIEST KNOWN USE 1958

BASIC INFRASTRUCTURE/MATERIALS Transmitter (including energy and/or power suppliers, oscillator, amplifier, and modulator), receiver, antennas, diplexer, rocket, waveguide (optional), satellite tracking app (optional), telemetry beacon (optional), transponder (optional), digital repeater (optional), magnetometer (optional), thruster (optional), solar panels (optional)

RELATED Amateur radio satellite [16.6], radio broadcast [17], packet radio network [26], microwave radio-relay [31], radio broadcast network [46], television broadcast network [36], cellular network [49], telephone [34], telautograph [36], cable television [48]

DESCRIPTION Communications satellites are entities launched into the Earth's atmosphere, usually by rocket or spacecraft, at which point they are set into orbit for the purposes of receiving/transmitting information between other satellites and/or to/from Earth stations using radio waves. The information transmitted by these satellites includes Cable and Broadcast Television [48, 47], Telephone [34], Telegraph [33], and electronic data of all kinds. Whereas amateur radio satellites have used HF, VHF, and UHF frequencies for their transmissions, these government or commercially owned satellites avoid the line-of-sight limitations of high-frequency radio wave transmissions (between three and thirty megahertz) largely by using super high (between three and thirty gigahertz) or extremely high (thirty to three hundred gigahertz) radio frequencies to relay signals around the curvature of the Earth, thereby enabling communication from one side of the planet to the other. The ranges of communications satellites may extend to deep space, and they may also be considered hybrid networks, insofar as they sometimes include or have included packet-switched networks such as ARPANET and Telenet, terrestrial radio networks such as Cellular Networks [49], and terrestrial telephone networks. These satellites may also be either passive (only reflecting signals from a source to a receiver) or active (amplifying a signal from a source before sending it to a receiver). They are also classified according to their orbit and the area they can cover: geostationary Earth orbit (GEO) satellites orbit the equator 35,000 kilometers above Earth and are considered ideal for radio, TV broadcasting, and long-distance communications; middle Earth orbit (MEO) satellites orbit the Earth between 2,000 and 35,786 kilometers up and are generally used for navigation systems such as GPS; low Earth orbit (LEO) satellites orbit the Earth 160 to 1,000 kilometers up and are considered ideal for data communication and remote sensing.

Science fiction writer Arthur C. Clarke is often cited as the inventor of satellite communication as he speculated about the possibility of using orbiting satellites to relay radio signals in an October 1945 issue of *Wireless World*. However, in the wake of the successful launch of the Russian satellite Sputnik in 1957 (a satellite that was only a beacon), the US launched the first communication satellite SCORE (Signal Communications by Orbiting Relay Equipment) in 1958. While SCORE's battery died after operating for just eight hours, it had the capacity for one voice or six teletype channels along with a recorder that had a four-minute capacity. Earth stations could command the satellite to transmit a stored message or put it into record mode to receive and store a new message—a capability that was demonstrated with the broadcast of a Christmas message from then US President Dwight D. Eisenhower that had been stored on the satellite. One of the next major communications satellites was Telstar, launched in 1962 and used for transatlantic transmission of telephone and television signals. The first US domestic satellite was WESTAR, launched in 1974 by Western Union. Among other uses, *The Wall Street Journal* used WESTAR to transmit pages in facsimile form to remote printing plants; these pages were received on page-sized photographic film that each took three to five minutes to transmit. This also marked the point at which Europe and Japan eclipsed the US in terms of satellite research and development. Today, communications satellites are within reach of small organizations and even individuals thanks to the creation of small, light, and relatively affordable LEO satellites such as CUBEsats. However, given their affordability and off-the-shelf parts, satellites have also become targets for hackers in recent years. One of the earlier (and possibly earliest) satellite hacks took place in 1986, when John MacDougall, then known as Captain Midnight, hijacked HBO's Galaxy 1 satellite signal as a way to register his objection to HBO's decision to start scrambling its satellite signal and charging viewers a $12.95 monthly fee. As of 2020, the US Air Force hosts Hack-a-Sat events in an attempt to preempt satellite hacks.

EXPERIMENTS From 1975 through 1977, US artists Kit Galloway and Sherrie Rabinowitz developed a series of projects called Aesthetic Research in Telecommunications, which included experiments with communications satellites they dubbed Satellite Arts Project. They were the first artists in the world not simply to use satellites for broadcasting but instead "focused equally on transmission delays over long distance networks, and performed a number of telecollaborative dance, music, and performance scores to determine what traditional genres could be supported, while exploring new genres that would emerge over time as intrinsic to these new ways of being-in-the-world." Using the CTS satellite (jointly operated by NASA, the Canadian Department of Communications, and the European Space Agency), their first live rehearsals, tests, and performances took place between July and November 1977. In September of the same year, US artists Liza Béar and Keith Sonnier co-produced and co-directed *Send/Receive Satellite Network*, which comprised a fifteen-hour two-way interactive transmission featuring artists in San Francisco and in New York City, including, among other features, a cross-country improvised soundtrack, an interactive discussion among participants, and a joint performance by dancers Margaret Fisher and Nancy Lewis, who echoed each other's moves from either side of the split screen display. This project also used the CTS satellite; while it only was in operation from 1976 to 1979, CTS was both experimental and intended to be used by the public. Also, according to Liza Béar, the satellite's "amazing 200-watt transmitter, the most powerful of its time . . . could be accessed by low-powered and relatively inexpensive satellite dishes within its footprint. That characteristic made it ideally suited for use by non-profits and community organizations." More recently, media artist and research Matthias Hurtl has been developing his project Drowning in Æther since 2014; the project focuses on the "audibility" of satellite signals as well as a range of artifacts associated with these signals. Thus, in Hurtl's words, it "aims to delve into the narratives of the invisible signals around us, highlighting the expansive, systemic, and geological attributes of the Technosphere. Drowning in AEther is an archive, a collection of sounds and recordings, images, as well as found footage, debris, and artifacts found during the research. It attempts to give an insight into the practice called 'signal hunting' and maybe inspire others to drown in the Æther too."

SOURCES Nathan J. Muller, *Wireless A to Z* (McGraw-Hill, 2003); Arthur C. Clarke, "Extra-Terrestrial Relays," *Wireless World* (October 1945); Hugh R. Slotten, *Beyond Sputnik and the Space Race* (Johns Hopkins University Press, 2022); James Martin, *Telecommunications and the Computer*, 2nd edition (Prentice Hall, 1976); Donald H. Martin, *Communication Satellites 1958–1995* (Aerospace Corporation, 1996); Bruce R. Elbert, *Introduction to Satellite Communication*, 3rd edition (Artech House, 2008); Dimov Stojce Ilcev, "Architecture of ORBCOMM Little LEO Global Satellite System for Mobile and Personal Communications," *Microwave Journal* (February 10, 2023); "Video Pirate Interrupts HBO," *The New York Times* (April 28, 1986); Christian Vasquez, "First in Space: Space-X and NASA Launch Satellite that Hackers Will Attempt to Infiltrate During DEF CON," *Cyberscoop* (June 5, 2023); Kit Galloway and Sherrie Rabinowitz, "Telecollaborative Art Projects of Electronic Cafe International Founders" website; Steve Durland, "Defining the Image as Place: A Conversation with Kit Galloway, Sherrie Rabinowitz, and Gene Youngblood," *Social Media Archaeology and Poetics*, ed. Judy Malloy (MIT Press, 2016); Liza Béar, "Send/Receive Satellite Network: Some History," *Send/Receive Satellite Network* website (June 30, 2020); Matthias Hurtl, "Drowning in Æther" website

[32] Photograph from September 11, 1977, of the New York earth station at the Battery Park City landfill used for Liza Béar and Keith Sonnier's *Send/Receive Satellite Network*; Sonnier appears on the far left.

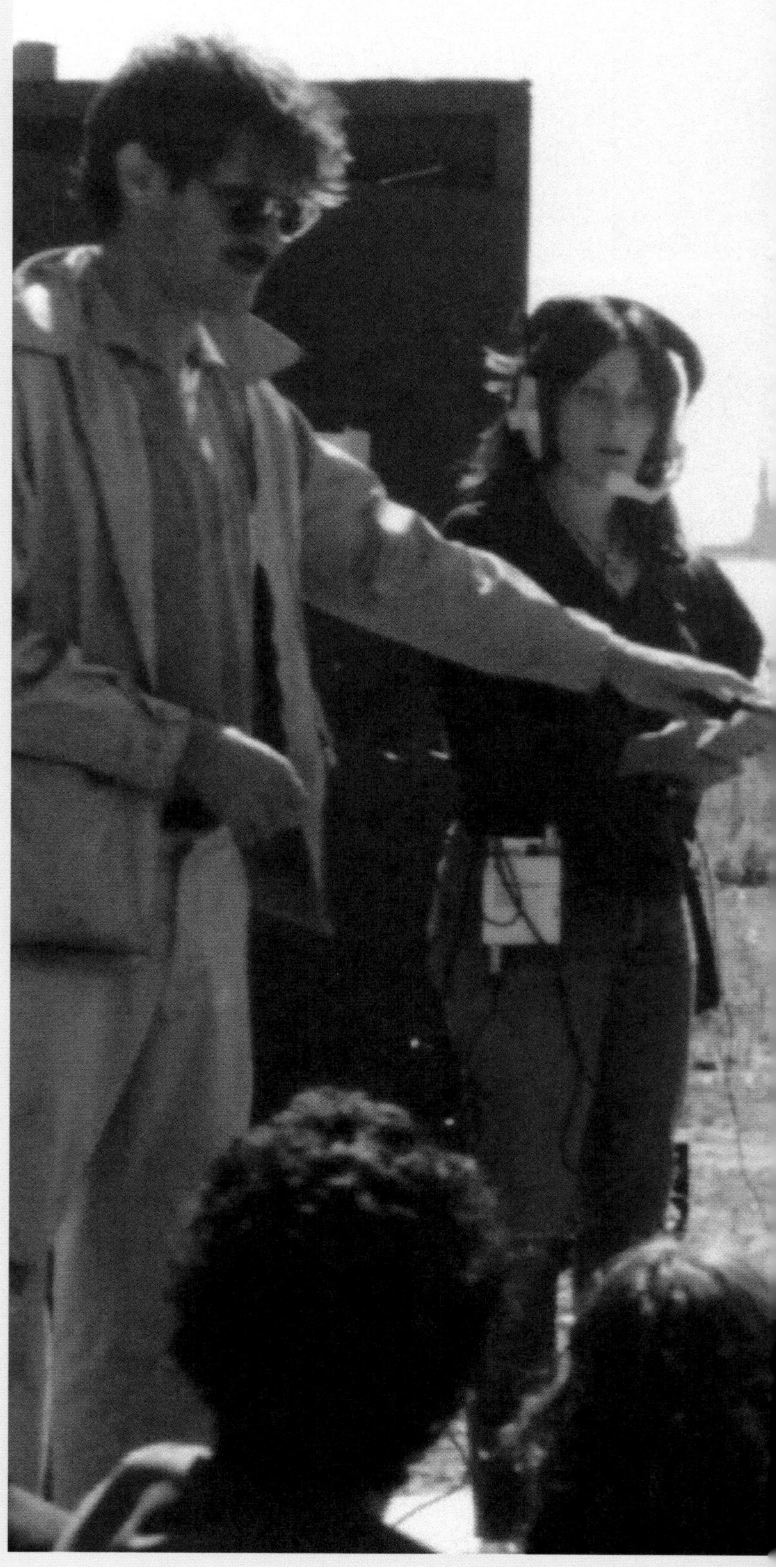

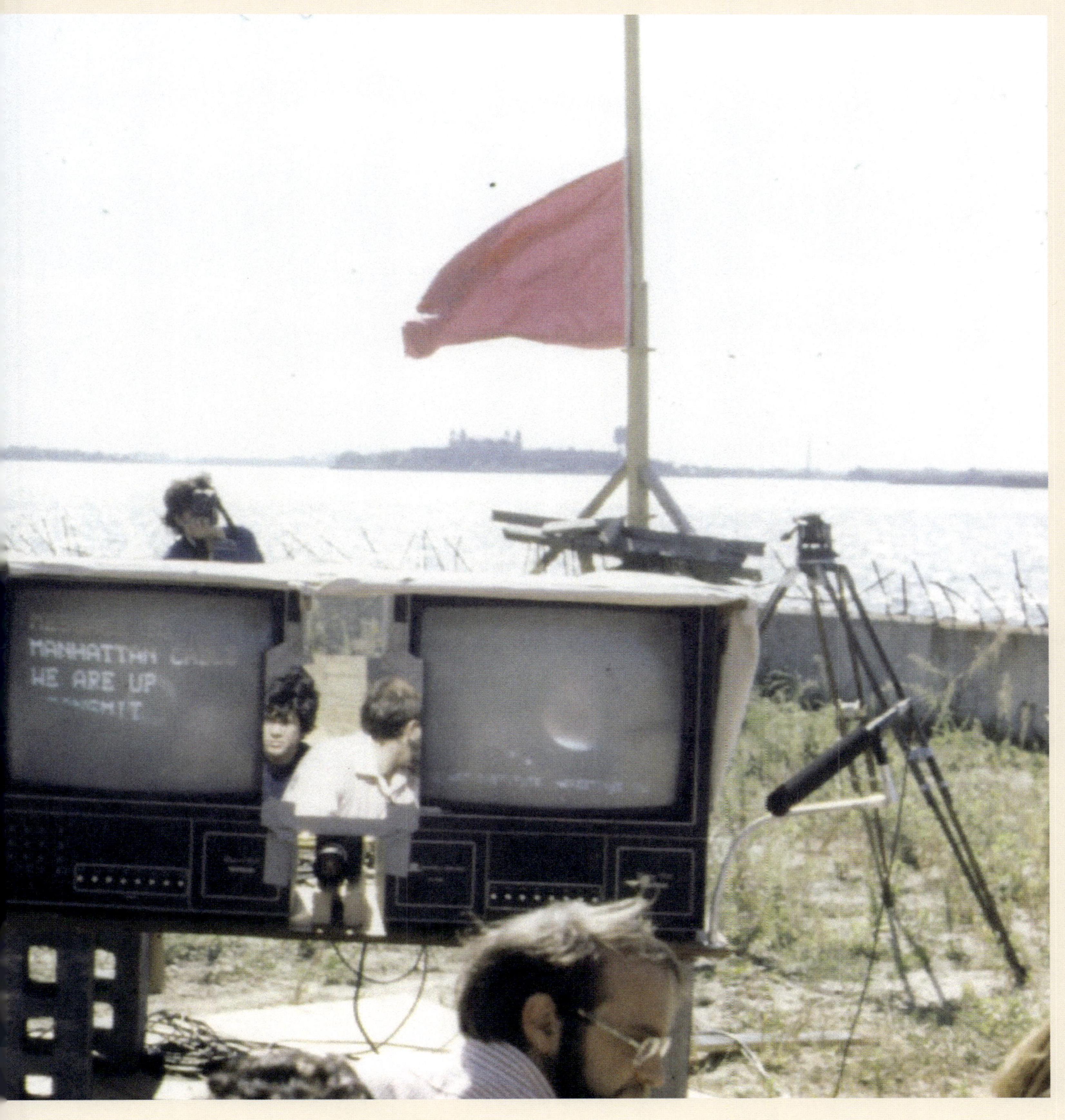
MANHATTAN
WE ARE UP

WIRED

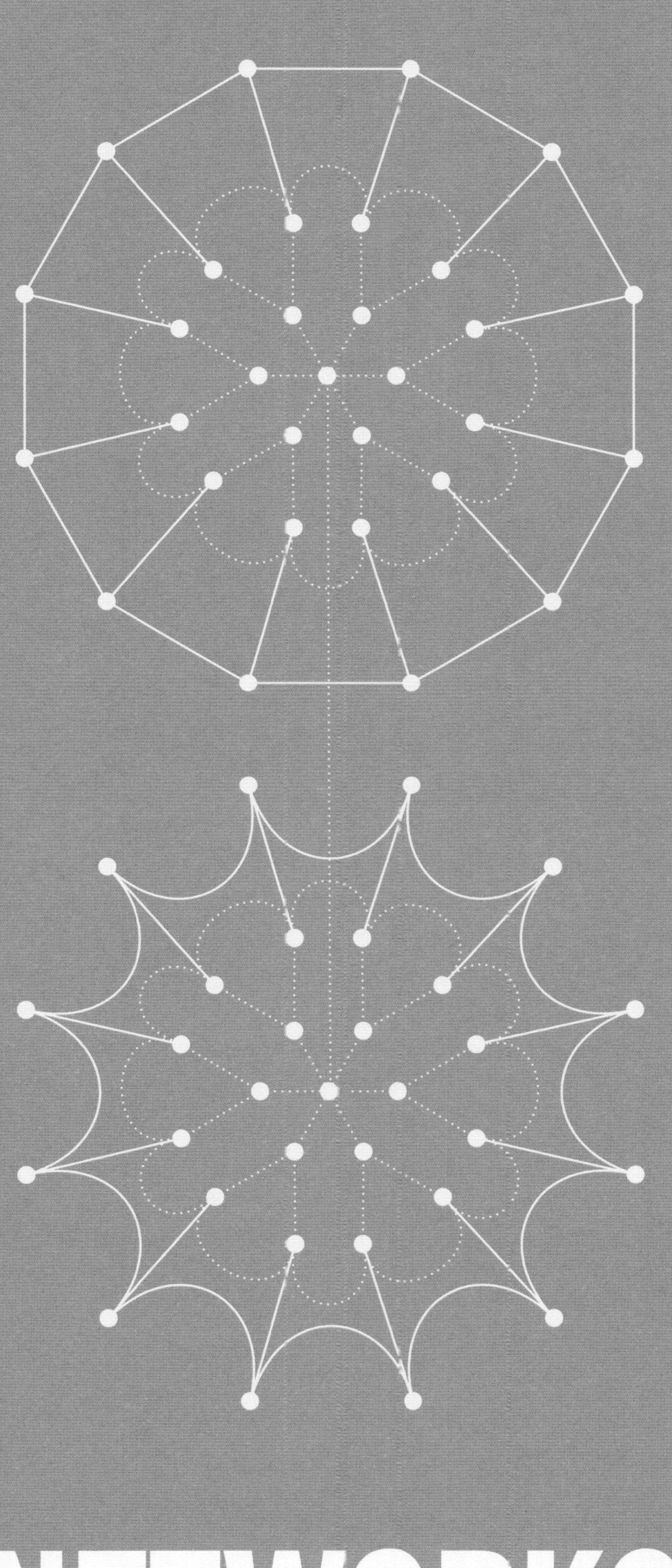

NETWORKS

ELECTRICAL WIRE NETWORKS

Similar to how wireless radio networks are grouped together first by underlying infrastructure (i.e., the manipulation of radio waves) and then by the different, discrete networks that rest on top of this infrastructure, this section groups together networks whose underlying infrastructure is the transmission of electricity through wires and then by the different discrete networks that rest on top of these wires. That said, given this emphasis on infrastructure, it would have been ideal to further organize electrical wire networks by type of wire if there weren't so much overlap between networks that operate over single wire, paired twisted wire, coaxial cable, and fiber optic cable (for example, one can build a telegraph or telephone network using any of the aforementioned wires and cables).

Most national electrical codes refer to electrical wires as "conductors," as the majority of metals are commonly used for conducting electricity. Moreover, even though the terms "wire" and "cable" are often used interchangeably, engineers often use the term "cable" to refer to bundles of insulated wires, whereas "wire" usually refers to just a single bare or covered (insulated) bar of flexible metal. Broadly speaking, the insulating material used was originally cloth, yarn, or paper, before rubber (or rubber-like material such as gutta-percha), polymer, varnish, plastic, or dialectric fluid (for example, mineral or castor oil) was employed. The wire itself could be copper, aluminum, steel, iron, nickel, silver, gold, or any combination of the foregoing.

The first wired networks were telegraph networks and, at the time, the wire used was referred to as "telegraph wire." While the construction of telegraph wire varied over time and by geography, generally it was made from iron for shorter distances, and from copper for distances over 150 miles. By the 1880s, the consensus was that "iron wire for telegraph use should be soft and capable of stretching 18 to 20 per cent before breaking . . . for general use the soft wire should be capable of being bent at right angles several times backward and forward without breaking, so that joints may be made securely." Once the telephone was invented in 1876, it soon became apparent—partly because of the growing number of electrical power lines producing interference over neighboring telegraph wires—that the method of using a single grounded wire was not adequate for transmitting voice. While the Bell Telephone Company began using two-wire circuits in 1878, most telegraph and telephone networks transitioned to using a twisted pair of cables (consisting of shielded copper wires) after Alexander Graham Bell patented the design in 1881. Now, while the specifications of telephone wire and cable vary enormously depending on their use and their location, copper telephone cables generally consist of anywhere from two to thousands of pairs of twisted, insulated cables, which are further insulated from other cables within a single bundle either by a gel-like substance or with air. Although it is gradually being replaced by fiber optic cable, twisted pair cabling has also long been used for computer networking, including Ethernet.

Coaxial cable was first used in the first transatlantic cable in 1858 but was not patented until 1880, by British physicist Oliver Heaviside. In 1929, engineers at Bell

Telephone Laboratories patented a design for a "Concentric Conducting System," which is the basis for modern coaxial cables. By the 1960s, coaxial cable included steel wires encased in another layer of copper running through the center. A coaxial cable consists of an outer tube-shaped conductor with an inner conductor (usually made out of solid, stranded, or braided copper) held at the center of the cable by insulators made out of a material such as rubber, ceramic, or plastic. They are used to transmit many telephone conversations simultaneously, as well as for television and radio.

Fiber optic cable consists of a bundle of optical fibers that carry flashes of light (from sources such as a laser [14] or LED) along a glass fiber; at the end of the fiber, the light is converted to an electrical signal with a photo-electric cell. When used in a cable, the fibers are made out of fused silica glass or plastic and are coated with several layers of protective material. Optical fibers were first demonstrated in 1953 by Dutch physicist Bram van Heel, but it was not until 1983 that optical fiber was successfully used for data transmission over a distance of 240 kilometers (150 miles). It is now used for telephone, television, and data communications.

Undersea cables were and are constructed differently. In the early 1850s, they consisted of four copper wires, a double covering of gutta-percha (a type of natural rubber derived from gutta-percha trees, found mostly in southeast Asia), yarn saturated with tar, which also covered the wires, and a final outer layer of ten galvanized iron wires. By 1858, the first transatlantic cable consisted of seven copper wires covered in three coats of gutta-percha and wound with tarred hemp, along with an outer layer of eighteen strands (each consisting of seven wires) that formed a sheath. The construction of undersea cables (in terms of number and type of loading coils, repeaters, and multiplexing techniques) varied from this point until the use of coaxial cable in 1956. Now, they are largely made out of optical fiber.

SOURCES *Galvanized Iron Wire as Employed in the Telegraph and Telephone* (Washburn and Moen Manufacturing Co., 1881); *Telephone and Telegraph Engineers' Handbook; a Convenient Reference Book for All Persons Interested in Telephone and Telegraph Systems, Location of Faults, Electricity, Magnetism, Electrical Measurements, and Batteries* (International Correspondence Schools, 1908); Ken Beauchamp, *History of Telegraphy* (Institution of Electrical Engineers, 2001); Alexander Graham Bell, "Telephone Circuit," Patent #US244,426A (July 19, 1881); Lloyd Espenschied and Herman A. Affel, "Concentric Conducting System," Patent #US1,835,031A (May 23, 1929); W. D. Wilkens, "Telephone Cable: Overview and Dielectric Challenges," *IEEE Electrical Insulation Magazine* 6:2 (March–April 1990); K.D. Mayne, "Heaviside's Communication Cable Designs," *Electronics and Power* (September 1977); A. Lemuel Albert, *Electrical Communication* (Wiley, 1950); H.H. Schenck, *The World's Submarine Telephone Cable Systems* (Dept. of Commerce, Office of Telecommunications, 1975); John Crisp, *Introduction to Fiber Optics* (Elsevier Science & Technology, 2005); Jeff Hecht, *City of Light: The Story of Fiber Optics* (Oxford University Press, 2004); Ivan Kaminow, Tingye Li, and Alan E. Willner, eds., *Optical Fiber Telecommunications* (Academic Press, 2013)

Electrical Telegraph [33]

COUNTRY OF ORIGIN Germany
CREATOR(S) Samuel Thomas von Sömmerring
EARLIEST KNOWN USE 1809

BASIC INFRASTRUCTURE/MATERIALS Electrical power source, telegraph wire, wire, utility poles (optional), leads (optional), pins (optional), keys (optional), water-filled tank (optional), stylus and ink or pen (optional), paper tape (optional), chemically treated paper (optional), magnetic needle (optional), galvonometer (optional), paper disc (optional), mercury (optional), insulator (optional)

RELATED Radiotelegraphy [16.1], electrical printing telegraph [33.1], image telegraph [33.2], fire alarm telegraph [33.3], pantelegraph [33.4], telephonic telegraph [33.5], barbed wire telegraph [40]

DESCRIPTION Electrical telegraphs include any telegraph device whose functioning works by electrochemicals, electromagnetism, magnetized needle, electrovoltaic effect, or electric shock. There were so many different, short-lived telegraph devices developed around the world throughout the eighteenth and nineteenth centuries (Ken Beauchamp lists no fewer than forty-eight different electrical telegraphs developed between 1753 and 1848) that this entry only discusses a handful of pivotal developments. While there were likewise many proposals for electrical telegraphs that relied on static electricity, it seems likely that these only amounted to proposals rather than functioning systems, since static electricity could not have been sufficient to transmit messages over any meaningful distance. In 1809, German anatomist and physiologist Samuel Thomas von Sömmerring successfully demonstrated an electrical telegraph that used thirty-five wires to represent letters of the alphabet and numerals. According to Russell Burns, each wire "terminated in a pin which projected through the base of a glass vessel filled with acidulated water. By connecting any pair of the wires to the extremities of a Volta pile, [von Sömmerring] was able to cause bubbles to rise from the appropriate pins." In other words, bubbles would emerge next to the relevant letter or numeral. Even though Sömmerring had successfully sent messages through 10,000 feet (3040 meters) of wire by 1812, his telegraph was not broadly adopted.

By the 1830s, the electromagnetic telegraph had become the dominant type of electrical telegraph. The Schilling Telegraph, demonstrated in 1832 by Russian military officer Baron Pavel Lvovitch Schilling, was the earliest electromagnetic telegraph and also one of the first needle telegraphs. To receive signals, the device used eight coiled wires (six of which were for signaling, with the remaining two reserved for sending and receiving); the wires acted as electromagnets that caused a needle, functioning as a magnet and hanging above the wires by silk thread, to move from one side of a disc of paper (colored white) to the other side (colored black), thereby producing two signals that could be turned into code; for example, the letter *A* was represented by the needle indicating black and then white. To transmit signals, the device used a keyboard with sixteen black and white keys, much like a piano keyboard. Each pair of black and white keys was attached to one of eight wires, and each key was also assigned a positive or negative voltage that in turn corresponded to the black or white side of the paper disc on the receiving end. The device was slated to be installed in St. Petersburg by Tsar Nicholas I, but the telegraph line was never realized because Schilling died in 1837.

Numerous other electromagnetic telegraph devices had successful demonstrations after Schilling's own in 1832. However, the two telegraphs produced by William Cooke/Charles Wheatstone and Samuel Morse in 1937, both of which were developed for commercial purposes, became dominant in many countries. English inventor William Fothergill Cooke and English scientist Charles Wheatstone first developed a five-needle telegraph that used six wires in 1837 for railway signaling across the two-kilometer stretch between Euston Station and Camden Town in London. An even more simple, inexpensive, and robust one-needle-and-two-wire system was developed for the British railway system in 1843, and it was at this point that the Cooke and Wheatstone telegraph also became available to the public. The one-needle system continued to be installed with new

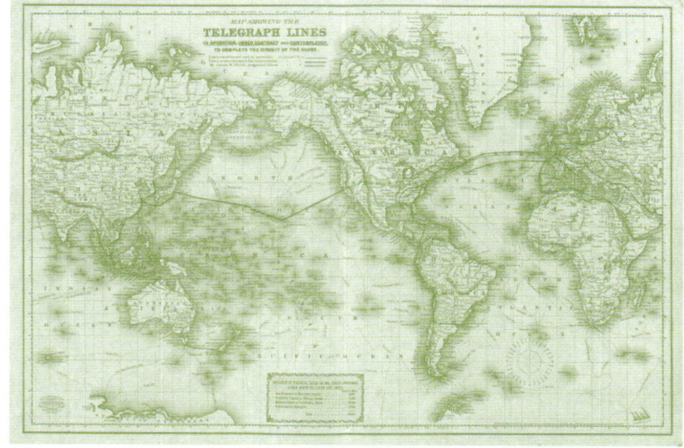

Map from 1871 "showing the telegraph lines in operation, under contract, and contemplated, to complete the circuit of the globe."

railway lines radiating out of London, and by 1869, after Cooke formed the Electric Telegraph Company and the company was nationalized to become part of the General Post Office, the one-needle telegraph become part of the nationwide infrastructure. The Cooke and Wheatstone telegraph was also used in Spain. This model of telegraph featured needles affixed to a surface similar to a clock-face; the needles would turn left or right, depending on the positive or negative direction of the current in the telegraph wires. The face of the five-needle telegraph featured a diamond-shaped grid with twenty different letters that the needles could point to, excluding the letters *C*, *J*, *Q*, *U*, *X*, and *Z*. The one-needle telegraph had a four-unit code that could account for the entire alphabet.

American Samuel Morse came up with an idea for a simple electromagnetic telegraph in 1832, which he then developed into an early prototype in 1835 while acting as professor of the literature of the arts and design at New York University. After receiving invaluable advice from Leonard Gale (professor of chemistry also at NYU) and Joseph Henry (scientist and expert on electromagnetics) to use finer wire and a more powerful battery, Morse and Gale successfully sent a message across ten miles (sixteen kilometers) of wire strung across NYU. Despite the fact that dozens of individuals across many different countries had proposed and sometimes successfully demonstrated models of the electrical telegraph before this point, Morse nonetheless declared himself "the first proposer and inventor of electromagnetic telegraphy [in 1832] . . . All telegraphs in Europe are, without one exception, invented later than mine." Between 1837 and 1843, Morse spent considerable time attempting to gain government funding for his telegraph. His lobbying resulted in the US Congress passing the Telegraph Bill in 1844 and allocating him $30,000 for the construction of a telegraph line between Washington, D.C., and Baltimore, Maryland. On May 27, 1844, Morse transmitted "What hath God wrought" on the DC–Baltimore line, after which the telegraph became available to the American public.

The system developed by Morse, Gale, and later Alfred Vail relied on a system of code, now known as Morse code, which indicated letters and numbers through a series of short (dots) and long (dashes) pulses of electricity. To send these short and long pulses, an operator used a telegraph key, or a metal plate with long and short metal bars or knobs representing dots and dashes, connected to a battery. The Morse telegraph receiver included an electromagnet that received the pulses; these pulses were then either translated into audible sounds that an operator could interpret, or recorded by an ink roller and stylus

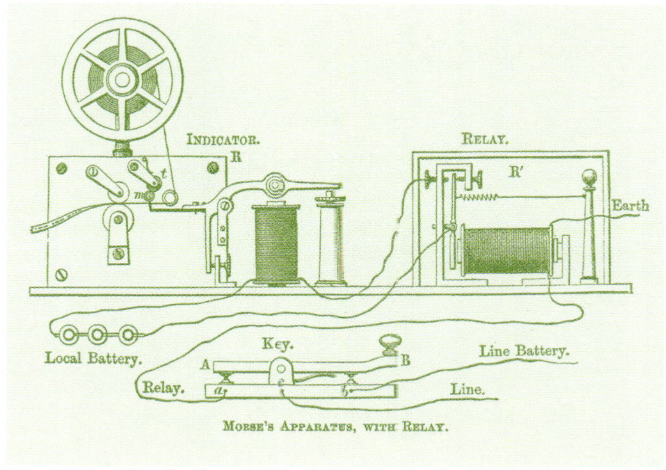

Diagram illustrating Samuel Morse's telegraph from 1837.

as short and long marks on a strip of paper powered by a clockwork motor.

Despite the fact that the Cook and Wheatstone and Morse systems are the best-known systems, it's worth noting that a pointer telegraph developed by German electrical engineer Werner von Siemens and German mechanic Johann Georg Halske in 1848 was used for a 670-kilometer underground link from Berlin to Frankfurt and then, starting in 1853, for a 9,000-kilometer Russian state telegraph network. Unlike the Cook and Wheatstone and Morse systems, the transmitter and receiver for the Siemens pointer telegraph were identical. Each had their own battery and, according to Siemens's recounting of their own history, "pointers at both stations rotated synchronously around the radially arranged letters. If someone at the transmitting station pressed a letter, the pointer stopped there—and the pointer at the receiver station likewise stopped on the same letter." By 1880, Great Britain, France, Germany, Belgium, the Netherlands, and Switzerland had the most developed telegraph infrastructure; Denmark, Austria and Italy followed close behind, while other non-European countries such as the USA were markedly underdeveloped.

In order for the telegraph to become a nationwide means of telecommunications, telegraph offices were set up, sometimes as part of railway stations or post offices, and were connected first by telegraph wires laid underground and then suspended overhead by utility poles. The offices were staffed by operators who manually sent, received, and decoded telegraphs; these operators were later replaced by teleprinters that automatically decoded telegrams. The messages themselves were often delivered by messengers limited to a three-mile radius from the telegraphic office; in Great Britain, these messengers were paid

[33] Design drawings for the Cooke and Wheatstone telegraph from 1837.

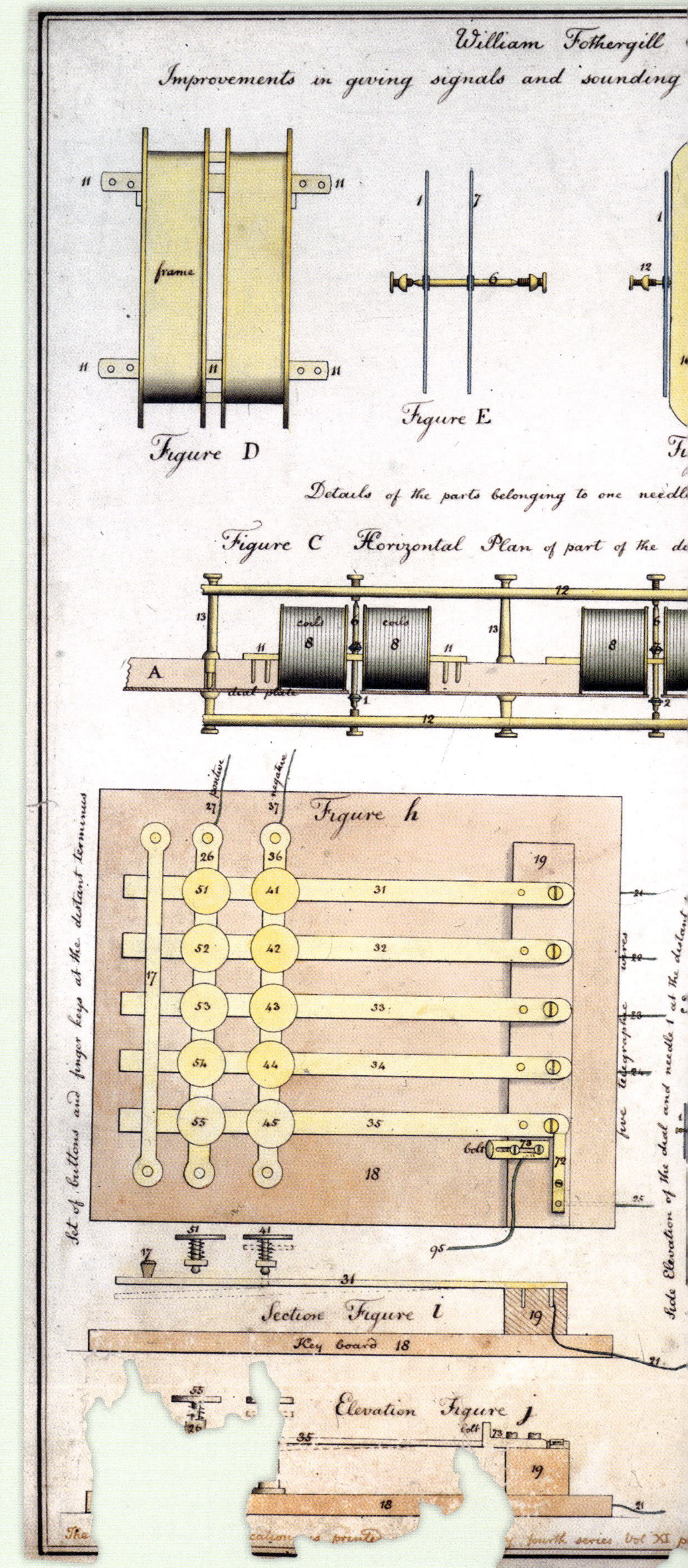

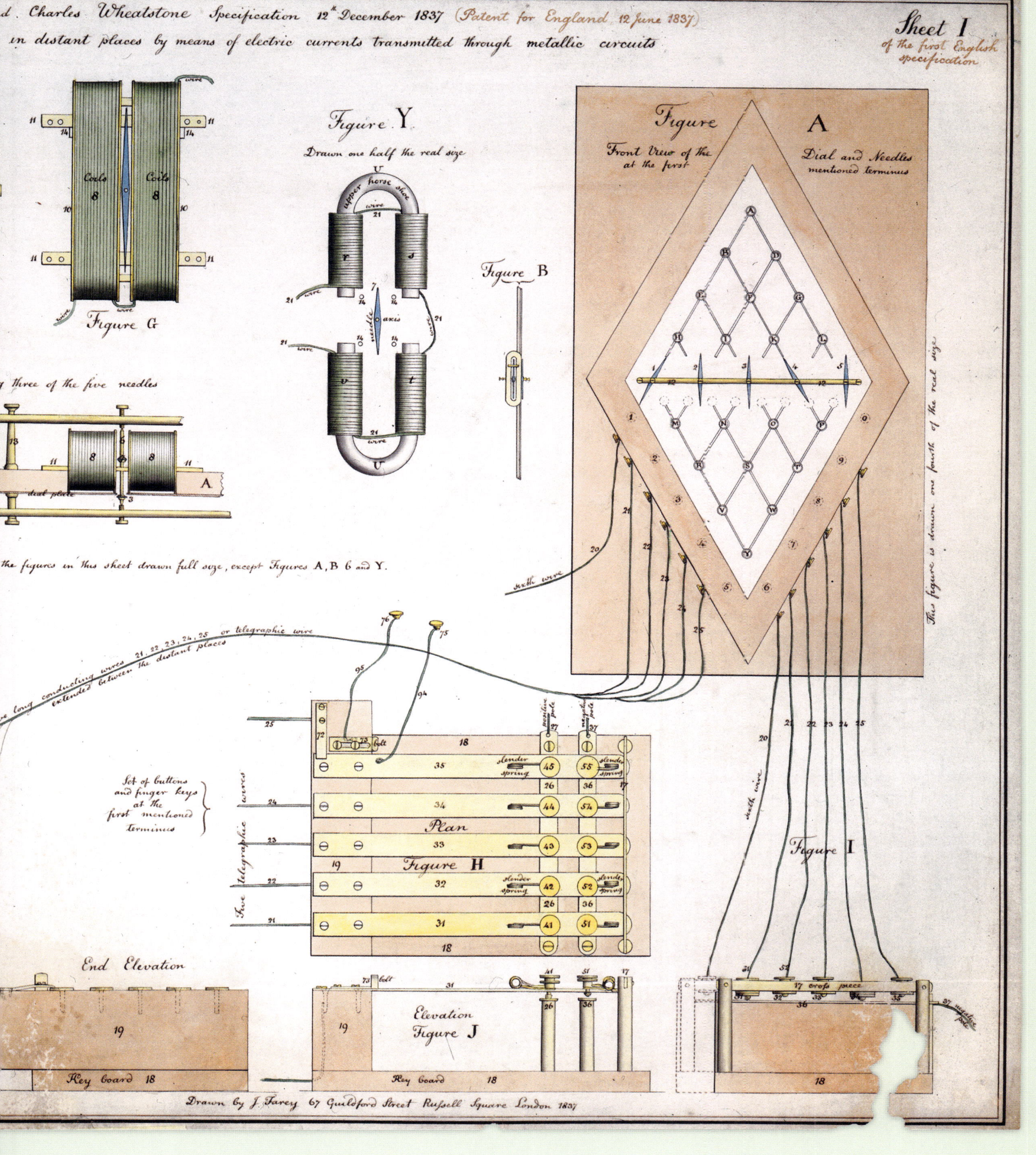

d. Charles Wheatstone Specification 12th December 1837 (Patent for England 12 june 1837)
in distant places by means of electric currents transmitted through metallic circuits
Sheet I
of the first English specification
Figure G
Coils
Figure Y
Drawn one half the real size
upper horse shoe
wire
needle
axis
Figure B
Figure A
Front View of the Dial and Needles at the first mentioned terminus
This figure is drawn one fourth of the real size
three of the five needles
dial plate
the figures in this sheet drawn full size, except Figures A, B 6 and Y.
sixth wire
long conducting wires 21, 22, 23, 24, 25 or telegraphic wire extended between the distant places
Set of buttons and finger keys at the first mentioned terminus
Five telegraphic wires
positive pole
negative pole
bolt
slender spring
Plan
Figure H
Figure I
End Elevation
Key board
Elevation
Figure J
cross piece
negative pole
Drawn by J. Farey 67 Guildford Street Russell Square London 1837

around a penny per message delivered, with the stipulation that they could deliver a maximum of thirty messages a day.

Because individuals were often charged by the character and by the word for messages they wished to transmit, the telegraph spawned a style of writing (similar to today's norms for text messaging) that became known as "telegram style." After the term "telegram" was coined in the USA in 1852 to refer to a "telegraphic dispatch," a wide variety of abbreviated styles of writing and even codes were developed. For example, by 1894 the *Adams Cable Codex* lists "Abash" as code for "Am quite ill. Please come at once" and "Abased" was used to mean "All well. Business O.K. Nothing here requiring your immediate return." There are even reports of the telegram being used for what we would call spam as early as 1864, when a London-based dental practice sent out a mass telegram informing receivers of their operating hours.

It is unclear when the electrical telegraph running on telegraph wires was last used, as those wires were gradually replaced by ones used for the telephone; telegrams continued to be sent over telephone wires well into the twentieth century.

EXPERIMENTS Once again, arguably nearly every (often short-lived) electrical telegraph from the nineteenth century was itself an experiment, though many didn't exist long enough for artistic experimentation to take place. However, even though functioning telegraph wire hasn't existed since the early twentieth century, many artists have continued its spirit of experimentation by using telegraphic codes, such as Morse code, as their medium. Working from the premise that knitting is essentially a binary operation involving many different combinations of two different kinds of stitches, history professor and textile artist Kristen Haring mapped the binary elements of Morse code onto the two types of stitch in order to produce Morse code textiles. For example, in 2007 Haring created *Subtle Distress*—a sweater that has "SOS" repeatedly encoded into its red knit. Haring writes, "I spell out words by switching between the two stitches that make up knitting (knit and purl) like a telegrapher switches an electrical current on and off to send Morse code. With patience, text can be deciphered from the shape of the stitches. This project complicates reading partly to emphasize that all communication depends on cultural codes. Understanding Morse code knitting requires combined knowledge of domains often set apart as masculine and technical, in the case of Morse code, and feminine and folksy, in the case of knitting." Haring's work is thus important not only for the way in which it experiments with different media for code and encoding but also for the way it unsettles long-established gender norms about who gets to be an expert about what.

SOURCES Ken Beauchamp, *History of Telegraphy* (Institution of Electrical Engineers, 2001); Russell W. Burns, *Communications: An International History of the Formative Years* (Institution of Electrical Engineers, 2004); Anton Huurdeman, *The Worldwide History of Telecommunications* (John Wiley & Sons, 2003); John Fahie, *A History of Electric Telegraphy, to the Year 1837* (E. & F.N. Spon, 1884); J.J. Hamel, "Historical Account of the Introduction of the Galvanic and Electromagnetic Telegraph," *Journal of the Society of Arts* (1859); Siemens AG, "Pointer Telegraph, 1847," Siemens.com; Jean-François Fava-Verde, "Victorian Telegrams: The Early Development of the Telegraphic Dispatch and Its Interplay with the Letter Post," *Notes and Records: The Royal Society Journal of the History of Science* 72:3 (September 2018); Roland Wenzlhuemer, "The Development of Telegraphy, 1870–1900: A European Perspective on a World History Challenge," *History Compass* 5:5 (2007); E.A. Adams, *Adams Cable Codex* (E.A. Adams, 1894); "Getting the Message, at Last: The Etiquette of Telecommunications," *The Economist* 385:8559 (December 15, 2007); Jacqueline Witkowski, "Knit for Defense, Purl for Control," *InVisible Culture: An Electronic Journal for Visual Culture* 22 (April 15, 2015); UNESCO, *News Agencies: Their Structure and Operation* (United Nations Educational, Scientific and Cultural Organization, 1953); Kristen Haring, "Morse Code Knitting," Wavefarm.org (2007)

Electrical Printing Telegraph [33.1]

COUNTRY OF ORIGIN Scotland
CREATOR(S) Alexander Bain
EARLIEST KNOWN USE 1843

BASIC INFRASTRUCTURE/MATERIALS Electrical power source, utility poles (optional), telegraph wire, copper wires, pendulums, styluses, metal frames, metal type, chemically treated paper

RELATED Radiotelegraphy [16.1], electrical telegraph [33], image telegraph [33.2], fire alarm telegraph [33.3], pantelegraph [33.4], telephonic telegraph [33.5], barbed wire telegraph [40]

DESCRIPTION The electrical printing telegraph was the first facsimile device that could print alphabetic

[33]

Kristen Haring's *Subtle Distress* (2007).

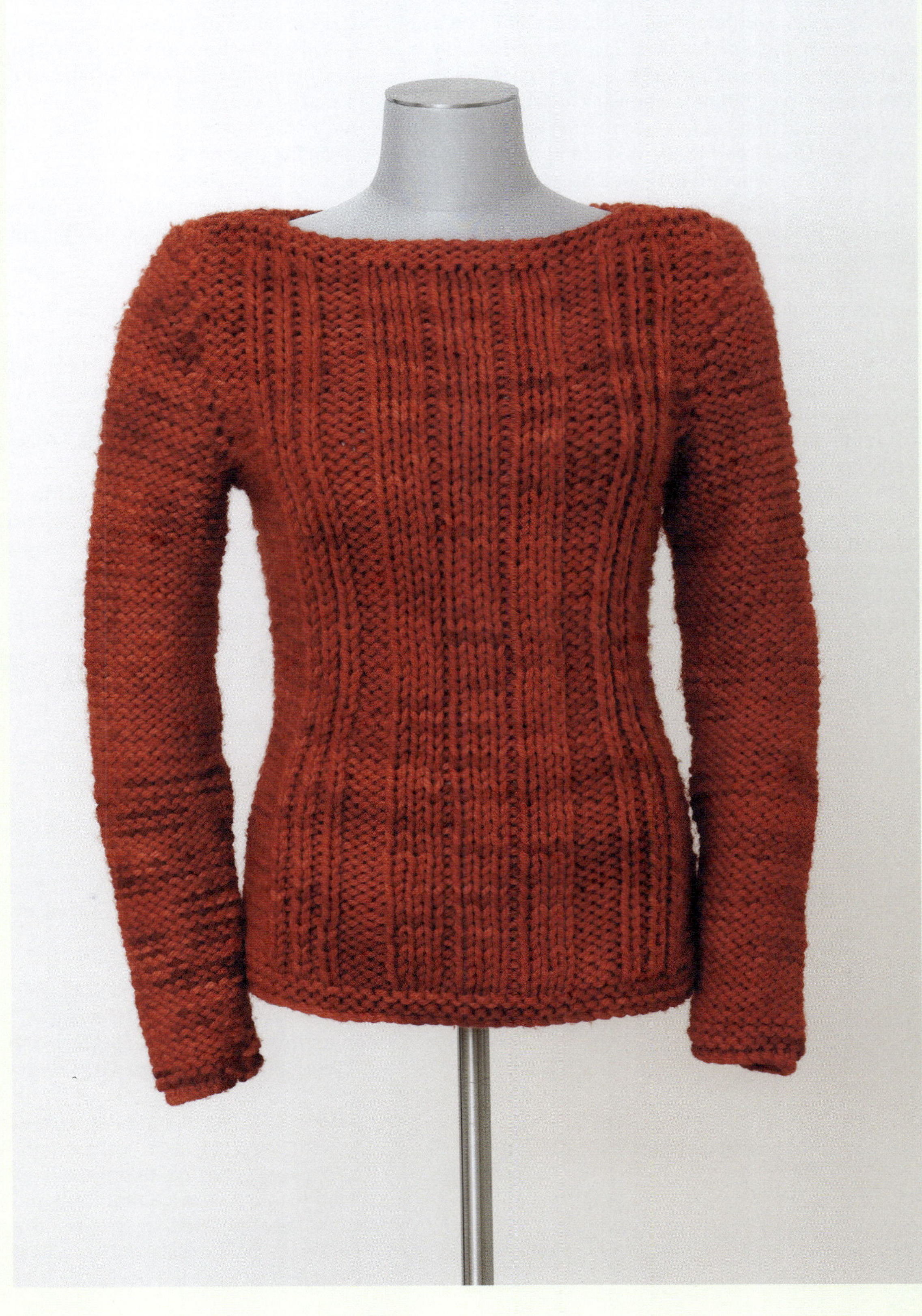

letters and numbers as well as two-dimensional images on paper using a suspended stylus; the latter feature means this device was arguably the first telautograph [36]. From 1843 to roughly 1850, Scottish clock and instrument maker Alexander Bain refined his electrical printing telegraph to the point where he could successfully send and receive black-and-white images over telegraph wire. Bain's device used a pendulum and a stylus that was suspended from a metal frame to scan the surface of a text or image. The transmitter and receiver were nearly identical and were synchronized by connecting each device's pendulum to a circuit that could hold one pendulum until the other reached the same position. As Russell Burns explains, with each swing of the pendulums, "the frame descended by a given constant amount so that the whole surface was scanned uniformly." Each transmitter consisted of five wires placed at right angles to a metal frame with metal type inside; the wires thus made contact with the raised surface of the metal type. On the other side of the metal frame was a stylus attached to a pendulum. Burns continues, "as the stylus moved across the frame, an electric circuit containing the stylus, the frame and type was continually made and broken according to the arrangement of the type." Each receiver also included a metal frame and five wires, but with the addition of chemically treated paper: "At the back of the paper there was a smooth metal plate that pressed the paper into contact with the ends of the parallel wires which filled the frame . . . By chemical action it was intended that the making and breaking of the current in the circuit should discolor the paper at the receiver to give a copy of the original surface."

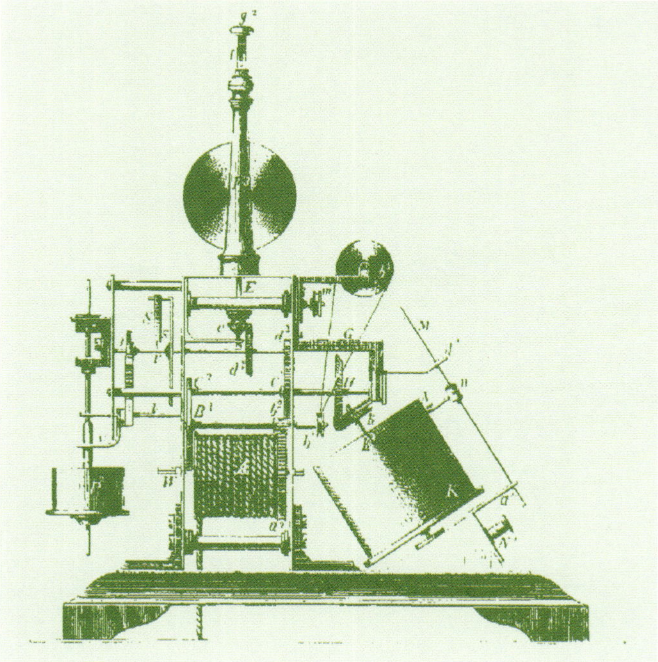

Design from 1850 for Alexander Bain's final design for his electrical printing telegraph.

It is unclear when the device fell out of use, but it was likely in the early 1850s, as Frederick Collier Bakewell publicly demonstrated a more convincing device called the image telegraph [33.2] in 1848, two years before Bain attempted to submit a patent for his improved design. While Bain's device was never adopted for regular service, he essentially created the method of raster scanning (the line-by-line display of an image) that became the basis of television as well as digital image storage and transmission. As Ivan Ruddock declares: "It is thus not an exaggeration to claim that Alexander Bain is the real father of television."

SOURCES "Mr. Bain's Electrical Printing Telegraph," *Mechanics' Magazine, Museum, Register, Journal, and Gazette* 1080 (April 20, 1844); Alexander Bain, "Bain's Patent Electro-Chemical Printing Telegraph," *Mechanics' Magazine, Museum, Register, Journal, and Gazette* (February 9, 1950); Russell W. Burns, *Communications: An International History of the Formative Years* (Institution of Electrical Engineers, 2004); Ivan S. Ruddock, "Alexander Bain: The Real Father of Television?," *Scottish Local History* 83 (Summer 2012)

Image Telegraph [33.2]

COUNTRY OF ORIGIN England
CREATOR(S) Frederick Collier Bakewell
EARLIEST KNOWN USE 1847

BASIC INFRASTRUCTURE/MATERIALS Electrical power source, utility poles (optional), telegraph wire, copper wires, pendulums, styluses, metal frames, non-conductive liquid, metal foil, cylinders, chemically treated paper

RELATED Radiotelegraphy [16.1], electrical telegraph [33], electrical printing telegraph [33.1], fire alarm telegraph [33.3], pantelegraph [33.4], telephonic telegraph [33.5], barbed wire telegraph [40]

DESCRIPTION The image telegraph—also technically a telautograph [36], and occasionally referred to as a "copying electrical telegraph"—was another short-lived facsimile device that transmitted images, especially handwriting, over telegraph wires. Physicist Frederick Collier Bakewell successfully gave a public demonstration of the device by transmitting an image

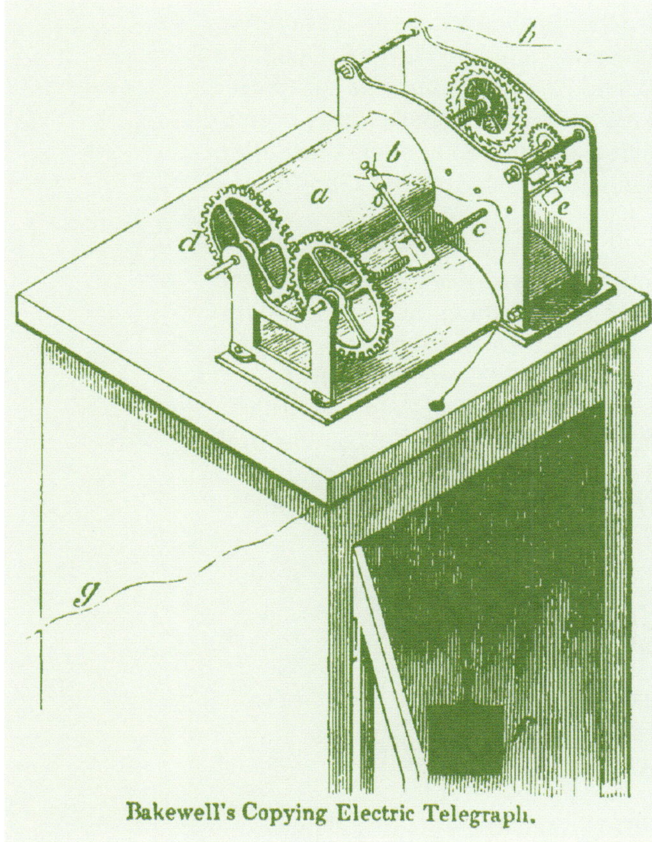

Illustration of Frederick Bakewell's image telegraph, here dubbed a Copying Electric Telegraph in the catalogue for the Great Exhibition of the Works of Industry of All Nations, 1851.

of handwriting from one end of London to another in 1847. Bakewell's design involved creating an image with nonconductive liquid such as shellac on metal or tin foil, wrapping the metal foil around a cylinder on the transmitter that was rotated by a pendulum, and then having a stylus rotate around the cylinder; the cylinder was also attached to a battery while the stylus was attached to a telegraph line. An electrical current would be created when the stylus touched the surface of the foil and would be broken every time the stylus touched the nonconductive liquid. The receiver also included a cylinder connected to a battery, but instead of metal foil it was wrapped with chemically treated white paper around which another stylus rotated to reproduce the image. As electricity ran through the stylus, a white image would appear against the blue background produced by the chemically treated paper. Bakewell also demonstrated his image telegraph at the first World's Fair that took place in London in 1851, called the Great Exhibition of the Works of Industry of All Nations. At this point, the image telegraph was described as being for "transmitting facsimiles of the handwriting of correspondents, so that their signatures may be identified. Its objects

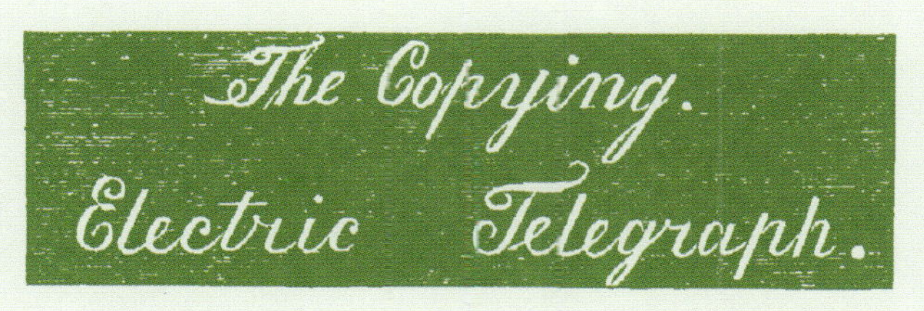

An image of handwritten text transmitted via Frederick Bakewell's image telegraph; the background in this image would have been dark blue.

are, authentication of communications, increased means of secrecy, rapidity of action, and economy, as it requires only a single wire." While the image telegraph was never adopted for service, Bakewell's design for what amounted to a "recording cylinder" became the basis of Thomas Edison's phonograph and was also later used in copy machines.

SOURCES Anton Huurdeman, *The Worldwide History of Telecommunications* (John Wiley & Sons, 2003); *Official Catalogue of the Great Exhibition of the Works of Industry of All Nations, 1851* (Spicer Brothers, 1851); Russell W. Burns, *Communications: An International History of the Formative Years* (Institution of Electrical Engineers, 2004)

Fire-Alarm Telegraph [33.3]

COUNTRY OF ORIGIN Germany
CREATOR(S) Werner von Siemens
EARLIEST KNOWN USE 1851

BASIC INFRASTRUCTURE/MATERIALS Electrical power source, utility poles (optional), telegraph wire, wire, relay magnet, clock gong, hammer, telegraph key, speaker (optional)

RELATED Radiotelegraphy [16.1], electrical telegraph [33], electrical printing telegraph [33.1], image telegraph [33.2], pantelegraph [33.4], telephonic telegraph [33.5], barbed wire telegraph [40]

DESCRIPTION While there is little remaining archival evidence, the first fire-alarm telegraph installed in a city was developed by German engineer Werner von Siemens in Berlin, Germany, in 1851. The better-known and more publicly accessible citywide fire-alarm telegraph was launched in Boston, Massachusetts, in 1852 by William Francis Channing and Moses G. Farmer. This system consisted of forty signal stations connected by forty-nine miles of wire. By 1861, once the system had been adopted by numerous cities across

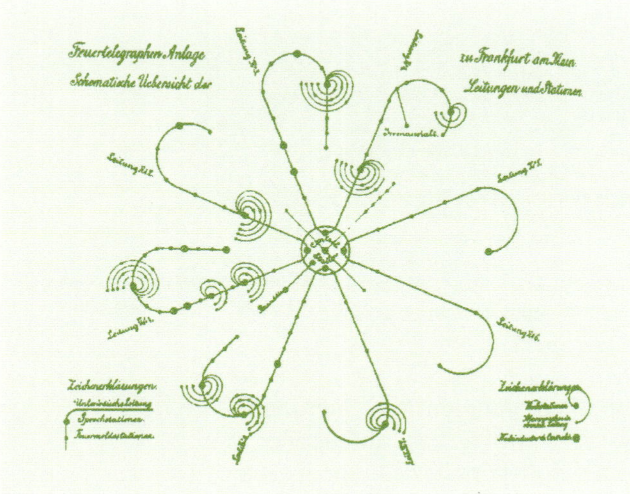

Image of Frankfurt's telegraph network before 1880 connecting individuals, via speaker stations (*Sprechstationen*), and the public, via alarm boxes (*Feuermeldestationen*).

the USA, an individual described in *The Scientific American* the fire-alarm in Providence, Rhode Island, working as follows:

> A single line of No. 8 galvanized wire, starting at an engine house at one extreme of the city, passes over and is frequently insulated upon the house tops, until it reaches the engine house at the other extreme end of said city, the line being about 4 1/4 miles long, and having at each end a "sulphate of copper" battery of 30 cups. The circuit embraces all the engine houses, the police office, the chief engineer's office, the room of the night watch, and several stores whose proprietors agree to ring, whenever required, several church bells in their neighborhood. In each of the places referred to, is a box containing a relay magnet, clock-gong, hammer and a [telegraph] key for operating—in all 17 boxes . . . as the electrical current is kept upon the line always, any person understanding the fire signals can give a proper alarm at any of the boxes by simply pressing the "key" the requisite number of times. At each pressure upon the said key the current is interrupted, and a simultaneous report of operations in the box used is made upon the other "gongs" in all the other boxes by their respective hammers. Generally, within two or three minutes after the signal is thus given, the church bells used for alarming the firemen are in motion and striking the number of the district in which the fire may be . . . Bells two miles apart often commence alarming at the same moment.

By 1903, fire-alarm telegraphs had been installed in cities around the world including Havana (Cuba), Honolulu (USA), and Kingston (Jamaica). Reportedly, many call boxes are still operational in large urban centers and are also considered a crucial part of emergency signaling because of their reliability and accessibility.

SOURCES Paul Roncallo, *History of the Fire Alarm and Police Telegraph* (M.T. Publishing, 2005); "Fire Alarm Telegraph," *The Scientific American* 5:13 (September 28, 1861); Jan Hua-Henning, "Opening the Red Box: The Fire Alarm Telegraph and Politics of Risk Response in Imperial Germany, 1873–1900," *Technology and Culture* 62:3 (2021); *Emergency Signaling* (Gamewell Fire Alarm Telegraph Company, 1916)

Pantelegraph [33.4]

COUNTRY OF ORIGIN Italy
CREATOR(S) Giovanni Caselli
EARLIEST KNOWN USE 1857

BASIC INFRASTRUCTURE/MATERIALS Electrical power source, utility poles (optional), telegraph wire, iron wires, pendulums, clock mechanisms, styluses, metal frames, nonconductive liquid, cylinders, chemically treated paper

RELATED Radiotelegraphy [16.1], electrical telegraph [33], electrical printing telegraph [33.1], image telegraph [33.2], fire alarm telegraph [33.3], telephonic telegraph [33.5], barbed wire telegraph [40]

DESCRIPTION The pantelegraph was a facsimile device (or, given its use of styluses, a telautograph [36]) capable of transmitting images and handwriting via telegraph wire over long distances. While the pantelegraph is often referred to as the first telefax or facsimile device, it was more accurately the first facsimile device widely adopted for both governmental and public use. The electrical printing telegraph and the image telegraph [33.2] were both viable means for transmitting images; but it's likely the pantelegraph succeeded commercially because its transmitter and receiver were more closely synchronized, and thus it could produce accurate reproductions of images more quickly.

Italian priest and physics professor Giovanni Caselli successfully demonstrated the pantelegraph in Paris in 1857. Similar to other telegraph-based facsimile devices of the time, the pantelegraph also used a regulating clock and a large, weighted pendulum that

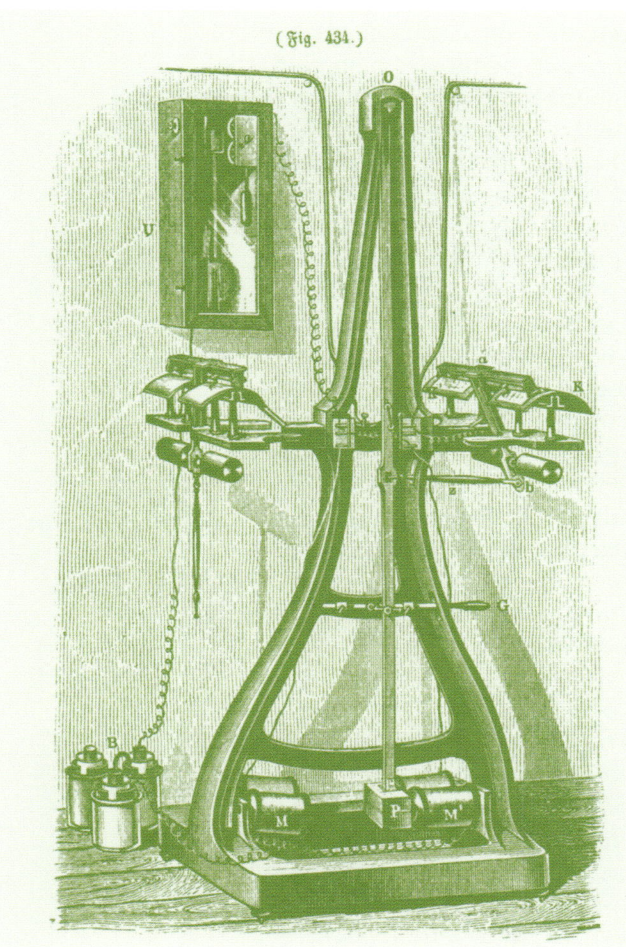

Engraved illustration of the pantelegraph from the 1876 publication *Die gesammten Naturwissenschaften*.

was powered by a battery; the clock and pendulum were also responsible for making and breaking the circuit to ensure that the transmitting and receiving styluses (attached to the pendulum and a cylinder) were synchronized. Each message was written with insulating ink on two separate metal plates, and the pendulum and stylus would scan one line of the image while it swung in one direction and scan another line as it swung in the opposite direction. The receiver included a chemically treated, damp sheet of paper on a metal plate and, as the stylus scanned the paper, the transmitted image would appear in dark blue. In 1861, French Emperor Louis Napoleon approved a test of the pantelegraph on a 140-kilometer (87 mile) section of telegraph wire between Paris and Amiens, France, and then again on an 800 kilometer (497 mile) section between Paris and Marseille. Reportedly, in a demonstration that took place between Paris and Lyon in 1865, the device transmitted roughly twenty-five words on a small sheet of paper in about 108 seconds. By 1867, roughly forty twenty-word telegrams could be transmitted in an hour via pantelegraph. The device was also used on a line between London and Liverpool in 1863 as well as a line between St. Petersburg and Moscow in 1864. While it is unclear when pantelegraph service ended, Huurdeman asserts than an economic crisis in England in 1864 halted the further development of the pantelegraph and the Franco-Prussian war in 1870–71 in France likewise led to its demise.

SOURCES Robert Sabine, *The History and Progress of the Electrical Telegraph* (Virtue, 1869); Anton Huurdeman, *The Worldwide History of Telecommunications* (John Wiley & Sons, 2003)

Telephonic Telegraph [33.5]

COUNTRY OF ORIGIN USA
CREATOR(S) Alexander Graham Bell and Thomas A. Watson
EARLIEST KNOWN USE 1876

BASIC INFRASTRUCTURE/MATERIALS Electrical power source, utility poles (optional), telegraph wire, drumhead, electromagnet, membrane, soft iron, steel springs

RELATED Radiotelegraphy [16.1], electrical telegraph [33], electrical printing telegraph [33.1], image telegraph [33.2], fire alarm telegraph [33.3], pantelegraph [33.4], barbed wire telegraph [40]

DESCRIPTION The telephonic telegraph—sometimes referred to as the articulating telephone, speaking telephone, or speaking telegraph—was capable of sending audible speech over telegraph wires. The telephonic telegraph lasted only a very short period of time before audible speech was sent via telephone [34] over telephone wires. Since it was such a short-lived network, even the name for it is barely known; moreover, the occasional use of the term "telephone" in this context creates terminological confusion given that, in the late nineteenth century, "telephone" just referred to long-distance voice transmissions, while for most of the twentieth century "telephone" referred to a particular device as well as the network of telephone wires on which the device depended; creating even more terminological confusion, in the twenty-first century, a "telephone call" could refer to voice transmissions via telephone wire, fiber optic cable, or even cellular [49].

In 1874, several years before Alexander Graham Bell and his assistant Thomas Watson produced a working model of the telephonic telegraph, Bell and electrical engineer Elisha Gray separately (but at roughly the same time) submitted patent designs for what they

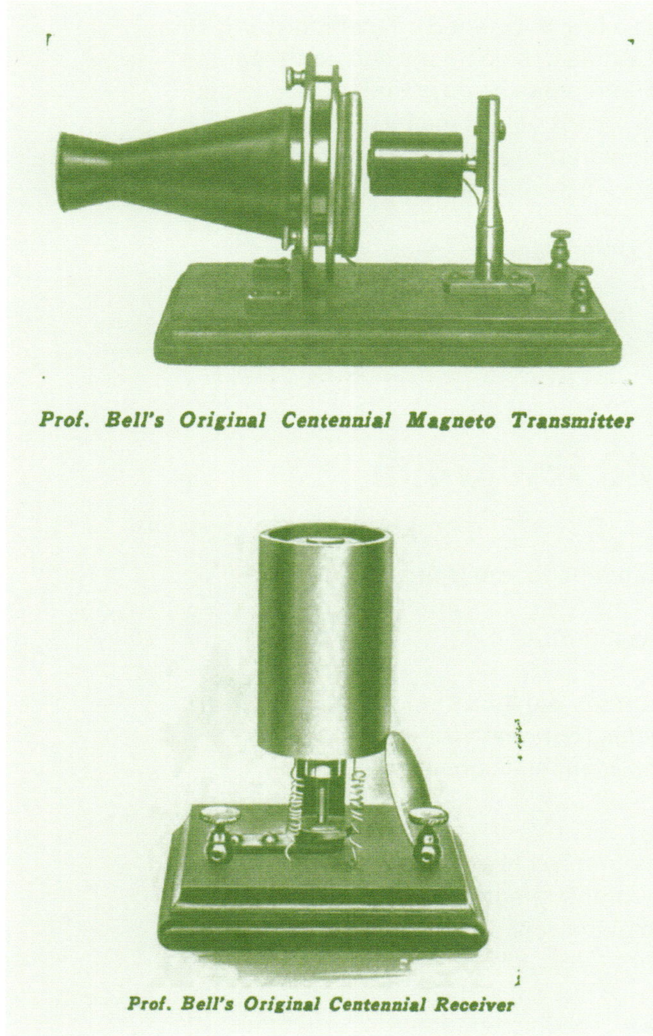

Alexander Graham Bell's transmitter and receiver for the first telephonic telegraph (1876–1877).

called a "harmonic telegraph"—a telegraph that could transmit musical tones and which was also intended to transmit speech. By 1876, both Bell and Gray were reportedly able to get their harmonic telegraph to transmit speech, but not over any significant distance and not with any clarity. However, Bell—who had been studying telegraphy since 1867—remained confident: he later wrote that "I used to tell my friends, that some day or other we should talk by telegraph."

This expectation came to fruition in October 1876, when Bell and Watson successfully demonstrated talking by telegraph over the two-mile distance between Boston and Cambridge (USA). According to Watson, the device basically sent an undulatory current "through the connecting wire to the distant receiver which, fortunately, was a mechanism that could transform that current back into an extremely faint echo of the sound of the vibrating spring that had generated it." After further instruction from Bell, Watson then mounted "a small drumhead of gold beater's skin over one of the receivers, join[ed] the center of the drumhead to the free end of the receiver spring and arrange[d] a mouthpiece over the drumhead to talk into. His idea was to force the steel spring to follow the vocal vibrations and generate a current of electricity that would vary in intensity as the air varies in density during the utterance of speech sounds." The first telephonic message delivered via telegraph wire was simply Bell saying, "Mr. Watson, please come here, I want you." It is unclear when audible speech ceased to be sent over telegraph wire. However, telephone wires consisted of a pair of copper wires twisted together by hand (or "hand-drawn," as the process was called) and Alexander Graham Bell's patent for "twisted-wire cabling" was approved in 1881; furthermore, Watson's chronology of the telephone states that 1884 was when telephone conversations began via hand-drawn copper wire strung overhead. Thus, it seems likely that 1884 marks the end of the use of telegraph wire.

SOURCES "Bell's Articulating Telephone," *Journal of the Society of Telegraph Engineers* (1876); Alexander Graham Bell, *The Multiple Telegraph: Alexander Graham Bell's Statement of Inventions Filed with the Honorable Commissioner of Patents in Conformity with Rule 53* (Franklin Press, 1876); Alexander Graham Bell, "Essay Written February 6, 1879" (Library of Congress); Thomas A. Watson, *The Birth and Babyhood of the Telephone* (American Telephone and Telegraph Company, 1940); Russell W. Burns, *Communications: An International History of the Formative Years* (Institution of Electrical Engineers, 2004); Alexander Graham Bell, "Telephone-circuit," US Patent 244,426 (1881)

Telephone [34]

COUNTRY OF ORIGIN USA
CREATOR(S) Alexander Graham Bell
EARLIEST KNOWN USE 1876

BASIC INFRASTRUCTURE/MATERIALS Electrical power source, utility poles (optional), electrical wire, transmitter, receiver, handset (optional), loudspeaker (optional), ringer or indicator (optional), operator (optional), switchboard (optional)

RELATED Amateur radio [16], two-way radio [20], telephonic telegraph [33.5], videophone [38], fence phones [41], telephonoscope [55], mundaneum [58]

DESCRIPTION The telephone refers both to the long-distance transmission of voice signals over electrical wires and to the device used to transmit and receive these signals. As networks of telephones were developed around the world, the terms "public switched telephone network" (PSTN) and later "plain old telephone service" (POTS) emerged to describe the full range of local, regional, and national telephone networks along with the wires, cables, microwave radio-relay links [31], satellites, switching centers, etc., that made possible this international network. The telephone was originally referred to as the "electric speaking telephone" as a way to distinguish it from "telephones operated electrically which were not available for conversation," as well as acoustic telegraphs and speaking telegraphs, both of which were often referred to as telephones before the twentieth century. The telephone, as we now commonly understand the term, quickly superceded the telegraph as the most popular and influential telecommunications network because, according to John Kingsbury in 1915, of its ability to provide direct communication "without the intervention of a third person," because it was much faster than communication via telegraph, and because "no expense is required either for its operation maintainence or repair. It needs no battery, and has no complicated machinery. It is unsurpassed for economy and simplicity."

While Anton Huurdeman documents how German schoolteacher Johann Philipp Reis was the first to demonstrate a device called the "telephon" capable of transmitting music (with plans for transmitting speech) over electrical wires, it is nonetheless still the case (after well over a century of debate) that, as I noted in the previous entry on the telephonic telegraph [33.5], Alexander Graham Bell is considered the inventor of the telephone as defined above. On February 14, 1876, only two hours before Elisha Gray applied for a patent for a device capable of transmitting speech, Alexander Graham Bell applied for a patent called Improvement in Telegraphy. Despite the importance of this patent, it is worth noting it does not mention the words "telephone" or "speech." Instead, Bell writes at the end of the patent that he desires "to secure by Letters Patent . . . The method of, and apparatus for, transmitting vocal or other sounds telegraphically, as herein described, by causing electrical undulations, similar in form to the vibrations of the air accompanying the said vocal or other sounds." The latter essentially involved using a diaphragm in the transmitter and receiver to convert sound energy into electrical energy and vice versa. Bell's assistant Thomas Watson lists 1876 as the year the first complete sentence was transmitted by telephone and also the year the first conversation was transmitted by overhead line over a two-mile distance from Boston to Cambridge.

After the Supreme Court confirmed Bell as the inventor of the telephone in 1879, "telephony became big business" (in the words of Huurdeman), particularly in the USA By 1880, telephone conversations were being transmitted over the forty-five-mile distance from Boston to Providence, Rhode Island, and there were supposedly 30,872 telephones in use in the USA Notably, the development of the telephone network was delayed in the UK because of the passing of the Telegraph Act in 1869, which meant that the post office was legally sanctioned to have a monopoly on all telegraphic communications, further ensured by the High Court's confirmation in 1889 that "the telephone was a telegraph within the meaning of the act." Still, by the end of the nineteenth century, telephone service, and even limited long-distance calling, was available in most cities across North America and Europe. Huurdeman notes that, at the time, the only other international telephone line that existed outside

How To Talk on the Telephone

When You Answer the Telephone

1. Pick up the handset. Hold the receiver part of it close against your ear and hold the mouthpiece about an inch in front of your mouth.

2. Say "Hello" or say your telephone number. Speak clearly, but don't shout. Talk into the telephone the way you would talk to someone face to face. Be polite and pleasant.

3. Sometimes the person who calls wants to speak to someone else at your house. Let's pretend you are Janie Allen, and Mr. Wright calls to talk to your Daddy. You and Mr. Wright know each other.

Mr. Wright says "Hello, Janie. May I speak to your Daddy?"

You say "Yes, Mr. Wright. I'll call him." Then you lay the handset down beside the base of the telephone. **Don't put it back in the cradle.** Next, you go find your Daddy and tell him that Mr. Wright wants him on the telephone. Don't stand close to the telephone and yell "Daddy!" That would hurt Mr. Wright's ear.

4. When someone calls you, let the person who called end the talk. Of course, if someone talks on, and on, and on, you may have to say "I'm sorry, but I have to stop now. Thank you for calling."

When You Call Someone on the Telephone

1. Try to be sure you are calling the right number.

2. When someone answers, tell your name right away.

3. Perhaps the person who answers the telephone is not the one you are calling. Let's pretend you are Howard Allen and you want to talk to Bill Wright. Bill's mother answers the telephone and you know her voice.

Mrs. Wright says "Sunnyside 5-3757."

You say "Hello, Mrs. Wright. This is Howard Allen. May I speak to Bill, please?"

4. If Mrs. Wright says "I'm sorry, Howard. Bill isn't in," don't say "Oh" and hang up. Say "Thank you, Mrs. Wright. I'll call again. Goodby."

5. When you call someone, you are supposed to close the conversation when you are through. Then you say "Goodby." And remember: don't talk on, and on, and **on!**

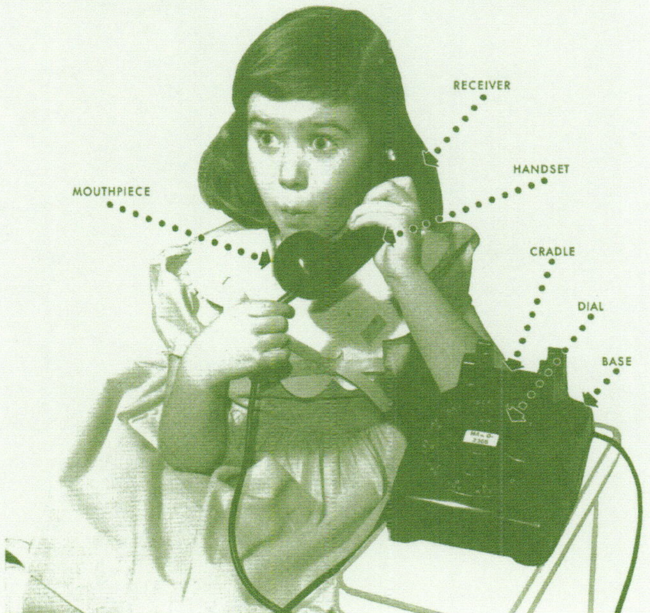

Instructions from the 1950s explaining proper telephone etiquette to children.

[34] John Giorno with his Dial-A-Poem answering machines in 1970, and newspaper clippings documenting the establishment of the hotline from 1969 to 1970.

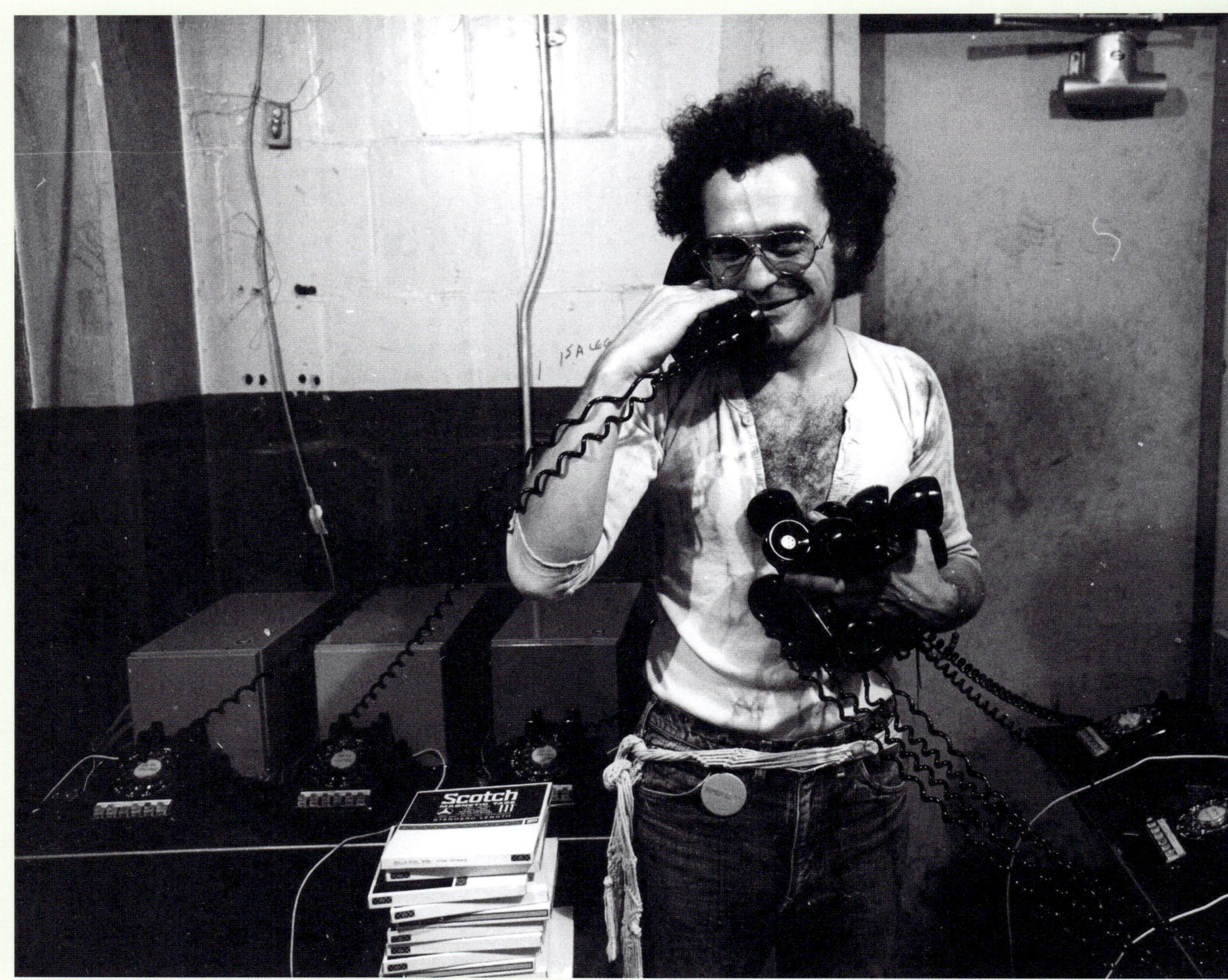

The Architectural League of New York presents

JOHN GIORNO'S DIAL - A - POEM

DIAL (212) 628-0400 NOW

POETS:	
BILL BERKSON	TAYLOR MEAD
WILLIAM BURROUGHS	RON PADGETT
JOHN CAGE	JOHN PERREAULT
ALLEN GINSBERG	PETER SCHJELDAHL
JOHN GIORNO	ANNE WALDMAN
DAVID HENDERSON	LEWIS WARSH
	EMMET WILLIAMS

6 Poets   101 Phone Lines   Changed Daily

The Architectural League of New York presents

JOHN GIORNO'S DIAL - A - POEM

DIAL (212) 628-0400 NOW

POETS:	TAYLOR MEAD
VITO ACCONCI	RON PADGETT
BILL BERKSON	JOHN PERREAULT
WILLIAM BURROUGHS	PETER SCHJELDAHL
JOHN CAGE	ANNE WALDMAN
ALLEN GINSBERG	LEWIS WARSH
JOHN GIORNO	JOHN WIENERS
DAVID HENDERSON	EMMET WILLIAMS

6 Poets   101 Phone Lines   Changed Daily

the village VOICE, *January 30, 1969*

Nygaard to Conduct

A chamber concert under the direction of Jens Nygaard will be presented on Wednesday, January 29, at 8.30 p. m. at Greenwich House Music School, 46 Barrow Street. The program will include Reger's "Canon and Fugue in the Olden Style for Two Solo Violins," Hindemith's English Horn Sonata, Schumann's "Four Duets for Two Sopranos and Piano," and Mozart's "Five Country Dances," K. 609. Livio Caroli will be soloist in the Hindemith. Admission is free.

Phone for Poetry

The Architectural League of New York has instituted "Dial-a-Poem," arranged by John Giorno. Two minute recorded poems by Bill Berkson, William Burroughs, Allen Ginsberg, John Giorno, David Henderson, Taylor Mead, Ron Padgett, John Perreault, Ed Sanders, Peter Schjeldahl, Anne Waldman, Lewis Warsh, and Emmett Williams can be heard by calling (212) MA 8-0400. The telephone number is connected by six lines to the same number of automatic-answering sets, each containing one taped poem. Poems will be changed daily and will be available at any time during the day for two months.

In Bloom

The Chinese witch-hazel has begun to bloom at the Brooklyn Botanic Garden. The delicate, spidery, yellow flowers of the fragrant shrub are appearing much earlier than usual this year and will probably last for at least a month even with snow.

Kossoys in Concert

The Kossoy Sisters, popular folksingers in Greenwich Village in the early '60s, will give their first New York concert in five years on Wednesday, February 5, at 8.30 p. m. at Washington Square Methodist Church, 135 West 4th Street. Admission is $2.

Brazilian Fete

The Brazilian Cultural Society will hold its 1969 Carnival celebration on Friday, January 31, at the Grand Ballroom of the Biltmore Hotel. For information call PL 7-4231.

THE VILLAGER, GREENWICH VILLAGE,

NEW YORK, THURSDAY, JANUARY 16, 1969

Weather, Time, Prayer . . . Now It's Dial-a-Poem

A year ago Marshall McLuhan, electric prophet, was roundly put down by *The New York Times Book Review* for using the poor old obsolete medium, print, to hand out his ear-and-touch over eye message. The Book Review suggested the right way to get McLuhan's message, since he wasn't due to be televised, was to call him up on the phone and ask for it, and they published his phone number.

Now some people generally engaged on the visual side, those in the Architectural League of New York, have picked up on the medium and by dialing 628-0400 you can dial-a-poem, and be massaged by Allen Ginsberg singing Hare Krishna, or messaged by the works of the 12 other contributing poets.

Among the 12 are William Burroughs, Taylor Mead, Ron Padgett, John Perreault, John Giorno, who thought up the idea and Ed Sanders of the Fugs. There are supposed to be six poets simultaneously over six lines that are changed daily, but the opening day, they provided only hare krishnas and busy signals.

Mr. Giorno was struck with the idea while dialing the weather one day, and conceives of it as "extending the poets' work through technology to a huge audience all over the world."

The poems will run 24 hours a day for those in need, and extend over the next two months.

The Architectural League of New York

presents

JOHN GIORNO'S DIAL-A-POEM

DIAL (212) 628-0400 NOW

POETS:	RON PADGETT
BILL BERKSON	JOHN PERREAULT
WILLIAM BURROUGHS	ED SANDERS
ALLEN GINSBERG	PETER SCHJELDAHL
JOHN GIORNO	ANNE WALDMAN
DAVID HENDERSON	LEWIS WARSH
TAYLOR MEAD	EMMET WILLIAMS

6 Poets   Over 6 Lines   Changed Daily

the village VOICE, *March 20, 1969*

JOHN GIORNO'S DIAL-A-POEM

(212) 628-0400

presents

FRANK O'HARA

on 10 lines for 24 hours

Tuesday, March 25

in conjunction with the opening of the Barnett Newman exhibition of paintings at Knoedler to benefit the Frank O'Hara Foundation.

of North America and Europe was a line built in 1894 by colonial powers to connect the West African countries Togo and Benin. By 1950, after substantial international investment in telephone network infrastructure (including the laying of transatlantic undersea cables), the number of telephones around the world was roughly 75 million.

The key developments in telephony that led to its stunning worldwide dominance were the transition to the use of pairs of twisted copper wires (patented by Bell in 1881 and briefly discussed in the introduction to this section) and the creation of switchboards in 1878, especially automatic switching in 1892. The first telephone lines were direct connections between users who would whistle into the wire to signal they wished to talk. As the telephone quickly became popular, telephone lines were instead connected to a central office or switchboard run by an operator who would manually connect telephone calls. The first commercial switchboard that operated in New Haven, Connecticut, was technically a "party line," or a single telephone line that served multiple households. Party lines were the norm in rural areas, and during wartime in the first half of the twentieth century, congestion on the lines prompted local telephone companies to produce numerous guides for "party line etiquette." The first automatic telephone exchange went into service in La Port, Indiana, in 1892, making it possible for subscribers to dial numbers instead of calling on the services of an operator. Eventually, operators were no longer needed, as the vast majority of telephone users were able to make local and later long-distance calls using automated pulse-dialing systems.

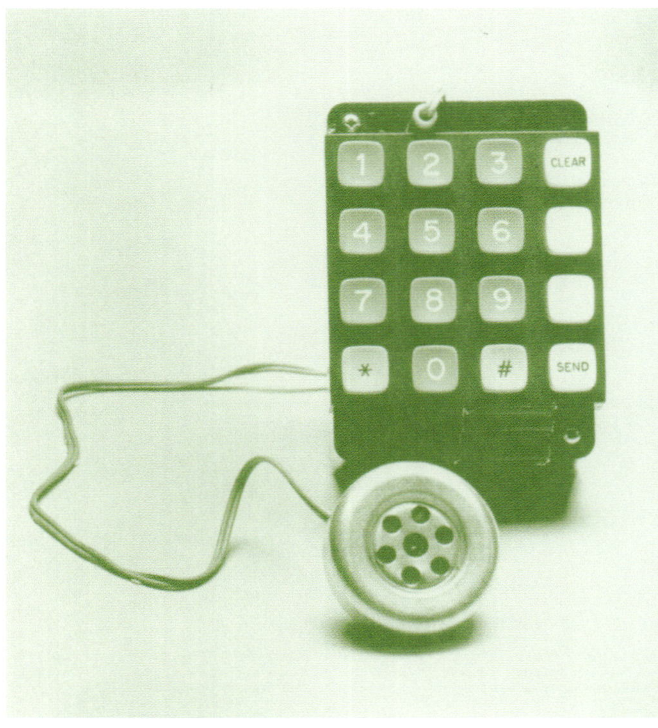

Apple co-founder Steve Wozniak's phone-phreaking blue box from 1972.

Telephone booths (coin-operated payphones available to members of the public, which were installed in private or self-contained booths) began to grow in ubiquity starting in the 1890s; the widespread availability of answering machines by the mid-1980s also significantly contributed to the dominance of the telephone network. While *Telephone Magazine* speculated in 1904 that a new invention called the "telegraphone" (a device created by Valdemar Pousen after his demonstration of magnetic recording in 1898) that could transmit electromagnetic sound recordings over electrical wire could be "especially suitable in connection with telephones in big offices, for recording the conversations which take place" and "in small offices . . . for giving and receiving notice in the case of the subscriber being absent," no commercially viable answering machine devices appeared until 1929. That year, *The New York Times* announced from London that "No longer will the telephone operator's 'No reply' face callers in this country. If there is nobody home, an invention just perfected at Elstree . . . functions in the absence of the subscriber, taking down the message provided the necessary attachment is connected to the telephone." In 1931, an advertisement appeared in *Popular Mechanics* for a "robot" that receives and answers telephone messages, and an "ansophone" was then announced in *Modern Mechanix* the following year. However, it was not until the breakup of the Bell system in the US in 1984 and the subsequent ability for customers to purchase their own telephone equipment that the answering machine went from being a piece of office equipment or a luxury item at home to an essential household appliance.

Since the advent of digital telephony (also in the mid-1980s), the PSTN has been largely superceded by internet protocol telephony and support for use of the PSTN is being phased out, even if its infrastructure is still clearly evident. In fact, the US Federal Communications Commission called for a phase-out of the PSTN in 2019, and the UK's PSTN will be switched off entirely in 2025.

EXPERIMENTS The telephone has been extensively used by and for artists throughout the twentieth century, especially as a means for either depersonalizing or mechanizing the art production process or, once telephone answering machines became affordable, to create one-to-many broadcasts of writing, poetry, and music. In 1896, seventeen years after Bell and Watson's successful telephone exchange, American

Thaddeus Cahill patented the first of three versions of a device he called the Telharmonium (or the Dynamophone). Because of a reference in Cahill's first patent to the device's ability to "synthesize," along with the way it was intended to electrically produce, as Simon Crab puts it, "a universal 'perfect instrument'; an instrument that could produce absolutely perfect tones, mechanically controlled with scientific certainty," the Telharmonium is considered the first music synthesizer. Crab continues: "The Telharmonium would allow the player to combine the sustain of a pipe organ with the expression of a piano, the musical intensity of a violin with polyphony of a string section and the timbre and power of wind instruments with the chord ability of an organ. Having corrected the 'defects' of these traditional instruments the superior Telharmonium would render them obsolete." Another key aspect of the Telharmonium was the fact that telephone receivers were needed to hear the music, which was in turn transmitted along dedicated telephone lines Cahill set up with the New York Telephone Company to run beside phone company cables. Using these dedicated lines, Cahill broadcast Telharmonium concerts to music halls, restaurants, hotels, and even private residences around New York City. Over a twenty-year period, Cahill designed three different versions of the Telharmonium that had two to four different keyboards, a clutch that worked like a foot pedal, acoustic horns up to six feet wide, a mainframe that was as long as sixty feet, ten switchboard panels and 2,000 switches; the device also weighed two hundred tons (about 18,150 kilograms) and cost around $200,000 (with inflation, that is equivalent to about $2.8 million USD in 2024). The Telharmonium models were eventually disassembled and sold for scrap after Cahill filed for bankruptcy in 1914 and after Cahill's brother died in 1958.

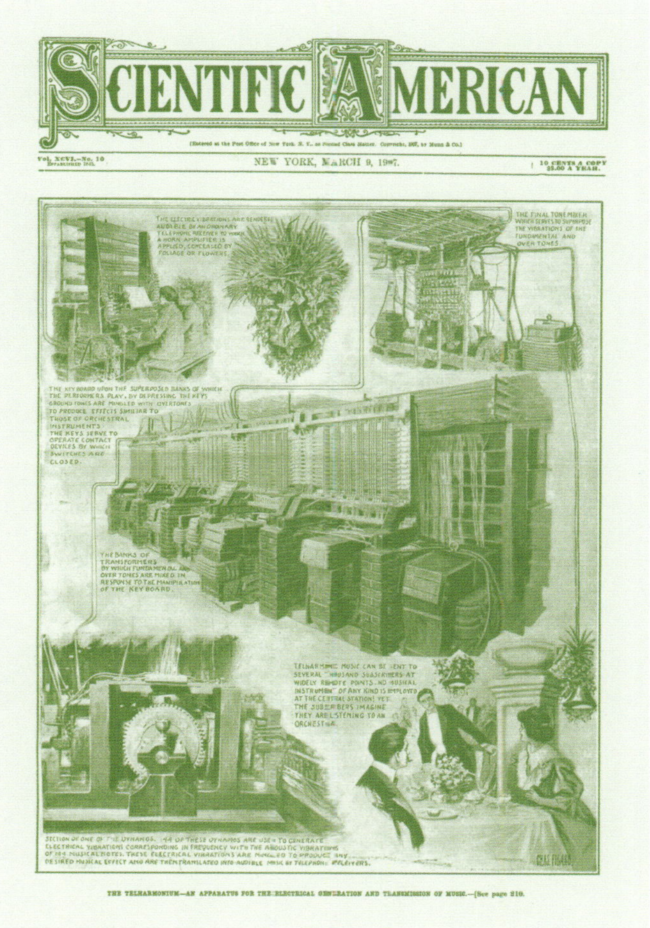

SCIENTIFIC AMERICAN

NEW YORK, MARCH 9, 1907.

Scientific American announces the Telharmonium on March 9, 1907, as "an apparatus for the electrical generation and transmission of music."

In his 1947 book *The New Vision and Abstract of an Artist*, Hungarian artist László Moholy-Nagy recounts how, in 1922, he "ordered by telephone from a sign factory five paintings in porcelain enamel." He describes the process as follows: "I had the factory's color chart before me and I sketched my paintings on graph paper. At the other end of the telephone the factory supervisor had the same kind of paper, divided into squares. He took down the dictated shapes in the correct position. (It was like playing chess by correspondence.) One of the pictures was delivered in three different sizes, so that I could study the subtle differences in the color relations caused by the enlargement and reduction. True, these pictures did not have the virtue of the 'individual touch,' but my action was directed exactly against this over-emphasis." It is perhaps worth mentioning, however, that Lucia Moholy later disputed the accuracy of his story about these so-called telephone pictures, claiming instead that Moholy-Nagy ordered the paintings in person at the factory.

In 1969, influenced by conceptual/performance artists Andy Warhol, Robert Rauschenberg, and Trisha Brown, American poet John Giorno launched his Dial-A-Poem phone service—a hotline that individuals could call to hear writers (including Allen Ginsberg, William Burroughs, and Anne Waldman) read poems recorded on one of ten reel-to-reel answering machines housed in New York City's Architectural League. In 1970, Dial-A-Poem moved to the Museum of Modern Art, at which point the service expanded to include Black Panther–affiliated and queer poets. The hotline received more than one million phone calls and had been investigated by the FBI for indecency and for inciting violence by the time it was forced to end in 1971. Giorno subsequently released LPs of the poetry readings; Ralf Webb, writing in one of the release's liner notes, puts it succinctly: "We used the telephone for poetry. They used it to spy on you." As

[34] *Scientific American* announces "the new Bell telephone" with a full front-page article on October 6, 1877.

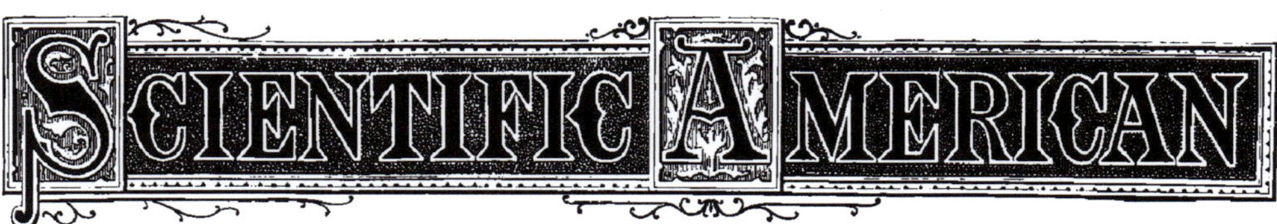

A WEEKLY JOURNAL OF PRACTICAL INFORMATION, ART, SCIENCE, MECHANICS, CHEMISTRY, AND MANUFACTURES.

Vol. XXXVII.—No. 14. [NEW SERIES.] NEW YORK, OCTOBER 6, 1877. [$3.20 per Annum. [POSTAGE PREPAID.]

THE NEW BELL TELEPHONE.

Professor Graham Bell's telephone has of late been somewhat simplified in construction and also arranged in more compact portable form. It consists now of but three metal portions and is contained in a casing of wood or light hard rubber, but five and five eighths inches in length and two and seven eighths inches in diameter at the enlarged end. It will be remembered that this telephone differs from all others in that it involves the use of no battery nor of any extraneous source of electricity whatever. The only current employed is that generated by the voice of the speaker himself.

The simplicity of the construction is clearly shown in Fig. 1 of our engravings, in which both sectional and exterior views of the device are given. Referring to the sectional view, A is a permanent magnet, held by the screw shown in the rear. Around one end of this magnet is wound a coil, B, of fine insulated copper wire (silk covered), the ends of which are attached to the larger wires, C, which extend to the rear and terminate in the binding screws, D. In front of the pole and coil, B, is a soft iron disk, E. Finally the whole is inclosed in a wooden casing having an aperture in front of the disk, and which, besides serving to protect the magnet, etc., acts somewhat as a resonator.

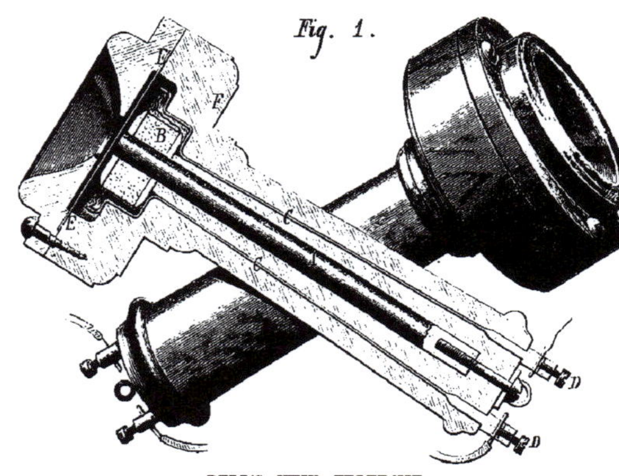

BELL'S NEW TELEPONE.

The principle of the apparatus we have already explained in some detail, but it may be summarized here as follows: The influence of the magnet induces all around it a magnetic field, and the iron diaphragm, E, is attracted towards the pole. Any alteration in the normal condition of the diaphragm, produces an alteration in the magnetic field, by strengthening or weakening it, and any such alteration of the magnetic field causes the induction of a current of electricity in the coil, B. The strength of this induced current is dependent upon the amplitude and rate of vibration of the disk, and these depend in turn upon the air disturbance made by the voice in speaking, or in any other similar source. Therefore, first, a wave of air throws the diaphragm into vibration; second, each movement produces a change in the magnetic field; and third, an induced

[*Continued on page* 212.]

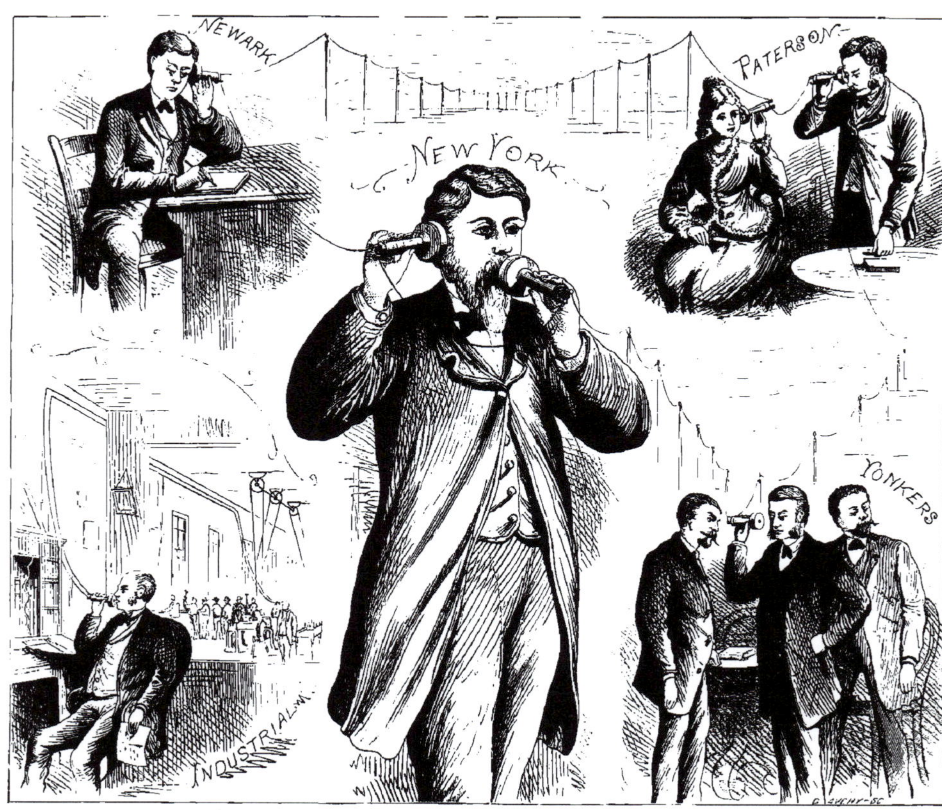

APPLICATIONS OF PROFESSOR BELL'S NEW TELEPHONE.

of 2024, a version of Dial-A-Poem can be accessed at +1(641) 793-8122, where listeners can hear one of about two hundred recordings of poems. The American band They Might Be Giants also had a "Dial-a-Song" service that operated from 1983 to 2006 and then again from 2015 to 2018. Originally called Dial-a-Machine, individuals could call +1(718) 387-6962 and connect with an answering machine that played back (usually on cassette tape) previews, demos, or released versions of songs. By the late 1980s, the service fielded more than one hundred calls a day.

Starting in 1955 and arguably continuing today as various practices of "frequency hacking," "phreaking" is the practice (often by self-identified hackers) of re-creating specific tones used by public switched telephone networks to explore the underlying workings of telephone switching, make free phone calls (in the era of long-distance charges), and participate in conference calls. American David Condon (a pseudonym) was likely the first to discover that a children's toy called a "Davy Crockett Cat and Canary Bird Call Flute" could generate a 1,000-hertz tone that he used in the mid-1950s for signalling between operators over long-distance telephone circuits. Condon orchestrated free long-distance calls with the help of a girlfriend who posed as an operator and two telephone booths: "We'd go out to Oak Ridge [California] and we'd get on a phone that wasn't monitorable, a pay station. You call an operator in a distant city, they don't have a call for you, and when the operator releases you, you ring and hand it to the girl! She knew what to say. I had written it down for her." In 1957, Joe Engressia (now widely cited as the "grandfather" of phreaking) discovered that a 2,600 hertz tone was used internally by AT&T to automatically signal that a long-distance call had ended, thereby leaving the line open and available for exploitation. Thereafter—particularly after phreakers figured out that various combinations of whistles could also be used to dial numbers— instructions for building devices called "blue boxes," "black boxes," and "red boxes" capable of automatically generating tones began to appear. After the October 1971 publication of "Secrets of the Little Blue Box" in *Esquire* magazine, phreaking became known around the world, and was most notoriously practiced by Apple co-founders Steve Jobs and Steve Wozniak.

SOURCES John E. Kingsbury, *The Telephone and Telephone Exchanges* (Longmans, Green, 1915); Anton Huurdeman, *The Worldwide History of Telecommunications* (John Wiley & Sons, 2003); Russell W. Burns, *Communications: An International History of the Formative Years* (Institution of Electrical Engineers, 2004); Alexander Graham Bell, "Improvements in Telegraphy," US Patent 174,465 (March 7, 1876); Thomas A. Watson, *The Birth and Babyhood of the Telephone* (American Telephone and Telegraph Company, 1927); Ronald Kline, *Consumers in the Country: Technology and Social Change in Rural America* (Johns Hopkins University Press, 2002); "The Use of the Telegraphone in Telephony," *Telephone Magazine* 23 (January–June 1904); "Answering Machine," *The New York Times* (October 10, 1929); "Robot That Answers Phone Takes Messages," *Popular Mechanics* (July 1931); "Device Answers Phone and Tells Caller When You Will Return to Office," *Modern Mechanix* (August 1932); AT&T and Wayne University College of Education, *The Telephone and How We Use It* (Bell Telephone System, 1961); Federal Communications Commission, Order 19-72A1 (August 2, 2019); "We're Retiring Our Copper Network," Openreach.com; László Moholy-Nagy, *The New Vision and Abstract of an Artist* (Wittenborn, Schultz, 1947); Simon Crab, "The 'Telharmonium' or 'Dynamophone' Thaddeus Cahill, USA 1897," 120years.net; Jay Williston, "Thaddeus Cahill's Teleharmonium," Synthmuseum.com; Lucia Moholy and László Moholy-Nagy, *Marginalien Zu Moholy-Nagy: Dokumentarische Ungereimtheiten* (Scherpe, 1972); "The Dial-a-Poem Poets, 1972," Ubu.com; Ralf Webb, "He's a Poet and the FBI Know It: How John Giorno's Dial-a-Poem Alarmed the Feds," *The Guardian* (October 18, 2021); "John Giorno's Dial-a-Poem Still Brings Poetry to the Masses," SFMOMA website (July 2020); "Dial-a-Song," *This Might Be a Wiki: The TMBG Knowledge Base* website; Peg Tyre, "Giant Steps," *New York Magazine* (February 2, 1989); Phil Lapsley, *Exploding the Phone: The Untold Story of the Teenagers and Outlaws Who Hacked Ma Bell* (Grove Press, 2013); Jason Scott, "Phone Phreaking," textfiles.com/phreak; Ron Rosenbaum, "Secrets of the Little Blue Box," *Esquire* (October 1971)

Wired Radio [35]

COUNTRY OF ORIGIN France
CREATOR(S) Clément Ader
EARLIEST KNOWN USE 1881

BASIC INFRASTRUCTURE/MATERIALS Electrical wire, power source, utility poles (optional), transmitters, receivers, microphones, speakers, antenna, telephone handset or wired radio set or intercom system, switching or exchange station (optional), filter (optional), transformer (optional), operator (optional)

RELATED Amateur radio [16], radio broadcast [17], telephone [34], fence phones [41]

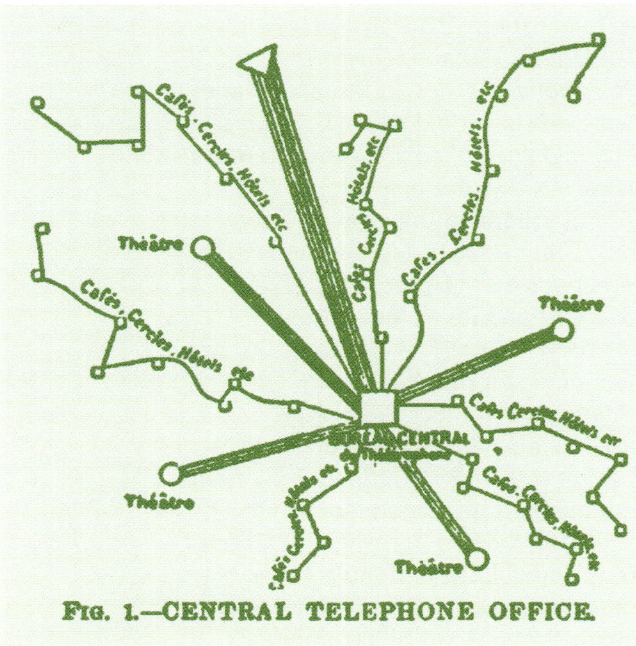

Rendering of the théâtrophone network radiating from the network's central office at No. 23 Louis-le-Grand Street in Paris that appeared in *Scientific American Supplement*, March 16, 1895.

DESCRIPTION Wired radio includes (and/or is also referred to as) wired wireless, telephone newspapers, circular telephones, théâtrophone, electrophone, carrier current radio, power line radio, and cable radio. While wired radio is not normally discussed alongside all of the aforementioned technologies (most likely because of firm disciplinary boundaries separating historians and engineers, along with a lack of technical detail about the underlying workings of these networks), I use the term here to describe voice-only broadcasts that take place over iron, copper, or coaxial cable. Transmissions may take place over or on top of telephone wires and telephone switching centers or they may take place entirely independent of telephone companies and services.

As Gabriele Balbi and Juraj Kittler point out, a lesser known history of the telephone is that it was "originally conceived as either a point-to-point or one-to-many medium." The first known instance of the use of electrical wires for voice-only broadcast was the launch of the théâtrophone (a type of circular telephone, according to Balbi and Kittler) in Paris, France, in 1881. At the International Electrical Exhibition, railroad engineer Clément Ader demonstrated a device that connected telephone switching stations to the Paris Opéra, located two kilometers away from the exhibition. The system was dismantled at the end of the exhibition and was not resurrected until 1889, as "the first permanent telephone-based entertainment service." According to Balbi, there were several public listening stations across Paris where individuals could insert a coin and listen to music for five minutes at a time. However, it was not until the launch of Tivadar Puskas's Telefon Hírmondó in 1893 in Budapest, Hungary, that listeners were given a detailed schedule of broadcast content that ranged from news reports to concerts, music programs, language lessons, literary reviews, the time signal, and advertising. Telefon Hírmondó was also unique both because it boasted as many as 6,200 subscribers and because it lasted until 1944.

The electrophone was another important pioneer in one-to-many telephone broadcasts. Launched by H.S.J. Booth and his Electrophone Company, Ltd., in London in 1895 and lasting until 1925, the service had at most 2,000 subscribers, including Queen Victoria. It broadcast concerts, performances, and even church services (where, according to J. Wright, microphones took "the form of a dummy bible [*sic*] lying in a natural position on the pulpit desk, or a hassock under the lectern"). Balbi points out the electrophone service was unique in that it was integrated with the national telephone system in such a way that subscribers could use their line as a two-way system, "calling 'the exchange using [their] normal telephone and asking for the 'Electrophone service.' They were then connected to the Electrophone operator who would offer them a selection of up to thirty programmes.'"

While numerous telephone broadcast services were attempted particularly across the USA shortly after the launch of the théâtrophone and the electrophone (including the 1906 launch of Tellevent in Michigan, the 1909 launch of Tele-musici in Delaware, the 1910 launch of Musolaphone in Chicago, and the 1911 launch of the

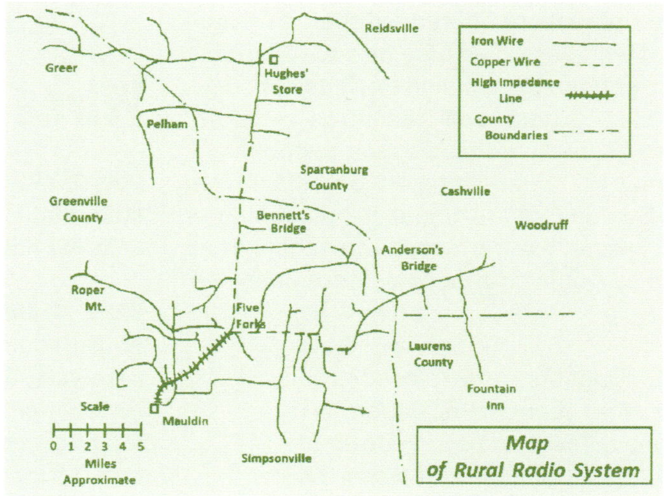

Map of South Carolina's "grapevine radio" system as it existed in 1936.

Telephone Herald in Newark, New Jersey), the most successful wired radio services seem to have been ones that operated independently of the telephone. Sometimes referred to as carrier current radio, cable radio, or power line communication, these systems transmitted audio at a different frequency than the telephone and did so over a variety of wires and cables. For example, Susan Opt notes that, similar to wired radio systems that were operational in New York City hotels such as the Waldorf Astoria and the Lexington Hotel, "Traveling down the dirt and gravel backroads of upstate South Caroline in the 1930s, one might have observed iron or copper wires dangling across treetops, attached to fencepots and cedar posts, and occasionally to telephone poles. These wires . . . served to connect nonelectrified rural communities to the radio airwaves passing overhead." This set up was called a "grapevine radio" and, according to interviews by Opt with original wired radio founders across South Carolina, was built from junk radio parts and loudspeakers. "Wherever the power line stopped, a small rebroadcasting station was established with lines running to the nonelectrified neighboring areas. These systems first appeared in the early 1930s and remained in operation until the Rural Electrification Authority Cooperatives brought electricity to the areas and residents could afford electric radios."

Wired radio was also used for college radio transmissions as early as 1936 at Brown University (USA). This was more commonly referred to as "carrier current broadcasting" in the context of college radio, such as when students David Borst and George Abrahams transmitted music and messages between Brown dorm rooms by low power AM radio frequency broadcast over the AC power lines. The latter marked the beginning of the first college radio station, initially called the Brown Network and later WBRU.

Wired radio systems were and still are also favored by government regimes intent on controlling the dissemination of information. For example, the Soviet Union began their wired radio broadcasts in 1924 via fifty loudspeakers installed across the city of Moscow. The system reportedly peaked in 1974 with 55 million public loudspeakers. It was not until 1962 that listeners had the ability to select more than one program to listen to. Germany, Singapore, Malta, Norway, and Italy, among others, also had extensive wired or cable radio networks in place. Italy's network in particular, Filodiffusione, remarkably lasted from 1958 until 2023. David Morton also points out that numerous so-called developing nations had wired radio networks installed by colonial governments and corporations; Ghana, for example, had a wired radio network built in 1935 that, with 60,000 subscribers, lasted until 1978. Wired

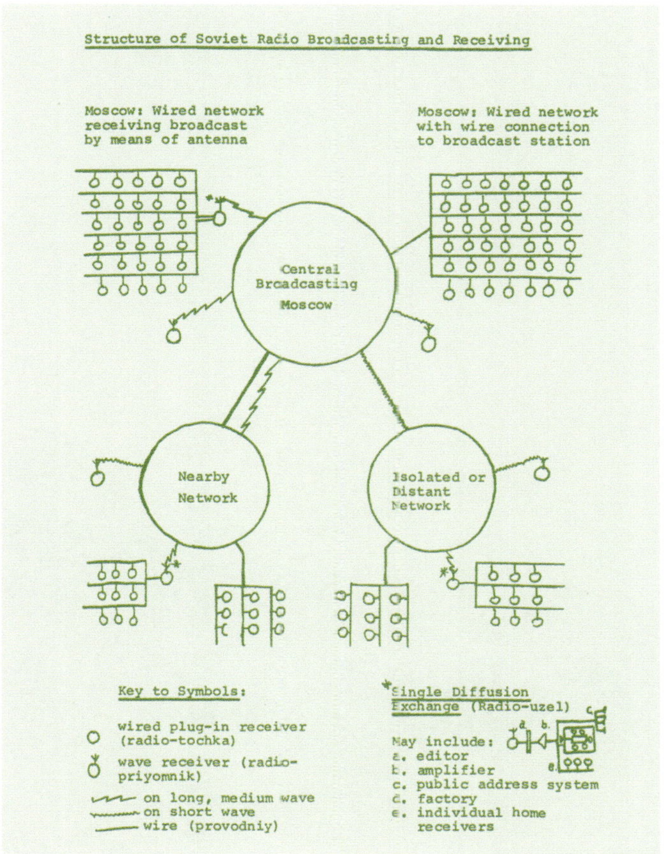

Map of the Soviet Union Wired Radio broadcasting network that appears in F. Gayle Durham's 1965 report "Radio and Television in the Soviet Union."

radio networks are rumored to still be in use in North Korea and, as of reports from 2022, authorities in Hanoi, Vietnam, have reinstated public loudspeakers to broadcast announcements.

SOURCES Gabriele Balbi and Juraj Kittler, "One-to-One and One-to-Many Dichotomy: Grand Theories, Periodization, and Historical Narratives in Communication Studies," *International Journal of Communication* 10 (2016); Gabriele Balbi, "Radio before Radio: Araldo Telefonico and the Invention of Italian Broadcasting," *Technology and Culture* 51:4 (October 2010); "The Théâtrophone," *Scientific American Supplement* (July 2, 1892); "The Queen and Her Electrophone," *Electrical Journal* 43 (April–October 1899); J. Wright, "The Electrophone," *The Electrical Engineer* 20 (July–December 1897); "The Telephone Newspaper," *The World's Work* 1 (November 1900–April 1901); "Theatrophone," *Michigan State Gazette* 1–2 (1905–1907); "Distributing Music Over Telephone Lines," *Telephony* 18 (1909); "College Radio Born at Brown," Imagine Brown 250+ website; Susan K. Opt, "The Development of Rural Wired Radio Systems in Upstate South Carolina," *Journal of Radio Studies* 1 (1992); Stephen Lovell, *Russia in*

[35] A 1950s-era cable radio device used by Italy's Filodiffusione system, with six different wired radio broadcast channels.

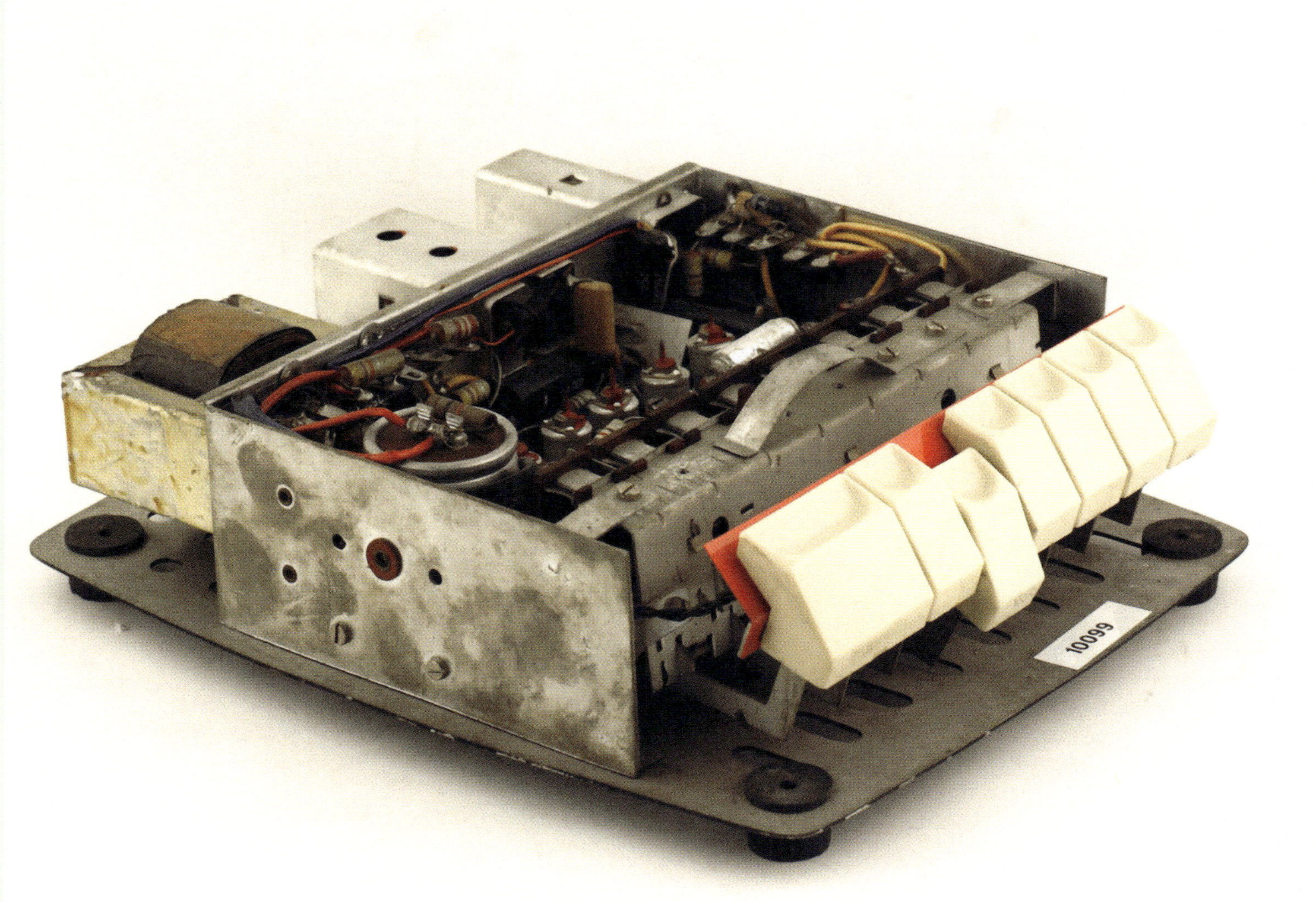

the *Microphone Age: A History of Soviet Radio, 1919 to 1970* (Oxford University Press, 2015); F. Gayle Durham, "Radio and Television in the Soviet Union," Research Program on Problems of International Communication and Security, MIT Center for International Studies (June 1965); "Early Soviet Radio Broadcasting," *Making It Up* blog (October 22, 2017); David L. Morton, *A History of Electronic Entertainment* (IEEE Press, 1999); Martyn Williams, "New Life for the Third Network," North Korea Tech website (April 29, 2021); Zachary Abuza, "Play It Loud: The Return of Hanoi's Loudspeakers Speaks Volumes," Radio Free Asia website (August 7, 2022)

Telautograph [36]

COUNTRY OF ORIGIN USA
CREATOR(S) Elisha Gray
EARLIEST KNOWN USE 1888

BASIC INFRASTRUCTURE/MATERIALS Electrical wires, power source, utility poles (optional), transmitters, receivers, wheels, drums, stylus, glass tube, ink, paper

RELATED Hellschreiber [16.4], radiofax [19], electric printing telegraph [33.1], image telegraph [33.2], pantelgraph [33.4], telefacsimile [37]

DESCRIPTION The telautograph, sometimes referred to as a writing telegraph or as a telewriter, was one of the earliest facsimile devices capable of accurately transmitting handwriting over telephone lines (or a two-wire circuit). Although it was a type of facsimile device (and also, generically, a telautograph much like the telegraph-based electrical printing telegraph [33.1], image telegraph [33.2], and pantelegraph [33.4]), in terms of how it worked, it was much closer to what we might today call a remote-controlled pen. Patented in 1888 by American engineer Elisha Gray as the first device explicitly named a telautograph, the device was unique in how it could reproduce handwriting (or any hand-drawn image) using a stylus that moved vertically as well as horizontally according to coded pulses over the electrical wires. In this sense, the telautograph is an important predecessor to any number of X-Y plotters used to print vector drawings. Gray stated in an interview five years after he launched the telautograph, "By my invention you can sit down in your office in Chicago, take a pencil in your hand, write a message to me, and as your pencil moves, a pencil here in my laboratory moves simultaneously, and forms the same letters and words in the same way . . . You may write in any language, use a code or cipher, no matter, a fac-simile is produced

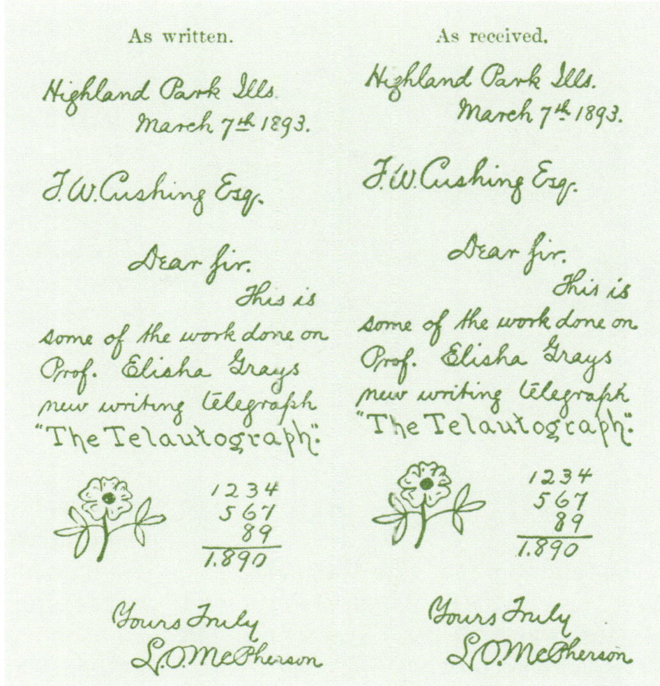

An example of a handwritten message transmitted and received on the telautograph, from 1893.

here. If you want to draw a picture it is the same, the picture is reproduced here." The telautograph was adopted by doctors to facilitate the exchange of prescriptions and by train stations in the US and Canada. The telautograph was later improved by Gray's assistant Foster Ritchie so that one could supposedly transmit handwriting or images at the same time while transmitting voice. Ritchie renamed the improved telautograph the "telewriter" and opened exchanges for subscribers (consisting mostly of stockbrokers and merchants) in London in 1910.

The Telautograph Corporation manufactured facsimile devices for a remarkably long time before being purchased by Danka Industries in 1993 and then Xerox in 1999. Similar "remote-controlled pens" have been used recently to register voters or for remote author signings. For example, in 2004 Canadian writer Margaret Atwood conceived of the LongPen, which was officially launched in New York City in 2006, supposedly allowing "an author to see the reader she is signing for, and vice-versa, using a videoconferencing system, and an image of the page to be signed." The device is now called the Syngraffi LongPen and is described as "a patented robotic apparatus that allows users to remotely sign hard-copy documents with wet-ink signatures, in real time."

SOURCES Elisha Gray, "Telautograph," Patent US386,815A (July 31, 1888); Anton Huurdeman, *The Worldwide History of Telecommunications* (John Wiley

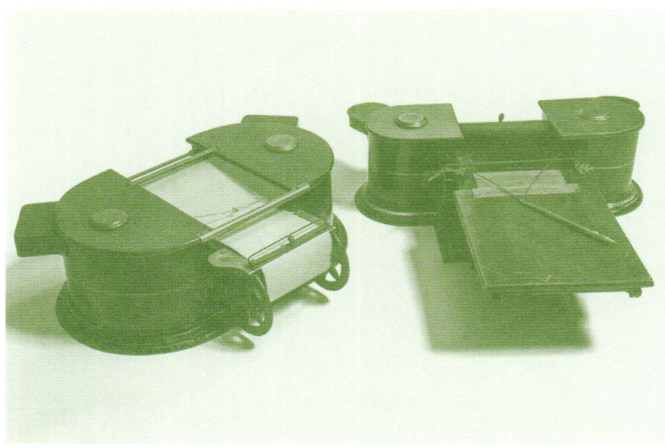

A commercial model of a telautograph transmitter and receiver from around 1890.

& Sons, 2003); "The Telautograph," *The Manufacturer and Builder: A Practical Journal of Industrial Progress* 25 (1893); "The Telautograph," *Popular Science Monthly* 43 (June 1893); "To Write Letters by Wire," *The Phonoscope: A Monthly Journal Devoted to Scientific and Amusement Inventions Appertaining to Sound & Sight* 1:8 (July 1897); "Writing by Wire," *The Pilbarra Goldfield News* (September 16, 1910); Sarah Lai Stirland, "Thousands of People Have Used Remote-Controlled Pens Over the Internet to Register to Vote," Techpresident.com (October 26, 2012); Oliver Burkeman, "Atwood Sign of the Times Draws Blank," *The Guardian* (March 6, 2006); "LongPen," Syngraffi.com

Telefacsimile [37]

COUNTRY OF ORIGIN Germany
CREATOR(S) Gustav Grzanna
EARLIEST KNOWN USE 1901

BASIC INFRASTRUCTURE/MATERIALS Electrical wires, power source, transmitters, receivers, drums or cylinders, paper and/or documents and/or photographs, utility poles (optional), (optional), stylus (optional), rheostat (optional), levers (optional), oscillograph (optional), tape perforator (optional), telegraph tape (optional), photographic film (optional)

RELATED Hellschreiber [16.4], radiofax [19], electric printing telegraph [33.1], image telegraph [33.2], pantelegraph [33.4], telephone [34], telautograph [36]

DESCRIPTION Telefacsimile here refers to techniques as well as devices for scanning, sending, and receiving handwriting, hand-drawn images, photographs, and eventually documents over telephone wires. As such, over the last 125 years of its existence, it includes techniques and devices (often with very little differentiation between the two) that were called anything from telephotography, phototelegraphy, picture telegraphy, Bélinographie or Belinography, facsimile telegraphy, facsimile, and fax to telautographs [36], Bélinographes, teleostereographs (or, alternatively, téléstéreographes), electrographs, telectrographs, copying telegraphs, Desk-Fax machines, telecopiers, fax machines, and more. Even more confusing is the fact that some devices/techniques worked first over telegraph wire and then telephone wire and sometimes were also used over wireless radio.

While German inventor Gustav Grzanna successfully demonstrated a Kopiertelegraph (or a copying telegraph) that worked over telephone wires in 1901, German physicist Arthur Korn's Bildtelegraph, first successfully demonstrated in 1904, used telegraph wire; nonetheless, it proved to be even more influential to the construction of subsequent telefacsimile devices than earlier facsimile devices that also transmitted over telegraph. Jonathan Dentler writes that Korn's device worked by using "a light to scan a picture attached to a rotating drum. The reflected light would be registered by a photovoltaic cell, which transmitted a current into a telegraph wire and carried it to the receiver, which would convert the signal back into light, exposing a negative attached to a synchronized rotating drum." Regular transmissions of photographs using Korn's device took place between newspapers located in Munich, Berlin, Paris, London, and Manchester from 1907 to the beginning of World War I in 1914, at which point the service was shut down. By 1921, however, engineers such as Marcus J. Martin believed that Korn developed "the first practical phototelegraphic system that was used for commercial purposes."

The first widely used and most influential telefacsimile device was Édouard Belin's telegraphoscope or teleostereograph or, alternatively, téléstéreographe, first demonstrated in 1907 over a Paris–Lyons–Bourdeaux–Paris telegraph circuit (transmitting an image of a small Alsatian chapel in twenty-two minutes) and then in 1914 over telephone wires (transmitting an image of four men celebrating the opening of the International Urban Exhibition that took place in Lyons, a process that took only four minutes) using a portable version about the size of a typewriter he called a Bélinographe. Belin's device was innovative for its time because he used dichromated gelatin to develop photographs that had a raised surface. Myriam Chermette clearly explains the process and, given the Bélinographe's importance to the history of telefacsimile, it is worth quoting at length: "By

[37] Tom Klinkowstein in the Mazzo Club, Amsterdam (The Netherlands), doing a live performance of *Telecommunications via Facsimile* with Robert Adrian in Vienna (Austria) along with faxes generated during the piece, 1980.

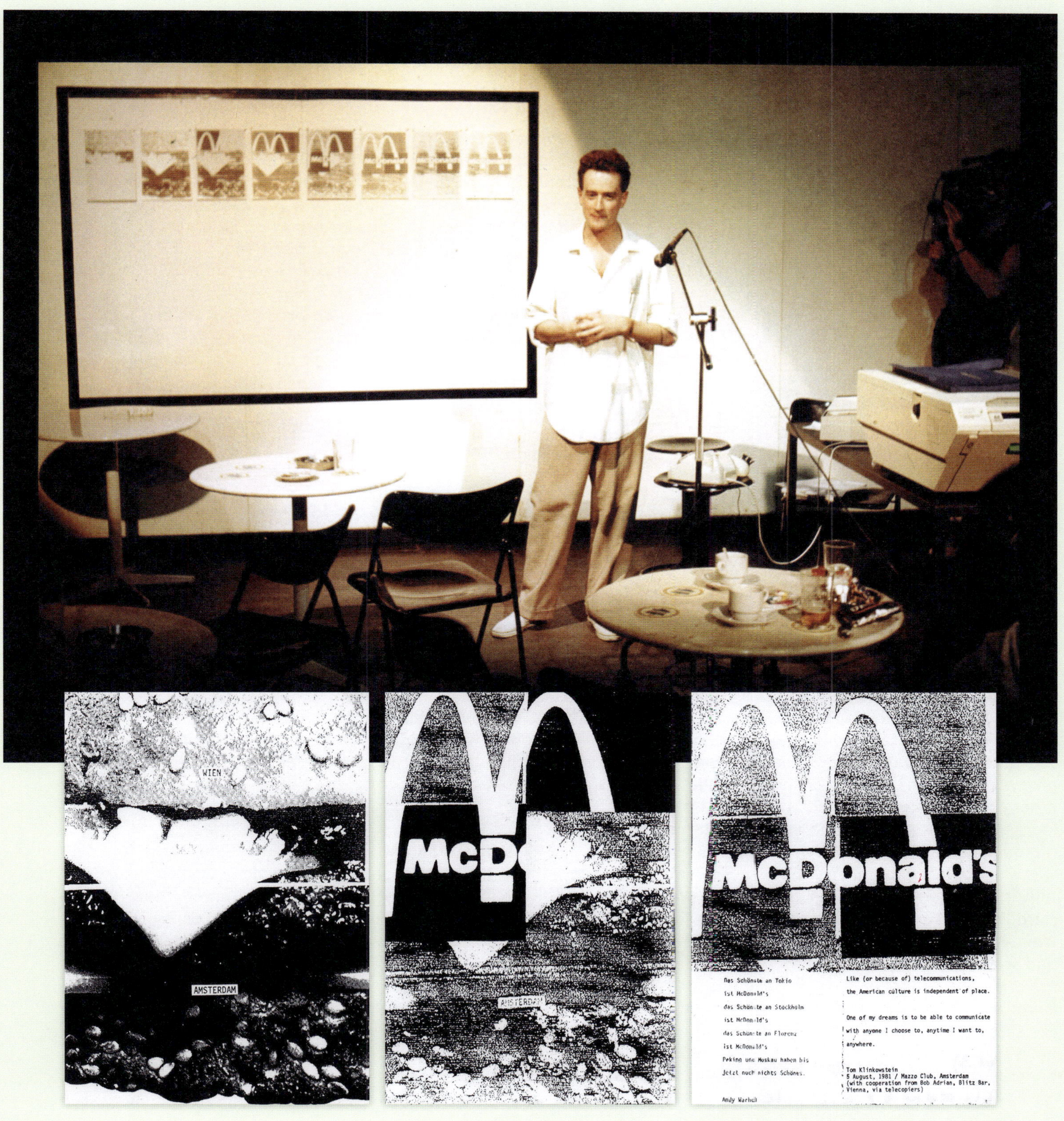

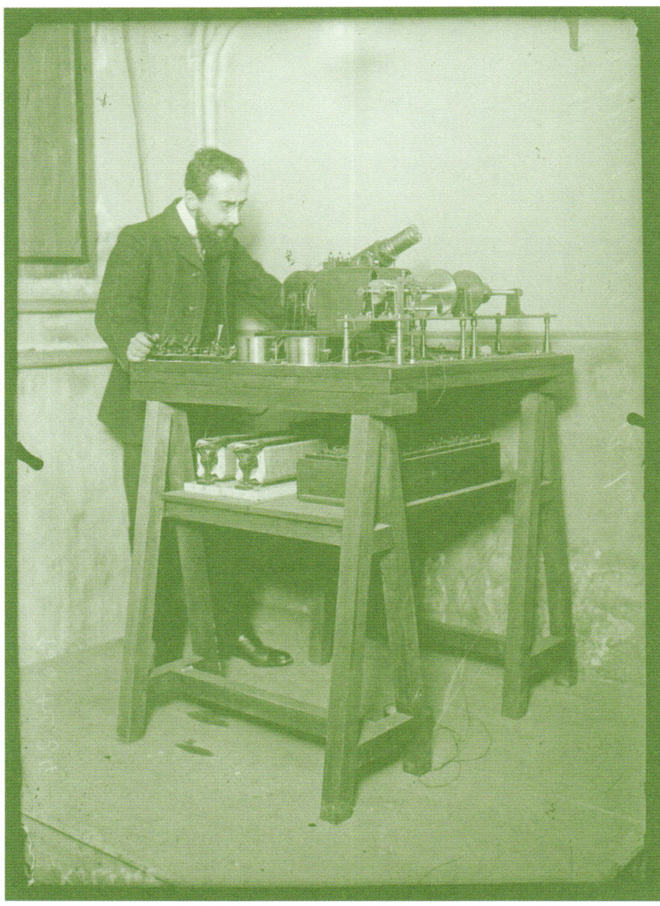

Édouard Belin standing with an early prototype of his Bélinographe from 1907.

increasing the thickness of the gelatin layer, the relief became more visible, with the elevated areas corresponding to the lighter portions of the image and the depressed areas to the dark portions. Mounted on the cylinder of the transmitting set . . . the image was scanned by a stylus, which imparted to a lever movements whose amplitude precisely corresponded to the varied height of the surface. A rheostat affixed to the arm of this lever transmitted a current of variable strength across the telephone line, with the current always proportional to the height of the relief. At the other end, the receiver was equipped with an oscillograph that translated the variations in strength, reproducing the image on a sensitive surface which was also rolled on a cylinder." By 1921, Belin successfully transmitted a picture of US President Warren G. Harding from Annapolis, Maryland, to Paris in roughly twenty minutes. By 1924, a small Bélinographe network had been established between newspapers located in Paris, Lyon, Strasbourg, Nice, Marseille, and Bordeaux, perhaps prompting its adoption by Great Britain in 1928. It was then used heavily by European news agencies throughout the 1930s and 1940s, until the process was largely replaced by radiofax [19].

The early 1920s also marked the beginning of telefacsimile networks dedicated to the transmission and sharing of photographs between newspapers, known as telephotographic networks or wirephoto networks (after the US-based Associated Press's Wirephoto service). Beginning in 1920, the AP, in conjunction with Western Union and AT&T, developed a small network between St. Louis and Cleveland using what they called an electrograph; however, as Martin reported in 1921, the total time needed to prepare and transmit the picture was about eighty minutes. Three years later in 1924, AT&T demonstrated its own system (at that time, known either as Type A or as Telephoto) by transmitting photographs of the Republican and Democratic National Conventions from Cleveland to New York; however, according to R.W. Burns, scanning and transmitting the photographs took about forty-five minutes. AT&T ended their service in 1934, prompting the AP to begin its own wirephoto service the next year in 1935 after acquiring (with United Press) what was then known as the Bell System (a later version of the Telephoto system, also referred to as Type B). The network was now capable of sending an image to twenty-five cities at once, according to Myriam Chermette, and by 1953, the AP Wirephoto service could transmit a photograph to any subscribing newspaper across the USA in under eight minutes.

While the nationwide transmission of photographs between newspapers was thus accomplished, transmitting photographs internationally—particularly across the Atlantic—was incredibly difficult. During the 1920s and 1930s, news agencies such as the AP utilized a transmission technique called the Bartlane process that, according to Jonathan L. Dentler, applied to photographs "a series of translations or compression and decompression techniques between analog (film, print, raster scan) and digital (telegraph punch tape, halftone matrix) forms." H.G. Bartholomew and M.D. McFarlane began working on the technique in 1920 and the process was ready for regular use by 1926. Dentler describes the process as follows:

> First, a photograph would be printed five times from the same negative onto small metal sheets with tonal variations. The metal prints were then placed on a series of rotating cylinders, each of which was traced by a needle with a current running through it. The needle was connected electrically to a tape perforator, which would punch holes in a telegraph tape according to the amount of current running through the needle from the metal plate. The plate's surface area conducted more or less electricity according to the light and shade of the image printed on it. The

perforated tape would then be delivered by messenger to Western Union and transmitted as an ordinary cable message across the Atlantic. On the other end, the tape was taken to a Bartlane device and run through the reproduction apparatus, inside which a light was projected through the holes in the tape as it rapidly unspooled, registering on a photographic film and building up the image with pixels.

As with the fate of the Bélinographe, the Bartlane process was phased out by the end of World War II with the introduction of "radiophoto" devices.

As for the development of personal or desk-sized facsimile machines capable of transmitting all kinds of documents as well as photographs, Western Union built on its patent for Teledeltos paper (paper that has an electro-sensitive coating on one side and carbon black on the other, such that when a current is applied to the paper, the coating is burned away to leave text in carbon) with the release of the Western Union Telefax machine in 1938, followed by the Desk-Fax in 1948. Anton Huurdeman points out that within just a few years of its release, roughly 40,000 Desk-Fax machines were in use across the US The Desk-Fax, then, helped spark the popularity of "fax machines" in the coming decades partly because it made it easier to transmit telegraphs, partly because it was compact and could easily fit on a desk, and partly because it was intended to be easily operated by inexperienced members of the public.

In 1964, Xerox released its LDX machine, declaring it was the first commercially viable fax machine because of its relatively low cost, ease of use, and speed of transmission. The LDX weighed 1,100 pounds and was soon followed by the forty-six-pound Magafax Telecopier, which Xerox advertised in a June 1966 issue of *Life* magazine as allowing people to "mail letters over the phone!" Throughout the 1970s and 1980s, numerous companies including the AP continued to release ever smaller, compact, and portable fax machines. However, since they were often very costly and/or involved monthly rental, Federal Express introduced Zapmail in 1984—a service whereby individuals could drop off documents at a nearby FedEx office, which would then fax them to an office nearest to the recipient to be printed and delivered as a hard copy. FedEx also set up Zapmail fax machines in offices, which would transmit documents over FedEx's own packet-switched network. The appeal of Zapmail was that documents would be delivered to anyone in the country within two hours (but of course it was only appealing if one didn't have access to one's own fax machine).

While fax continues to be used in healthcare as a secure method of transmitting patient information and prescription requests, it likely reached its zenith as a powerful and ubiquitous technology in the late 1980s and early 1990s, before the rise of internet-based alternatives. Notably, fax was used by students in Hong Kong to send news about the 1989 Tiananmen Square student protests and massacre back to those in mainland China, many of whom, because of strict government censorship, knew little to nothing about what had transpired. Writes King-Wa Fu, "Our plan was simple but ambitious: create a daily news digest about what was happening in Beijing; obtain the list of every fax number in the country that was printed in the Yellow Pages; send the digest out to all of these fax machines . . . In pre-internet 1989, we used phone-in talk shows and newspaper stories to recruit Hong Kong companies with underused fax machines. We managed to get a pool of more than 500 corporate partners to join the *crowd-faxing* campaign. We eventually even got some second-hand machines to use in our college dormitory."

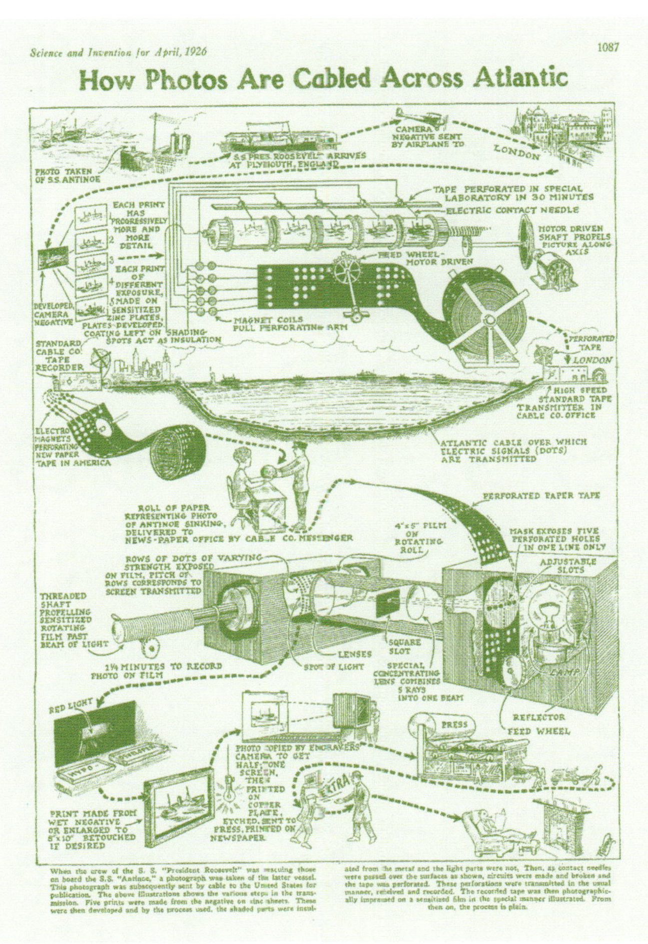

Science and Invention for April, 1926 1087

How Photos Are Cabled Across Atlantic

When the crew of the S. S. "President Roosevelt" was rescuing those on board the S.S. "Antinoe," a photograph was taken of the latter vessel. This photograph was subsequently sent by cable to the United States for publication. The above illustrations shows the various steps in the transmission. Five prints were made from the negative on zinc sheets. These were then developed and by the process used, the shaded parts were insulated from the metal and the light parts were not. Then, as contact needles were passed over the surfaces as shown, circuits were made and broken and the tape was perforated. These perforations were transmitted in the usual manner, received and recorded. The recorded tape was then photographically impressed on a sensitized film in the special manner illustrated. From then on, the process is plain.

H. Winfield Secor's illustration "How Photos Are Cabled Across Atlantic," an illustration of the Bartlane process, from *Science and Invention*.

[37] Fax piece by Spanish artist Marisa González from her series “Women Violence,” 1975–1976.

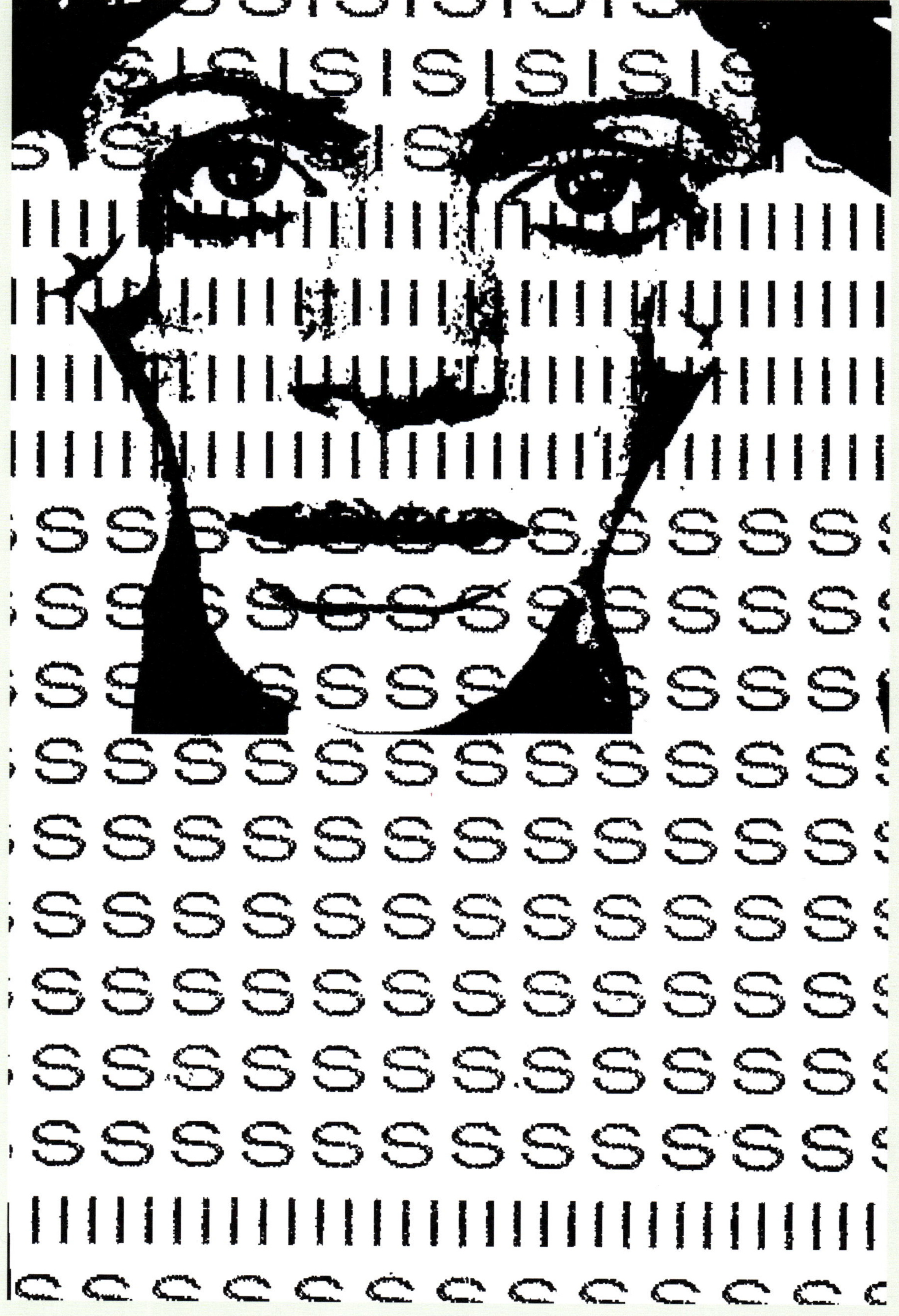

EXPERIMENTS João Ribas, curator of the exhibit *FAX*, which took place at the Drawing Center in New York City from April to July 2009, explains that telefacsimile has been so popular with artists partly because "the fax machine allowed for the creation of participatory, bi-directional, and collaborative networks over vast geographical areas in real time." The result of the creation of these long-distance, participatory networks was a move away from viewing art as an object to viewing it as a means to create more open-ended, creative means of communication. However, as American artist Tom Klinkowstein discovered in the late 1970s, the logistics—not just of the transmissions themselves, but of making them collaborative and participatory—were daunting. In the case of his live performance *Telecommunications via Facsimile* that took place in the Mazzo nightclub in Amsterdam in 1981, simply preparing for the event took hundreds of hours. After confirming Robert Adrian's participation in Vienna (Austria), Klinkowstein and Adrian had to arrange for nightclub venues in both cities, plan for access to the conferencing system ARTEX that ran on the time-sharing network [50] IPSA for planning purposes (a less expensive alternative to making long-distance phone calls), procure sponsorship from a Dutch state telecommunications company for the use of the fax machine, which cost roughly $2,000 in the early 1980s (equivalent to about $7,400 in 2024) as well as additional phone lines, and more. Klinkowstein describes the evening of the event as follows:

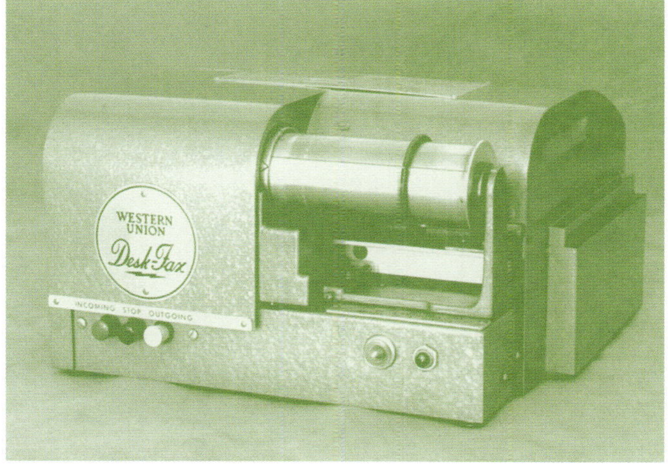

Western Union's Desk-Fax machine from 1948.

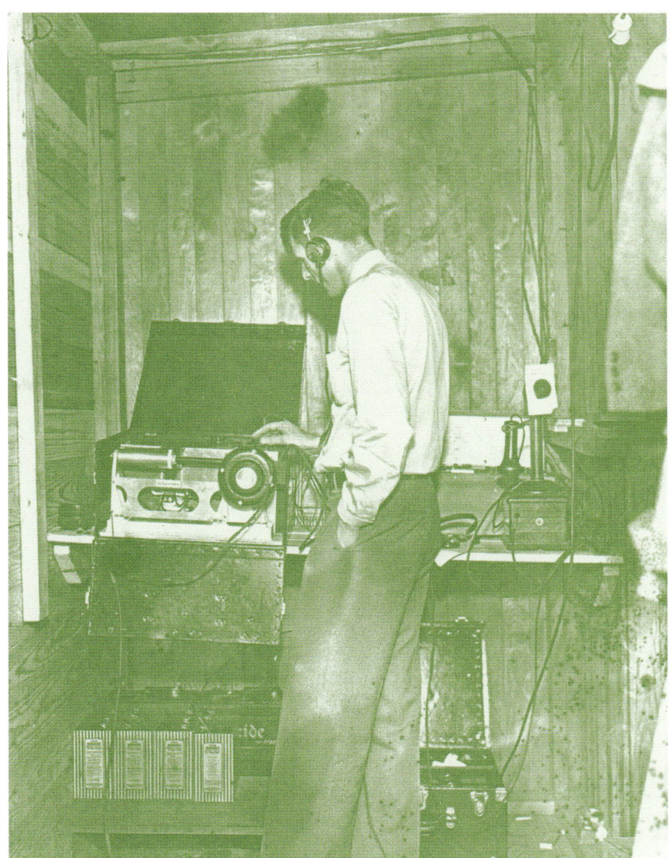
Chief Engineer Harold Carlson checks a portable Associated Press Wirephoto machine on May 2, 1936 at the Kentucky Derby in Louisville, Kentucky.

> Two telephone lines were used, one for voice communication and one for transmitting and receiving text. Both lines were amplified. creating a sound track. This also enabled the audience to hear the conversation between Bob and me and the whining, beeping sounds made as the telecopies were transmitted and received. In addition the audience could hear me explaining the technology in Dutch and English and Bob's explanations in German. Bob began by transmitting the first image from Vienna, an 8-½-by-11-inch partial image that would eventually be pieced together with successive images to form our completed 8-½-by-38-inch collage. After receiving Bob's image I photocopied it, together with my addition to the piece, and sent it back to him. What the audience saw unfolding was an image of an oversized cheeseburger, followed by the familiar McDonald's Restaurant's golden arches logo. Each step required about two minutes for transmission, with half a minute or so in between for photocopying. The last transmission from Vienna was a German translation of a quote by Andy Warhol: "The most beautiful thing in Tokyo is McDonald's. The most beautiful thing in Stockholm is McDonald's. The most beautiful thing in Florence is McDonald's. Peking and Moscow don't have anything beautiful yet." My statement followed: "Like (or because of) telecommunications, the American culture is independent of place. One of my dreams is to be able to communicate with anyone I choose to, anytime I want to, anywhere."

After the event was complete, Klinkowstein and Adrian encouraged audience members in both cities to transmit messages to each other.

An additional telecommunications event involving fax machines was *The World in 24 Hours*, orchestrated by Robert Adrian in 1982, which connected artists in sixteen cities for twenty-four hours using telefacsimile, slow-scan TV [23], the IPSA time-sharing network, and telephones [34]. Each city had one hour in which to create and transmit content using any of the aforementioned technologies/networks.

Telefacsimile also became an important part of the exchange of mail art in the 1980s. In 1984, Lisa Sellyeh, Peter Sepp, and Mary Misner organized *pARTiciFAX* at four different nodes across the province of Ontario (Canada) to facilitate the exchange of mail art. Artists also used telefacsimile in conjunction with cable television [48]. For example, Retrato Suposto—Rosto Roto (translated roughly as Presumed Portrait—Foul Face) was a mini network that Mario Ramiro and Eduardo Kac created in 1988, connecting Ramiro in São Paulo (Brazil) with Kac in nearby Rio de Janeiro via two private fax machines and the public medium of live TV transmission. As Ramiro writes, "From the studios of TV Cultura in São Paulo, I sent to the Rio de Janeiro studio a repertoire of images—eyes, mouths, noses, ears—for the potential creation of a portrait, which I called Retrato Suposto. Minutes later, Kac sent back a montage of those images, which he called Rosto Roto. The whole process took place on

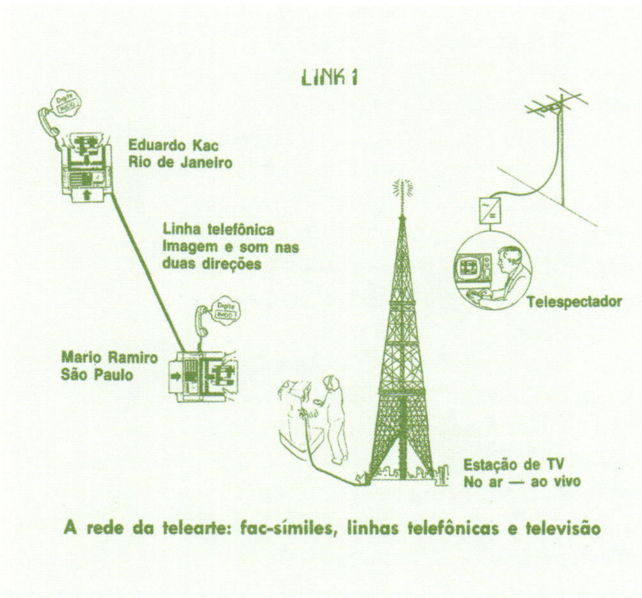

Illustration of the network created by Ramiro and Kac for "Retrato Suposto–Rosto Roto," involving telefacsimile, television, and telephone.

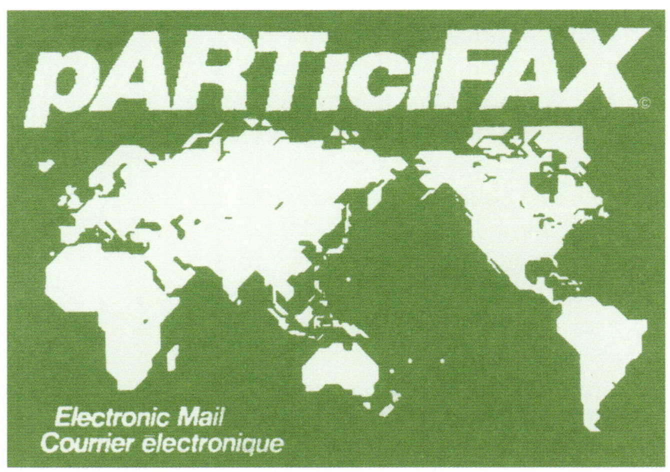

A facsimile of the poster for pARTiciFAX.

the air." However, the point of the project was not the transmission of images themselves but rather the exploration of the affordances of personal and public telecommunications media.

SOURCES "A Long-Distance Writer," *Modern Electrics* (May 1908); Jonathan L. Dentler, "Techniques of Transmission: Wire Service Photography and the Digital Image," *Images on the Move: Materiality—Networks—Formats* (Transcript Verlag, 2021); "Recent Work in the Telegraphic Transmission of Pictures," *Nature* (January 13, 1910); Marcus J. Martin, *The Electrical Transmission of Photographs* (Sir Isaac Pitman & Sons, 1921); "The Thorne-Baker Tele-Photographic Apparatus," *Scientific American* (May 21, 1910); Ernie Smith, "Pushing Photos Through Wires," Tedium.com (November 19, 2021); Anton Huurdeman, *The Worldwide History of Telecommunications* (John Wiley & Sons, 2003); Russell W. Burns, *Communications: An International History of the Formative Years* (Institution of Electrical Engineers, 2004); Myriam Chermette, "The Remote Transmission of Images," trans. James Gussen, *Études photographiques* 29 (May 24, 2012); UNESCO, *News Agencies: Their Structure and Operation* (United Nations Educational, Scientific and Cultural Organization,1953); Grosvenor Hotchkiss, "Electrosensitive Recording Paper for Facsimile Telegraph Apparatus and Graphic Chart Instruments," *Western Union Technical Review* 3:1 (March 1949); G.H. Ridings, "An Improved Desk-Fax Transceiver," *Western Union Technical Review* 6:3 (July 1952); Phil Goldstein, "How the Xerox Magnafax Telecopier Helped Make 'Fax' a Verb," *Biztech Magazine* (September 27, 2016); Calvin Sims, "Federal Express to End Electronic Mail Service," *The New York Times* (September 30, 1986); Lucas Mearian, "The Fax Is Still King in Healthcare—and It's Not Going Away Anytime Soon," Computerworld.com (May 22, 2023);

[37] Faxed and televised images exchanged by Ramiro and Kac as part of "Retrato Suposto–Rosto Roto" from 1988.

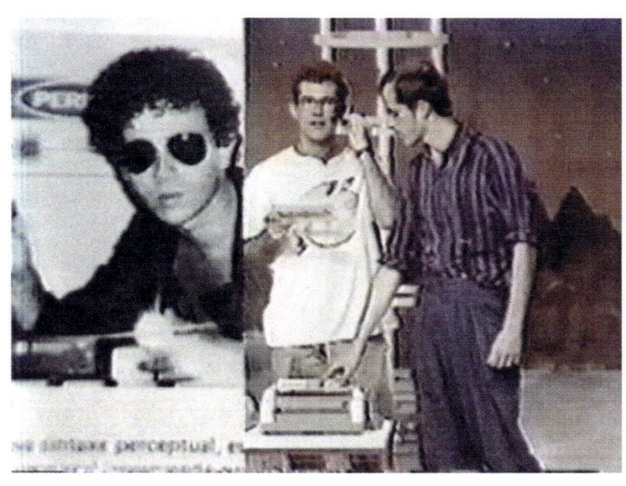
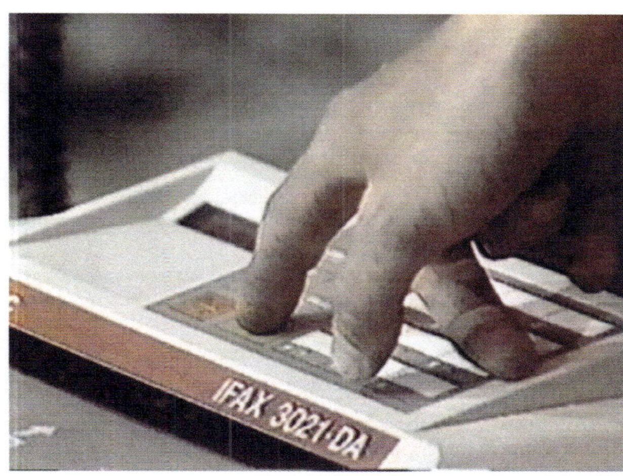

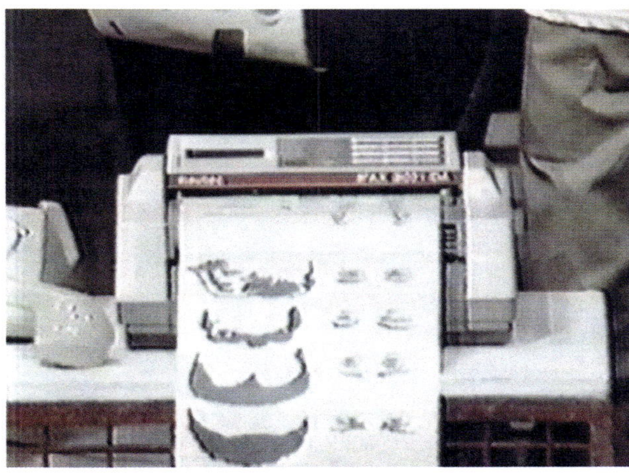
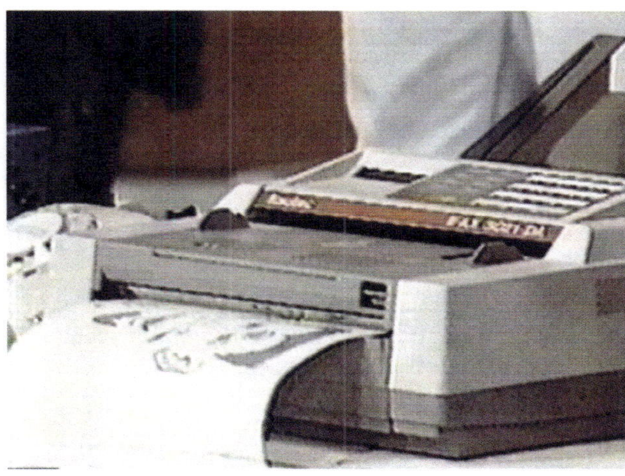
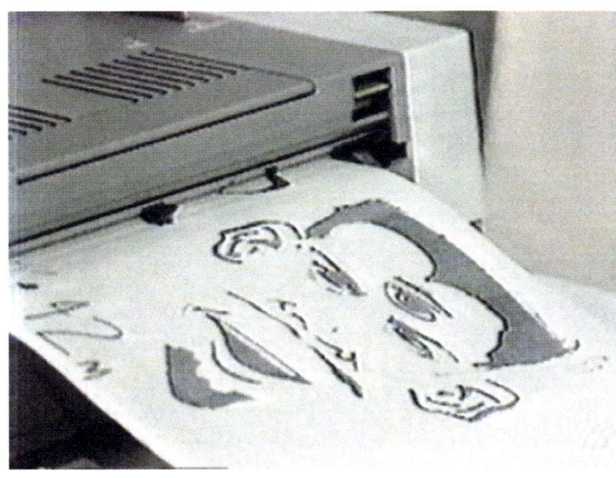
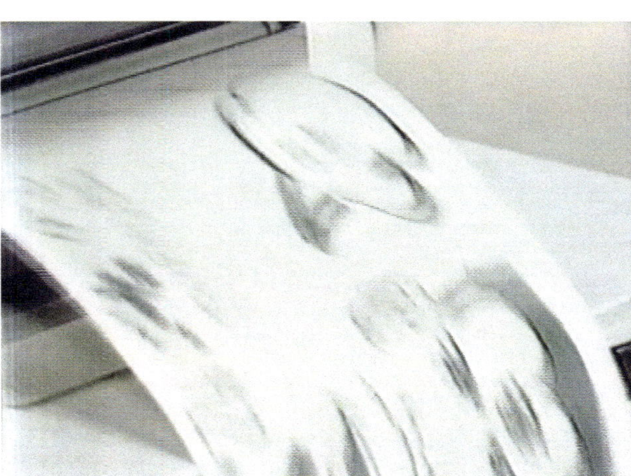

[38] A rare photograph of Bell Laboratories' two-way television viewing booth that appears in their 1930 pamphlet *Two-Way Television*.

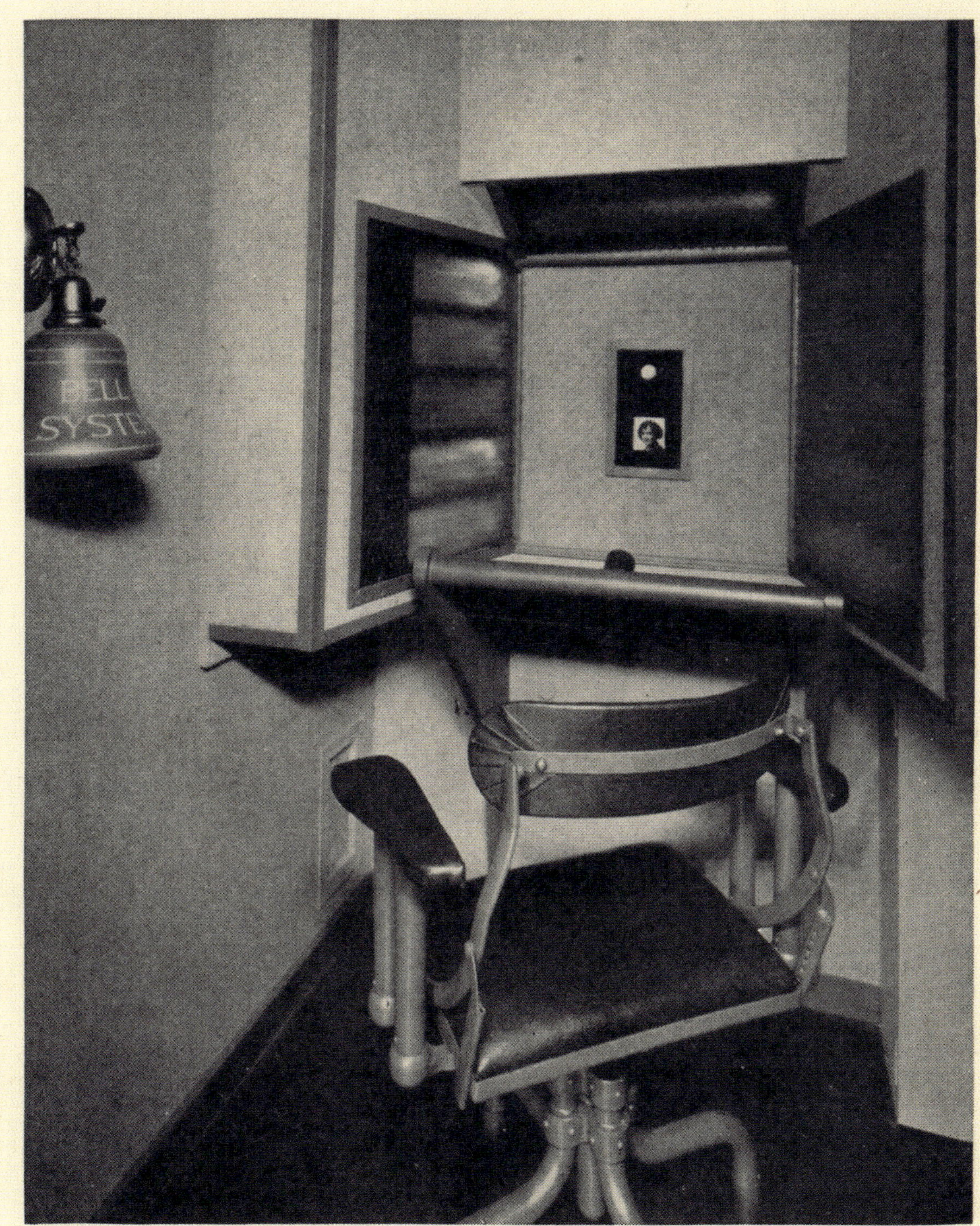

INTERIOR of television booth showing position of incoming image and, just above it, the hole through which the scanning beam is projected.

King-Wa Fu, "Students in Hong Kong Used Fax Machines to Fight Chinese Censorship of Tiananmen Square," Quartz.com (October 29, 2019); João Ribas, "FAX," from *FAX* (Drawing Center and Independent Curators International, 2009); Judy Malloy, "Interview with Tom Klinkowstein," Social Media Archaeology and Poetics: Web Archive (September 2017); Tom Klinkowstein, "Works with Telecommunication Tom Klinkowstein, 1979–1992," PDF (self-published); Tom Klinkowstein with Eve de Grywin and Dana Turner, "Portraying Suburban America in a Global Context Using Telecommunications," *Leonardo* 19:2 (1986); Eric Gidney, "Art and Telecommunications: 10 Years on," *Leonardo* 24:2 (1991); Chuck Welch, "Mail Art in Cyberspace," Crosses.net (2023); Mario Ramiro, "Between Form and Force: Connecting Architectonic, Telematic and Thermal Spaces," *Leonardo* 31:4 (1998)

Videophone [38]

COUNTRY OF ORIGIN USA
CREATOR(S) Herbert E. Ives / AT&T's Bell Laboratories
EARLIEST KNOWN USE 1927

BASIC INFRASTRUCTURE/MATERIALS Electrical wires, power source, videophone device, telephone

RELATED Amateur television [16.3], telephone [34], telefacsimile [37], cable television [48], telephonoscope [55]

DESCRIPTION Videophone is a stand-alone device that allows for two-way transmission of live images (or at least images transmitted so quickly that they appear to be live), often but not always including audio, over coaxial cable or telephone wires. The term "videophone" largely came into being in the 1980s with the advent of affordable devices capable of two-way transmission that were essentially very low-resolution televisions with built-in modems and cameras. One such device, originally dubbed an ikonophone and subsequently referred to as a Picturephone, included two-way television, videotelephony, and videoconferencing, and can thus be seen as the progenitor of contemporary Web conferencing platforms such as Google Meet, Skype, and Zoom. The fact that two-way television is included as a type of videophone indicates that there is significant overlap between the history of cable television [48] and the history of the videophone. Moreover, since video is, properly speaking, the rapid transmission of still images, there is also significant overlap between cable television, the videophone, and telefacsimile [37].

In 1927, Bell Laboratories demonstrated their ikonophone with a two-way transmission (picture and voice) of Secretary of Commerce Herbert Hoover giving an address over telephone wire and over radio from Washington, D.C., to New York. According to Ivy Roberts, Hoover's "words came over the loudspeaker to the New York crowd, while they watched him speak on a big screen" and the audio the audience heard was of Hoover speaking with AT&T president Walter Gifford. Roberts summarizes the process as follows: "A photoelectric cell at the transmitting end detects light and converts it into electrical current; on the other end, a lamp receives that pulse and turns the energy back into visible light; a spinning disc placed in front of the light synchronizes with a mechanism at the transmitting end to reconstruct the pulses into a perceivable image; a lens located in front of the disc focuses the light so that the viewer can better see 'the light flashing before their eyes.'" By 1930, AT&T demonstrated two-way television (in the sense that both speakers were able to hear each other and see images of each other). As *The New York Times* described this near-magical event, "special television booths have been developed about the same size as an ordinary telephone booth. Upon entering the booth the person to be 'televised' sits in a swivel chair and faces a frame in which he will see the person at the other end of the line to whom he will speak. The face is illuminated by a mild glow of blue light which is reflected from the face to the photoelectric cells, known as 'radio eyes' . . When the speaker turns in the chair and faces the apparatus he sees on the glass screen the words, 'Ikonophone—Watch this space for the television image.' Then this sign lifts like a magic curtain and in its place the animated picture appears of the person at the other terminal."

Subsequently, live image transmission devices were installed in post offices in Berlin and Leipzig, Germany,

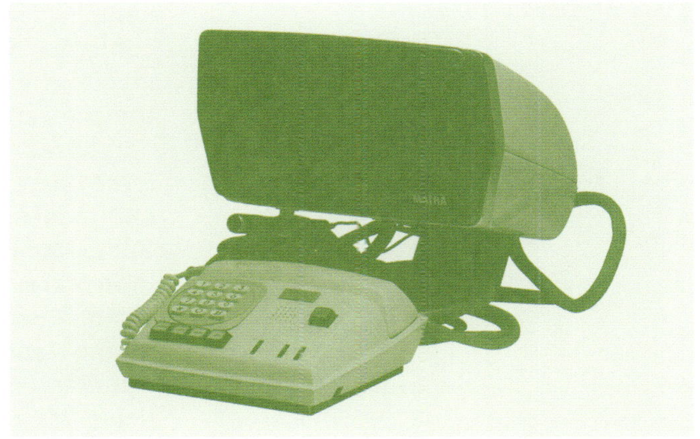

Videophone made by the French electronics manufacturer Matra in 1970.

in 1936 (called a Gegensehn-Fernsprechanlagen, or a visual telephone system), linked by coaxial cable. By 1938, the Third Reich regime had extended the network to Hamburg, Nuremburg, and Munich and made the booths available to the public.

Very little to no progress was made on the videophone until well after World War II. While AT&T reportedly promoted their then trademarked Picturephone at the World's Fair in 1939, a complete system (known as the Picturephone Mod I) was not implemented until 1959; it was demonstrated at the World's Fair in New York in 1964. Jon Gertner describes what visitors would have experienced: they "would enter one of seven booths and sit before what was called a 'picture unit.' The device was a long oval tube, measuring about one foot wide and seven inches high and about a foot in depth. Set within the oval face was a small camera and a rectangular video screen, measuring four and three-eighths inches by five and three-quarter inches. The picture unit was cabled to a touch-tone telephone handset with a line of buttons to control the screen. If you wanted to make a Picturephone call at the fair—or more precisely, if you wanted to talk with the Picturephone users at other booths—you simply pressed a button marked 'V' for video; after that you could either talk through the handset or through a speakerphone on the picture unit." Several months later, Picturephone booths were set up in New York, Washington, D.C., and Chicago. Gertner writes that a three-minute conversation would have cost about $16 (or $160 in 2024). AT&T continued to try to expand Picturephone service across the US (including releasing a Picturephone Mod II) but they shuttered the project in 1974 after investing about $1 billion (about $6.25 billion in 2024).

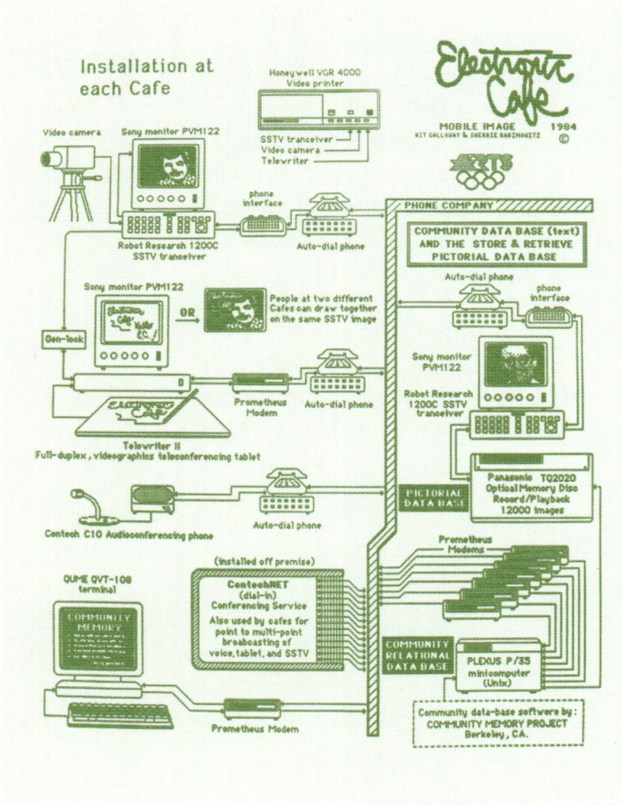

Diagram by Kit Galloway and Sherrie Rabinowitz from 1984 illustrating the then cutting-edge hardware, software, and networking setup at each Electronic Café, including a "pictorial database" that users could later add or access images with videophones.

While similar Picturephone services were launched in France, Russia, Sweden, and the UK, the emergence of digital telephone networks coupled with Japan's development beginning in the mid-1970s of what became videophones (produced by Mitsubishi, Panasonic, and Luma) were responsible for the worldwide popularity of the device. Now, with the phasing out both of the public switched telephone network and the Integrated Services Digital Network (or ISDN, communication standards developed in Bell Laboratories that make possible digital transmission of voice, video, and data over the public switched telephone network), videophones are increasingly difficult to use for long-distance communication. However, as Lori Emerson and libi striegl documented in 2020, they may still be easily connected to each other and the distance of transmission is almost entirely dependent on the length of the telephone cable connecting the two videophones.

EXPERIMENTS In the summer of 1984, Kit Galloway and Sherrie Rabinowitz (who worked under the name Mobile Image) initiated an ambitious project called *Electronic Café* (*EC*). *EC* consisted of five cafés located across Los Angeles, each of which housed elaborate networks of telecommunications devices for members of the public to experiment with; each setup included teleconferencing equipment, audio conferencing equipment, telephones, video cameras, monitors, videoprinters, slow-scan television [23] transceivers, computer conferencing services, and databases. In 1987, Galloway and Rabinowitz launched *Electronic Café International* (*ECI*)—a collection of sixty nodes around the world (with a home base in Santa Monica, CA) that produced and participated in virtual events. The events relied heavily upon videophones (or "vid-phones," as they called them) for the transmission of still images. Vid-phones were featured in numerous *ECI* virtual events particularly throughout 1991, ranging from international "link-ups" with Timothy Leary; online meetings with children from the USA, Japan, and South Africa; and a videophone celebration of American composer

[38] Diagram from 1992 illustrating the complex networking, including the Picturephone, involved in Van Gogh TV's *Piazza Virtuale*.

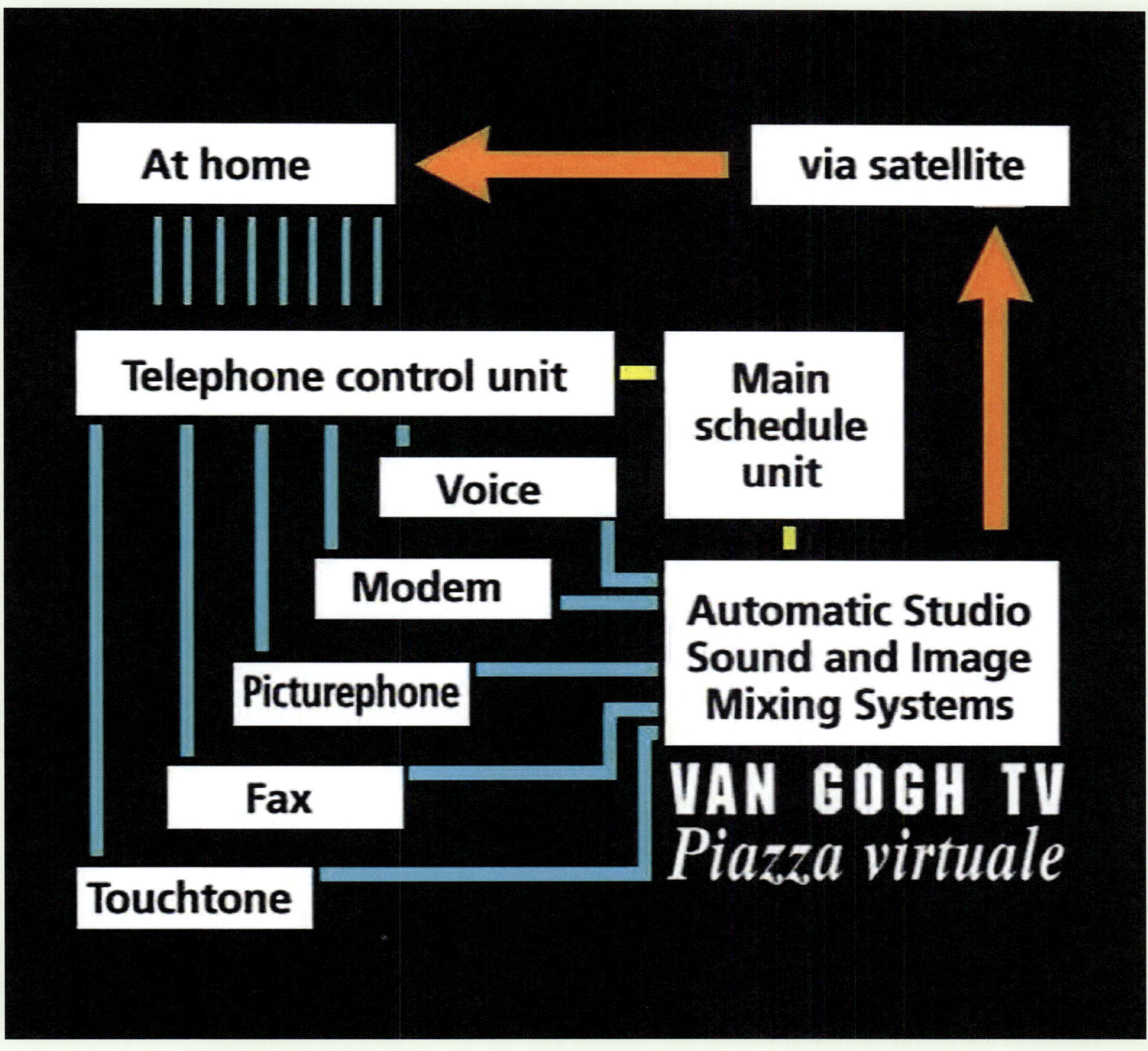

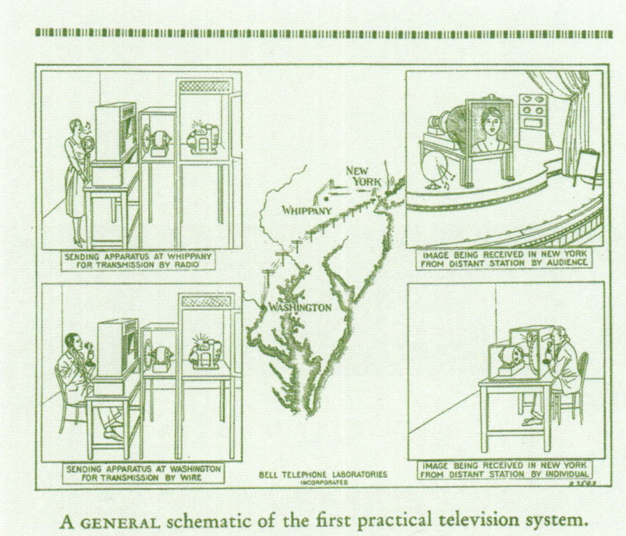

A schematic from the 1930 Bell Laboratories pamphlet *Two-Way Television* explaining their first two-way television transmission, which took place in 1927 from Washington, D.C., and Whippany, NY, to New York City.

Pauline Oliveros's *Four Decades of Composing and Community*, which, according to Joe Catalano, involved fifty artists and roughly twenty sound and video technicians in six cities and three time zones.

In 1986, roughly contemporaneous to the *Electronic Café*, Van Gogh TV was formed by artists and hackers from Austria and Germany (Mike Hentz, Karel Dudesek, Benjamin Heidersberger, and Salvatore Vanasco). Working primarily in interactive TV, they are perhaps most well-known for their 1992 project *Piazza Virtuale* that lasted for one hundred days during the contemporary art exhibition documenta IX that took place in Kassel, Germany. The most surprising, even shocking, aspect of *Piazza Virtuale* was that, with the help of Picturephones (along with fax, slow-scan television, and satellites), visitors experienced television as an interactive medium rather than a broadcast medium. Tilman Baumgärtel evocatively describes the project as follows: "There were no presenters, no announcements, no explanations. In fact, no show at all. Instead, you could call a telephone number that was displayed on the TV screen. And if you were lucky and got through, you were suddenly on air and could speak to the world via TV. Up to four callers found themselves in a strange, random community, could chat with each other or give a speech to mankind. Many callers were so startled that they hung up immediately. Others managed little more than 'Hallo.' Some tried to make conversation with the other callers. Others made farting noises until they were thrown off the line." There were also other miniature, so-called Piazettas set up in Hamburg, Göttingen, Cologne, and Zurich that, with the help of the Integrated Services Digital Nework, could transmit live images to Kassel using Panasonic WG-R2 picture telephones imported from the US.

SOURCES Russell W. Burns, *Communications: An International History of the Formative Years* (Institution of Electrical Engineers, 2004); Ivy Roberts, *Visions of Electric Media: Television in the Victorian and Machine Ages* (Amsterdam University Press, 2019); "2-Way Television in Phoning Tested," *The New York Times* (April 10, 1930); "Talk, Hear, SEE on This Phone," *Popular Science Monthly* (July 1930); Jon Gertner, *The Idea Factory: Bell Labs and the Great Age of American Innovation* (Penguin Books, 2013); Damon Darlin, "How the Future Looked in 1964: The Picturephone," *The New York Times* (June 26, 2014); *Bell Laboratories Record*, 47:5 (May/June 1969); Lori Emerson and libi striegl, "Slow Networks Experiment 4: Videotelephony," Loriemerson.net (December 18, 2020); Joe Catalano, "Electronic Midwifery," *Leonardo Music Journal* 3:29 (1993); Philip Glahn and Cary Levine, "The Future Is Present: *Electronic Café* and the Politics of Technological Fantasy," *Art Journal* 78:3 (September 30, 2019); Kit Galloway and Sherrie Rabinowitz, "Welcome

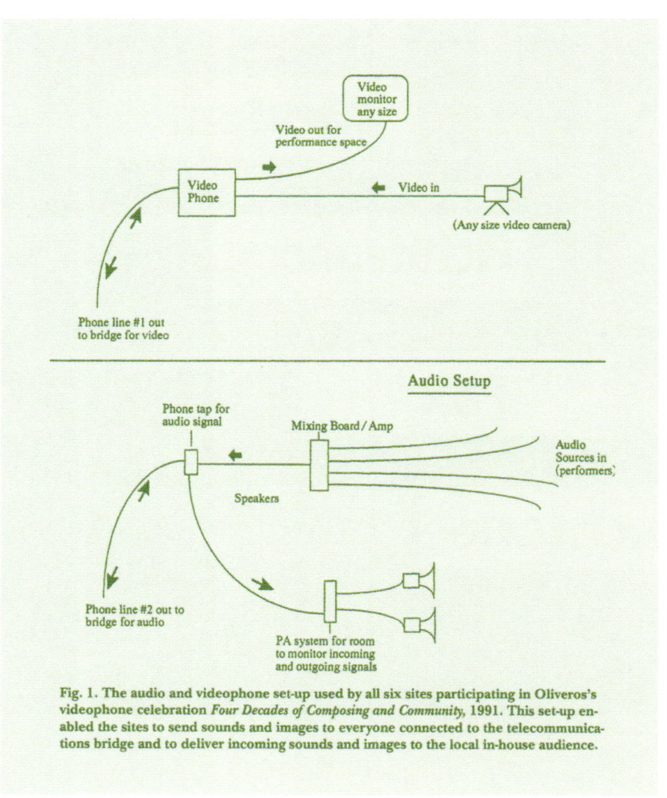

Diagram by Joe Catalano demonstrating the audio and videophone setup used for Pauline Oliveros's *Four Decades of Composing and Community* from 1991.

to Electronic Café International," from *CyberArts: Exploring Art & Technology*, ed. Linda Jacobson (A. Miller Freedman,1992); Tilman Baumgärtel, *Van Gogh TV's "Piazza Virtuale": The Invention of Social Media at Documenta IX in 1992* (Transcript Verlag, 2022); Tilman Baumgärtel et al., "Van Gogh TV's "Piazza Virtuale"—Report-in-Progress and Preliminary Case Study," *International Journal for Digital Art History* 5 (2020)

Telex [39]

COUNTRY OF ORIGIN USA
CREATOR(S) American Telegraph and Telephone Company
EARLIEST KNOWN USE 1931

BASIC INFRASTRUCTURE/MATERIALS Electrical wires, power source, transmitters, receivers, paper and/or documents and/or photographs, teletypewriters, utility poles (optional), tape perforator (optional), telegraph tape (optional), operator (optional), switchboard (optional)

RELATED Radiotelegraphy [16.1], radioteletype [16.2], Hellschreiber [16.4], communications satellite [32], telephone [34], telefacsimile [37]

DESCRIPTION Telex—also known as telephone typewriter or teletypewriter service or TWX—can refer to the teleprinter machines, the messages, the service, or the network of teleprinters that spans the globe. Considered in its totality, however, telex is primarily for text-based communication and uses the circuits of the public switched Telephone Network [34] (PSTN) or sometimes private lines, the sending and receiving devices are teletypewriters or teleprinter devices, and the sending and receiving signals are derived from telegraphic signaling, in that the presence or absence of a level of electric current indicates a mark and space (or dot and dash), which are either recorded on paper tape or automatically converted to characters. Thus, while telex might use the circuits of the PSTN, the network itself operates very differently from the telephone network in its use of a five-bit digital code and a separate system of assigning telex numbers. It has remained a popular communication network for nearly the last hundred years because it was (and sometimes still is) an inexpensive way to transmit legally recognized messages securely and simply.

The first documented and publicly available telex network was AT&T's Teletypewriter Exchange Service called TWX, which was launched in 1931 with 16,000 teletype machines. After this point,

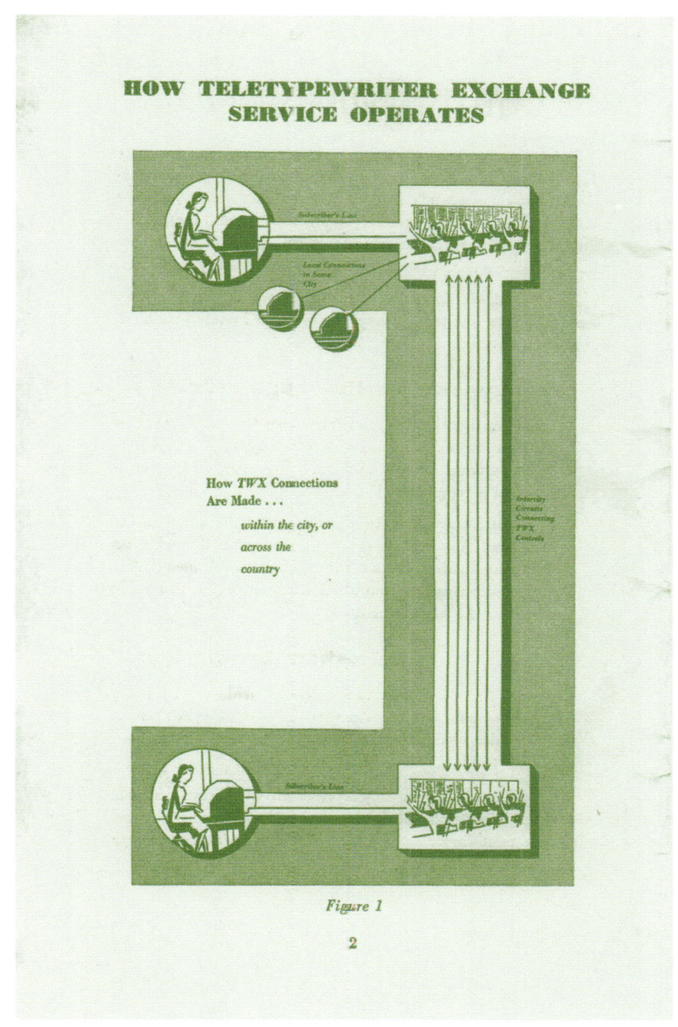

Diagram illustrating how telex connections are made to and from subscribers via telex operators, from AT&T's 1949 *How to Use Teletypewriter Exchange System (TWX)*.

still more machines were installed in companies, banks, and newspapers across the company. As Anton Huurdeman describes it, "To make a call, the customer looked up the number in the nationwide TWX directory and called the operator to be connected . . . Once connected, the two subscribers could type their messages and replies." A year later, in 1932, Western Union inaugurated its own telex network called Timed Wire Service, using its telegraph lines for one-way customer use. Given that sending a telex was cheaper and, in many ways, easier than placing a long-distance phone call, it was quickly adopted by the press, travel agencies, airlines, governments, and embassies. As the decades progressed, Western Union continued to develop its telex network, launching an automatic telex service between New York City and 12,000 subscribers in Canada in 1958 (making use of the telex network launched by Canadian Pacific and Canadian National Telegraph Companies in 1957). In 1969, Western Union purchased AT&T's TWX network,

[39] Telex operators at work in an AT&T TWX central exchange office, from AT&T's 1949 *How to Use Teletypewriter Exchange System (TWX)*.

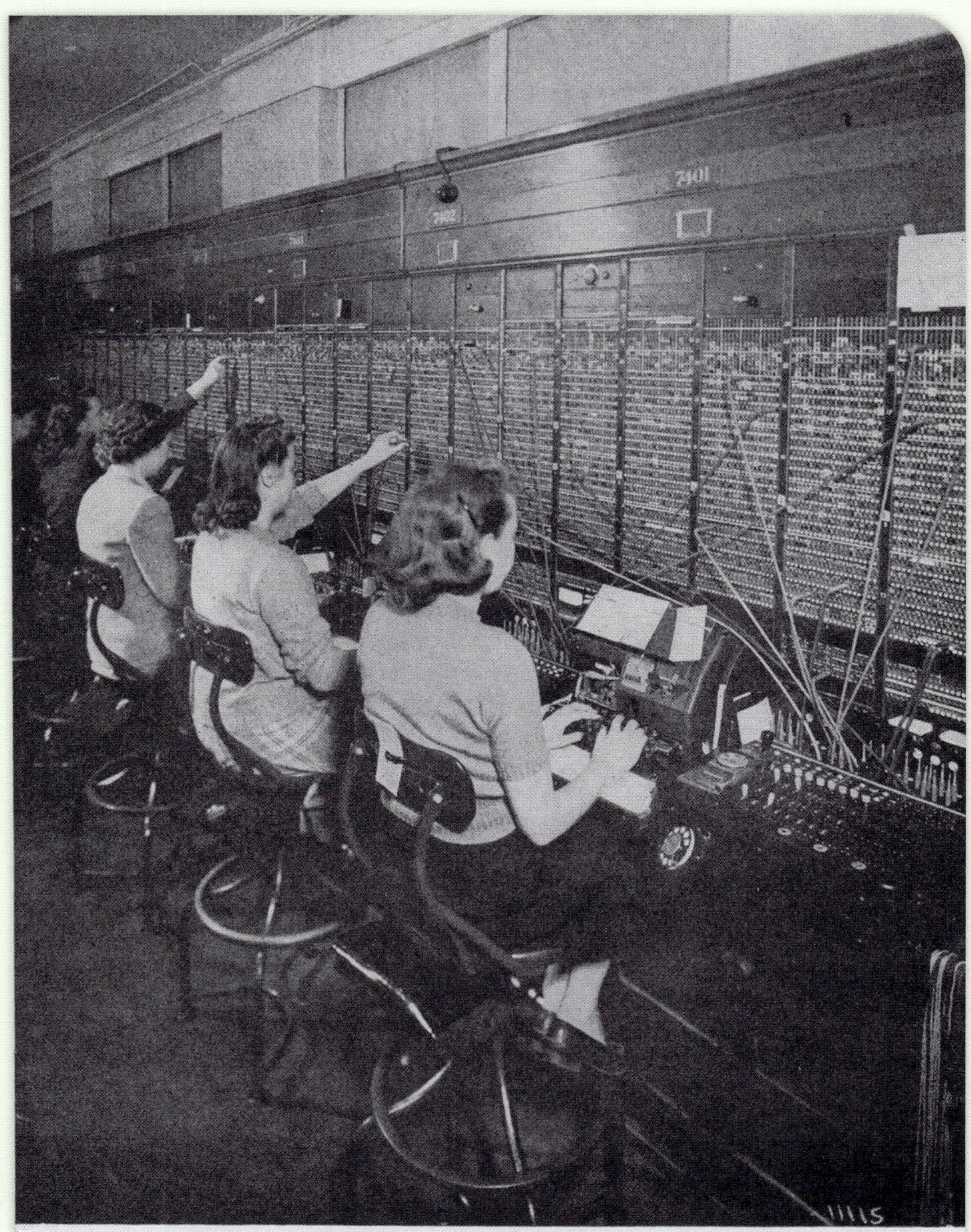

A Teletypewriter Central Office

Figure 2

3

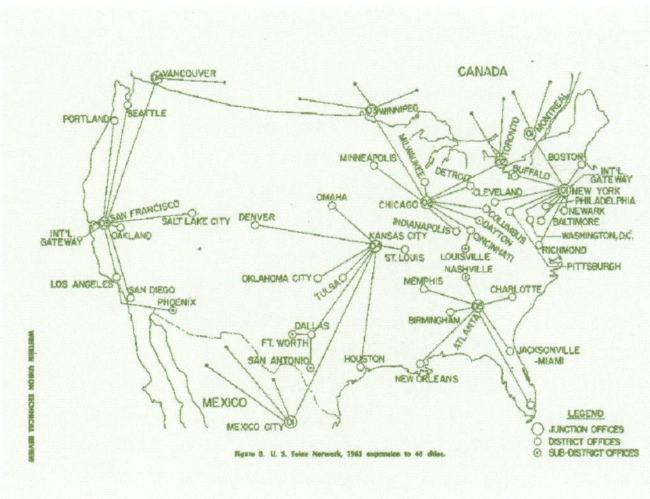

Map illustrating the US's telex network as it existed in 1961.

making them the de facto provider of telex to the USA until they ended the service in 2006.

The first telex networks in Europe were set up between Berlin and Hamburg, Germany, from 1933 to 1935 for about forty private subscribers. The German Reichspost adopted the system in 1935, and by 1940, the German telex network had roughly 10,000 subscribers. Networks were subsequently also set up in England in 1935, in France in 1946, in Japan (to select subscribers in Tokyo and Osaka) in 1956, and in Australia in 1959. According to Anton Huurdeman, by 1970 telex services were also available between North America and Europe and throughout most countries in the world. At its peak in the late 1980s, telex had roughly 1.7 million subscribers worldwide. It wasn't long, however, before its use was eclipsed by the popularity of telefacsimile [37] and computer-network based communication such as email. Still, the technology hasn't altogether disappeared; contemporarily, individuals in the Balkans (and elsewhere) still send telegrams to express condolences. One may also use the service International Telegram (or iTelegram) to send or receive telegrams around the world. The company operates the former Western Union telex network, which they took over in 2006, and claims that "telegrams remain a popular way to send important messages," since "people use them for cancelling contracts and sending legal notifications because a time-stamped copy of the message is retained in our files for 7 years and can be legally verified."

EXPERIMENTS One of the best-known instances of artists using telex for the production and distribution of art was the conceptual art collective N.E. Thing Co. formed by Iain and Ingrid Baxter in 1967 in Vancouver (Canada). Until their disintegration in 1978, the pair used N.E. Thing Co. as a way to appropriate the discourse of capitalism as a way to explore, in Derek Knight's words, "how the codes or symbols of the corporate world could be appropriated to serve both artistic and commercial ends." Among their many site-specific performances, multimedia projects, photographs, and "commercial ventures," the group also frequently used telex as part of their projects, likely because—as Knight points out—the telex symbolized "corporate legitimacy," and in the late 1960s and early 1970s it was a deeply engrained means for conducting business. N.E. Thing Co. created numerous telex and telecopier works between 1968 and 1970, several of which depicted triangle-shaped networks of nodes across maps of Canada and the USA, while the telexes themselves also inscribed these shapes as they were transmitted from place to place. In 1970, Joan Lowndes wrote in the *Vancouver Sun* that, with a telecopier and a telex installed in their home, N.E. Thing Co. are "concerned at the degree to which we are US dominated and [believe] that communications represent a field in which we can win international prominence." In other words, their appropriation of the telex was also an intervention in the Canadian media

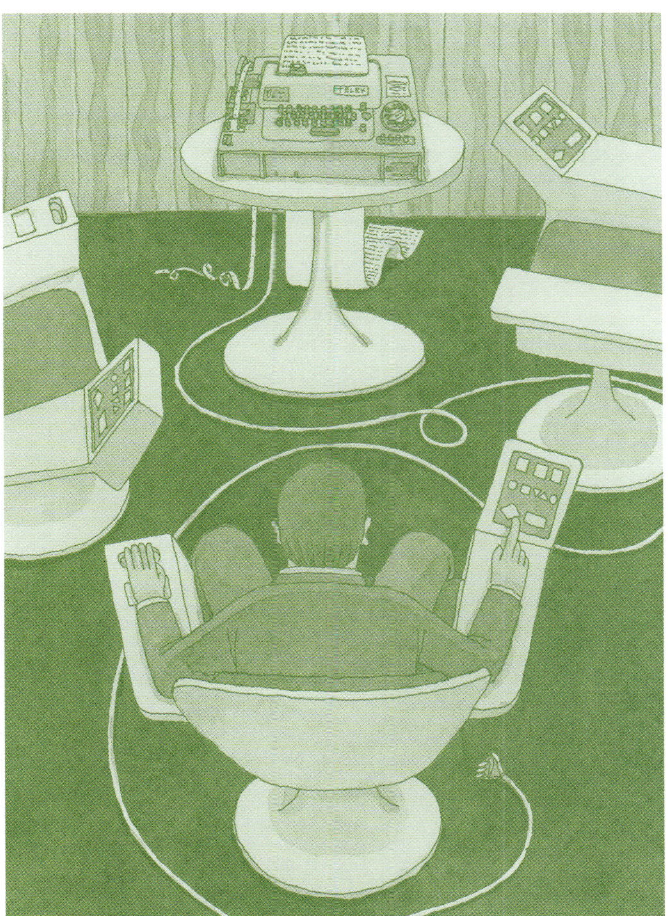

Illustration of the Project Cybersyn control room by Mattias Adolfsson from *The New Yorker* (October 6, 2014).

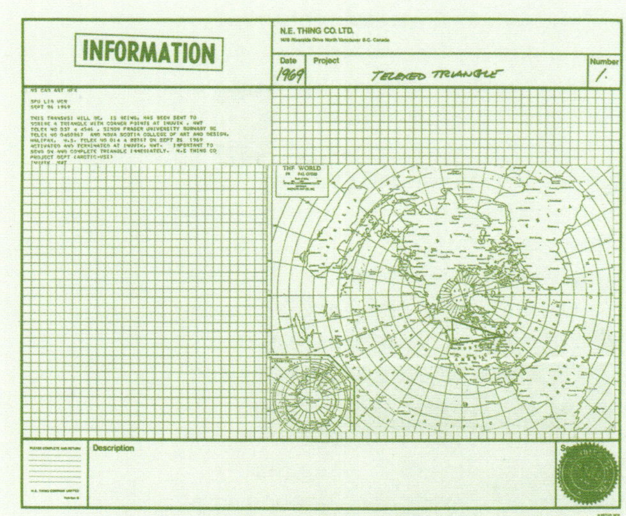

N.E. Thing Co.'s project "Telexed Triangle" (1969), which was sent "to scribe a triangle with corner points" from Inuvik (Canada), Simon Fraser University in Vancouver (Canada), Nova Scotia College of Art and Design in Halifax (Canada), and back to Inuvik.

communications landscape, which was (and continues to be) dominated by content from the US For the duo, the most powerful affordance of the telex was the fact that, as Joan Lowndes wrote, "It's an open channel. No one can stop the telex from working because it's a twenty-four-hour-a-day communication hookup. As soon as you dial the number you're really into that office." This quality was truly revolutionary for 1970, many decades before internet access became affordable and accessible.

Contemporaneous to N.E. Thing Co.'s experiments with telex was the Chilean project called Project Cybersyn, which lasted from 1971 to 1973. Salvador Allende's Popular Unity government brought in British cyberneticist Stafford Beer to propose, in Eden Medina's words, "a real-time control system capable of collecting economic data throughout the nation, transmitting it to the government, and combining it in ways that could assist government decision-making." When Beer arrived in Chile in 1971, there were only fifty computers in the entire country, prompting Beer to adopt a "cloud-based" solution. He proposed they use the more than four hundred telex machines they discovered in storage from a previous government regime, installing them on factory floors to communicate production data (about available resources, goods produced, hours worked, etc.) to the telex machine in the National Computer Corporation in Santiago, at which point programmers would translate the telexed data into punch cards that were then fed to the IBM 360/50 mainframe computer. The system played a key role in the events of October 1972, when 40,000 truck drivers went on a month-long strike and the government used the telex machines to organize alternative transportation. Beer believed that, during the strike, roughly 2,000 messages were transmitted a day over the network. Project Cybersyn came to an end with the military coup and Allende's death on September 11, 1973.

SOURCES Patrice A. Carré, "From the Telegraph To The Telex: A History of Technology, Early Networks and Issues in France in the 19th and 20th Centuries," *FLUX* 11 (1993); Anton Huurdeman, *The Worldwide History of Telecommunications* (John Wiley & Sons, 2003); A.E. Joel Jr., *History of Science and Engineering in the Bell System: Switching Technology (1925–1975)* (Bell Telephone Laboratories, 1982); G.A. Locke, "Telephone-Typewriter PBX Systems," *Bell Laboratories Record* 9:5 (January 1931); *How to Use Teletypewriter Exchange System (TWX)* (American Telephone and Telegraph Company, 1949); Philip Easterlin, "Telex in Private Wire Systems," *Western Union Technical Review* (October 1960); Philip Easterlin, "Telex in the USA," *Western Union Technical Review* 16:1 (January 1962); Edward E. Kleinschmidt, *Printing Telegraph . . . A New Era Begins* (publisher unknown, 1967); Mark Griffith, "Call of the Telex: 'I'm Not Dead Yet,'" *Salon* (December 5, 2000); "Customer Service," Itelegram.com (2024); Derek Knight, *N.E. Thing Co: The Ubiquitous Concept* (Oakville Galleries, 1995); Joan Lowndes, "'Easel' Is a Telex," in *You Are Now in the Middle of an N.E. Thing Co. Landscape* (UBC Fine Arts Gallery, 1993); Eden Medina, *Cybernetic Revolutionaries: Technology and Politics in Allende's Chile* (MIT Press, 2014); Alexei Barrionuevo, "Before '73 Coup, Chile Tried to Find the Right Software for Socialism," *The New York Times* (March 28, 2008); Evgeny Morozov, "The Planning Machine," *The New Yorker* (October 6, 2014)

BARBED WIRE NETWORKS

Barbed wire was originally proposed as an inexpensive and potentially painful material that could be used to create a fence and thus act as a deterrent to keep livestock within a confined area and/or to keep out intruders. Alan Krell documents numerous designs for wire that featured barbs throughout the nineteenth century, including one proposed by French inventor Léonce Eugène Grassin-Baledans in 1860 for a "Grating of wire-work for fences and other purposes." The first patent in the US for a wire fence featuring barbs was given to Lucien B. Smith from Kent, Ohio, in 1867. Illinois farmer Joseph Glidden submitted a patent for an improved version of barbed wire in 1874, which has since become the dominant design. As Reviel Netz puts it, after this point the physical control of wide-open spaces was largely complete. Many farmers objected to the cruelty built into barbed wire, the way in which the fencing meant cattle drives were no longer possible, and the fact that its advent marked the end of seemingly free and open public land; notably, they formed anti-barbed-wire associations and pleaded with legislators and government officials to enact laws limiting or regulating the use of the wire. Nonetheless, as the price of wire fell from twenty cents per pound in 1874 to two cents a pound by 1893, few ranchers could afford any other type of fencing material. By the 1890s, the barbed wire industry had become wealthy and powerful enough that it effectively quelled all opposition to the wire. The availability of inexpensive barbed wire, especially across the western US in the late nineteenth century, enabled farmers to keep larger herds of livestock than had been possible up to that point. It also played a significant role in "settling" the American west, by violently asserting individual ownership over land that was already occupied by Native Americans.

Appropriately nicknamed "the devil's rope," barbed wire is made from steel (later coated in zinc, a zinc-aluminum alloy, or a kind of polymer coating such as polyvinyl chloride) and single or double barbs placed roughly four to six inches apart. To erect a fence, one only needs barbed wire, posts, and materials to affix the wire to the posts. Although this section focuses on its use as a cooperative, noncommercial form of telecommunications network, it is also worth noting other more hostile uses of barbed wire: in trench warfare, or as a security measure atop walls or buildings.

SOURCES Alan Krell, *The Devil's Rope: A Cultural History of Barbed Wire* (Reaktion Books, 2002); Léonce Eugène Grassin-Baledans, "Grating of Wire-Work for Fences and Other Purposes," France Patent 45,827; Lucien B. Smith, "Wire Fence," US Patent 66,182A (June 25, 1867); Joseph Glidden, "Improvement in Wire Fences," US Patent 157,124A (October 27, 1873); Reviel Netz, *Barbed Wire: An Ecology of Modernity* (Wesleyan University Press, 2009)

Barbed Wire Telegraph [40]

COUNTRY OF ORIGIN USA
CREATOR(S) Unknown
EARLIEST KNOWN USE Roughly 1880s

BASIC INFRASTRUCTURE/MATERIALS Copper wire, barbed wire, posts, fasteners (such as nails or staples), insulators (such as porcelain knobs, glass bottles, leather, corn cobs, cow horns), telegraph keys

RELATED Electrical telegraph [33], wired radio [35], fence phones [41]

DESCRIPTION Since barbed wire did not become affordable until the 1890s, when the telephone [34] had begun to overtake the telegraph as a more popular telecommunications network, it seems likely that barbed wire was not commonly used for telegraphy. Nonetheless, in 1883, *Scientific American* reported on an experiment that was undertaken along a thirty-mile stretch of railway line between Milwaukee and Brandon, Wisconsin. Engineers attempted to use the barbed wire fencing on either side of the line to send telegraphic messages. Apparently, winter weather made the fence telegraph impractical; however, the article ends by observing that "there are thousands of wire fences along the Western lines, and it has been contended they should be used for this purpose." A little over twenty years later, *The New York Times* reported in 1904 on two young telegraph operators in Wilmington, Massachusetts, who used two coats of paint to first insulate every point on a nearby fence where barbed wire had been fastened. Next, they stated that "with a few hundred feet of covered wire we obtained from the railroad station agent we connected the ends of the fence at roads and other places where it was broken, burying the connecting wire slightly in the soil . . . Before this wire was laid it was also subjected to a heavy coating of paint, so that the moisture in the soil might not penetrate the covering and 'ground' the circuit." The telegraphers then placed a short section of iron rail six feet underground at each end of the line to act as a ground plate and "use the earth as a return wire in the usual way." Finally, rather than purchase batteries, they built their own batteries with half-gallon paint cans and salvaged scraps of zinc and copper they used to create plates. The account ends with the young men declaring they used the line for three years and became "expert telegraphers."

SOURCES "Wire Fence Telegraphing," *Scientific American* 49:26 (December 29, 1883); "Sending Telegrams Along a Barbed Wire Fence," *The New York Times* (March 20, 1904)

Fence Phones [41]

COUNTRY OF ORIGIN USA
CREATOR(S) Unknown
EARLIEST KNOWN USE Roughly 1893

BASIC INFRASTRUCTURE/MATERIALS Copper wire, barbed wire, posts, fasteners (such as nails or staples), insulators (such as porcelain knobs, glass bottles, leather, corn cobs, cow horns), battery-powered telephone handsets

RELATED Telephone [34], barbed wire telegraph [40]

DESCRIPTION A fence phone, also referred to as a barbed wire fence phone or squirrel line, is the use of "smooth" (presumably copper) wire running from a house to nearby barbed wire fencing to create an informal, ad hoc, cooperative, noncommercial, local telephone network [34]. Two key developments in the 1890s led to its adoption primarily by farmers, ranchers, and those living in rural or isolated areas especially in the US and Canada: the widespread availability and inexpensiveness of barbed wire in the 1890s, and the erosion of Alexander Graham Bell's patent monopoly in 1893 and 1894, which, according to Robert MacDougall, led to the sudden explosion of eighty to ninety independent telephone companies manufacturing telephone sets that could be used outside of the burgeoning Bell telephone system. According to Ronald Kline, the sudden explosion of independent telephone companies also set into motion the independent telephone movement, led by everyday people who were largely only interested in obtaining greater community access to a basic utility. Not only had Bell largely neglected to provide those in rural areas with telephone service in favor of focusing on those in urban areas, but early Bell telephone owners were also intent on controlling telephone usage. Writes MacDougall, "Bell's early managers sought to limit frivolous telephoning, especially undignified activities like courting or gossiping over the telephone, and to control certain groups of users, like women, children, and servants, who were thought to be particular offenders." By contrast, according to Kline, the independent telephone companies recognized that it would be too expensive to build lines in rural areas and instead openly "advised farm people to buy their own telephone equipment, build their own lines, and create cooperatives to bring phones to the countryside."

In need of a practical way to overcome social isolation and communicate emergencies, weather, and crop prices (and chafing under attempts to curtail free

[41] Poster from the 1880s advertising Glidden barbed wire and its crucial role in imposing physical control on wide open spaces. In this poster, the wire is celebrated for its ability to provide "safety to passengers and property" and because it "lasts twice as long as any other kind of fence," "sparks do not set it on fire," and "floods do not sweep it away."

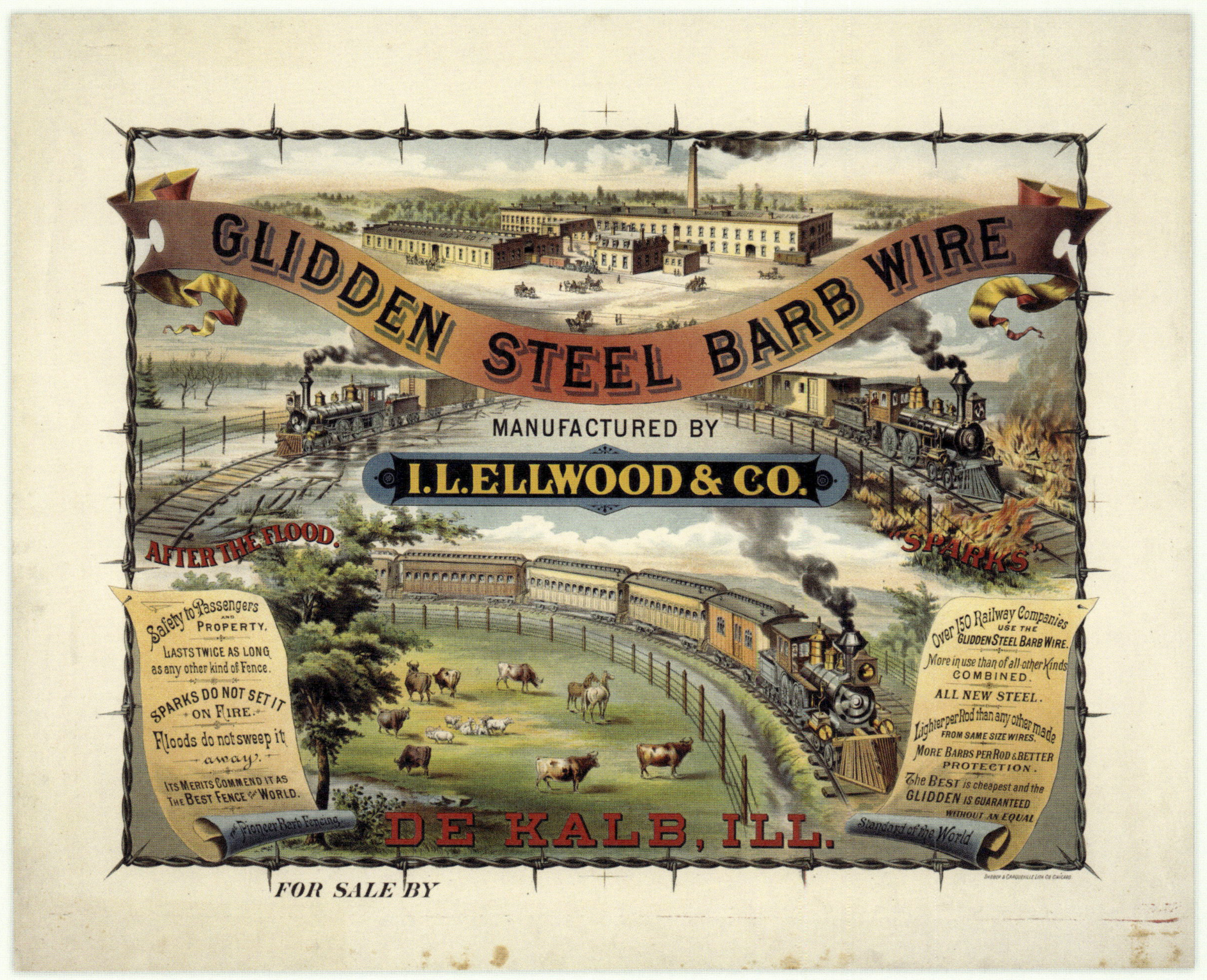

[41] Philip Peters and David Rueter's "Barbed Wire Fence Telephone" installed in the ARC Gallery in Chicago in 2014.

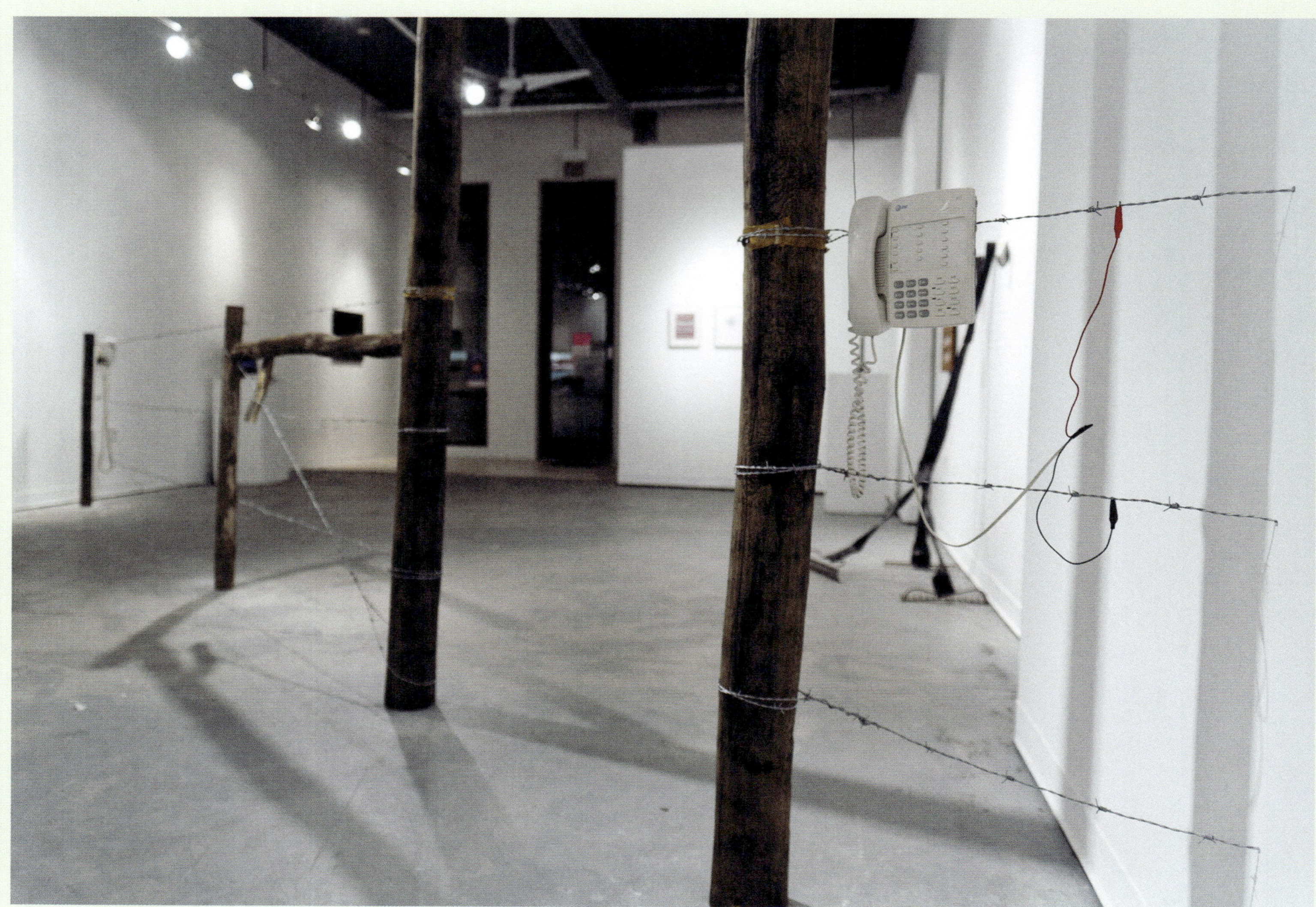

speech), ranchers and farmers began to take advantage of the growing ubiquity of both telephone sets and barbed wire fencing. They would hook up telephones to wire strung from their homes to a nearby fence—at the time, telephones had their own battery which produced a DC current that could carry a voice signal. Turning a crank on the phone would generate an AC current to produce a ring at the end of the line. Bob Holmes elaborates on the process: "the barbed wire networks had no central exchange, no operators—and no monthly bill. Instead of ringing through the exchange to a single address, every call made every phone on the system ring. Soon each household had its own personal ringtone . . . but anyone could pick up . . . Talk was free, and so people soon began to 'hang out' on the phone." The fence phone lines could also be used to broadcast urgent information to everyone on the line. Reportedly, the quality of the signal traveling over the heavy wire was excellent, but weather would frequently cause short circuits, which locals attempted to fix with anything that could serve as an insulator (such as leather straps, corn cobs, cow horns, or glass bottles).

There are newspaper reports of ranchers and farmers using fence phones in US states such as California, Texas, New Mexico, Colorado, Kansas, Iowa, Nebraska, Indiana, Minnesota, Ohio, Pennsylvania, New York, Montana, South Dakota, and also parts of Canada. For example, a 1902 issue of the Chicago-based magazine *Telephony* reported on a barbed wire fence telephone network that operated between Broomfield and Golden, Colorado, over a distance of twenty-five miles and which cost roughly ten dollars to build. The line was used for a "woman operator" to notify a worker at the end of the line "when to send down a head of water and how much." The author notes one "peculiar feature of this system is that only the operator can begin the talk. When it is decided to send down water the operator calls up the man at the headgate and gives him specific instructions, which he must follow. If he has anything to say he must say it then or hold his peace till he is called up again, for it is not a circuit system and only the Broomfield office can call up. This gives the lady the advantage of being able to shut off the other fellow at will and of getting in the last word." The fence phone systems also seemed to thrive in areas known for having cooperatives, especially related to farming. The model of a cooperative network particularly thrived throughout the 1920s, as farmers experienced economic depression some years before the Great Depression. For example, according to David Sicilia, farmers in Montana created the Montana East Line Telephone Association, to which they each contributed twenty-five dollars plus several dollars a year for maintenance along with telephone sets, batteries, wire, and insulators.

Anecdotally, fence phones were still being used throughout the 1970s and perhaps even later. C.F. Eckhardt describes calling his parents who lived in rural Texas and still used a fence phone; their number was simply 37, designated on the small local network by three long rings and one short ring.

EXPERIMENTS In 2014, artists Philip Peters and David Rueter installed their *Barbed Wire Fence Telephone* in the ARC Gallery in Chicago. Their installation hung telephones they had found at thrift stores from a barbed wire fence; the phones were connected to the barbed wire with simple copper wire and alligator clips. Visitors were then invited to communicate with each other using the fence phones.

SOURCES Alan Krell, *The Devil's Rope: A Cultural History of Barbed Wire* (Reaktion Books, 2002); David B. Sicilia, "How the West Was Wired," Inc.com (June 15, 1997); Early W. Hayter, *Free Range and Fencing*, vol. 3 (Kansas State Teachers College of Emporia Department of English, 1960); Robert MacDougall, *The People's Network: The Political Economy of the Telephone in the Gilded Age* (University of Pennsylvania Press, 2014); Ronald Kline, *Consumers in the Country: Technology and Social Change in Rural America* (Johns Hopkins University Press, 2002); "A Cheap Telephone System for Farmers," *Scientific American* 82:13 (March 31, 1900); "Bloomfield's Barbed Wire System," *Telephony: An Illustrated Monthly Telephone Journal* 4:6 (December 1902); Bob Holmes, "Wired Wild West: Cowpokes Chatted on Fence-wire Phones," *New Scientist* (December 17, 2013); C. F. Eckhardt, "Before Maw Bell: Rural Telephone Systems in the West," Texasescapes.com (2008); Phil Peters, *Barbed Wire Fence Telephone*, https://philipbpeters.com (2014)

HYBRID

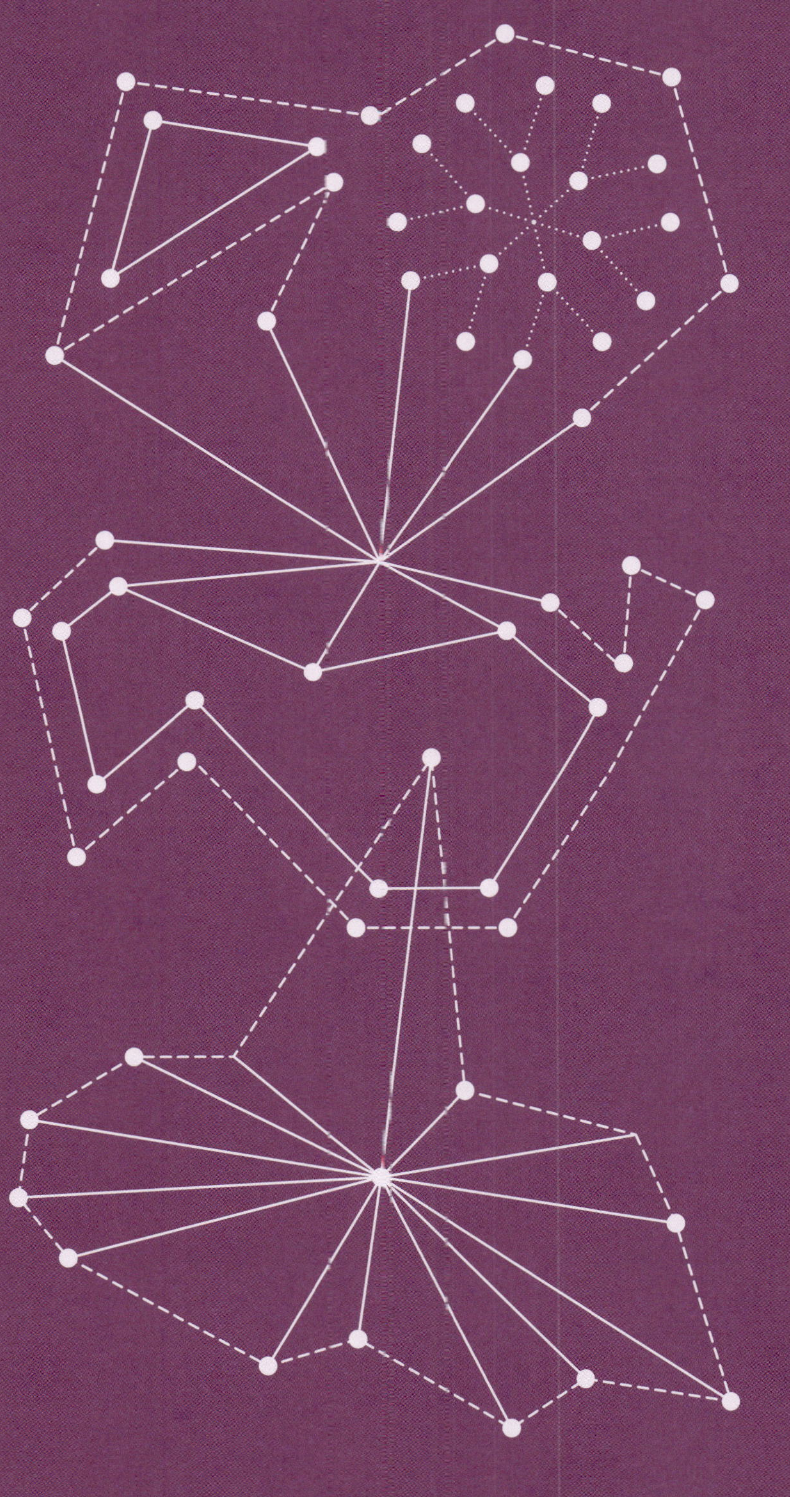

NETWORKS

HYBRID NETWORKS

This section contains networks that involve more than one transmission medium—for example, libraries [42], books [43], and postal systems [44] involve any combination of paper, buildings, roads, cars, trucks, planes, and more. Hybrid networks may therefore be wired and/or wireless. It is also worth noting that if one defines an internet simply as a network of networks, many of these hybrid networks could also be defined as such; however, I have attempted to adhere to the conventional definition of an internet as a network specifically of computer networks.

Library [42]

COUNTRY OF ORIGIN Present-day Iraq (originally Sumer, Ancient Mesopotamia)
CREATOR(S) Unknown
EARLIEST KNOWN USE 2600 BCE

BASIC INFRASTRUCTURE/MATERIALS Documents or literature (in any medium), physical or virtual space, classification system (optional), water/land/air transportation routes (optional)

RELATED Book [43], postal system [44], pigeon post [44.1], sneakernet [45], mundaneum [58], memex [60]

DESCRIPTION The first libraries were collections of clay tablets created by the Sumerians in the Mesopotamia region of the Fertile Crescent. The environment of the region was plentiful in clay, which led to the creation of not only clay tablets but also a form of writing consisting of wedges and lines that was well suited to this medium. Today, libraries are networks in multiple senses: with the aid of citational networks and classification systems like the Dewey decimal system that group materials together according to categories such as subject matter, these materials are in conversation with each other even if the communication takes place over months and years. If an internet is a network of networks, libraries have also been creating their own internets since the early twentieth century by using the postal system [44] and other forms of transportation infrastructure for sharing physical materials and by using the contemporary internet for sharing digital materials.

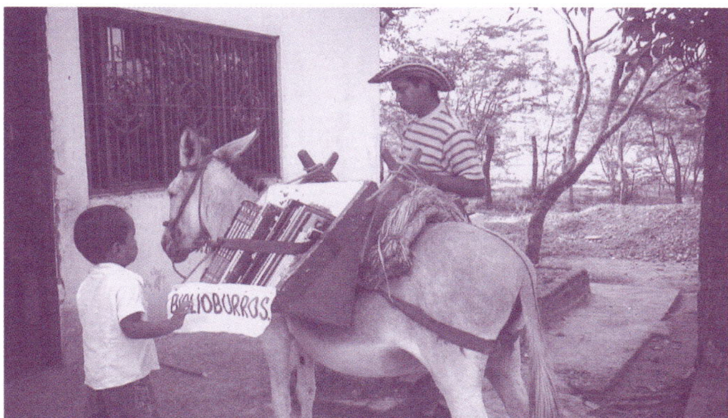

Biblioburros, traveling libraries using donkeys, are used in rural areas of Colombia.

EXPERIMENTS A recent example of artists engaging with the library itself as a platform for creative expression is Brazilian artist Mariana Lanari's *Moving Thinking* at the library of Amsterdam's Stedelijk Museum, where she "transformed the current classification of the books in the reading room—which are usually ranked alphabetically, by artist—by changing the order of the books, and replacing some with publications from the storage depot."

There are also many instances of experiments with the structure of the library itself as well as with library borrowing practices. One example is the traveling library, which dates back at least to nineteenth century Scotland and was used for educational purposes as well as for ships, lighthouses, and hospitals. Another example is the bookmobile, which, starting in the early twentieth century and continuing into the present, works to create more intangible networks of information sharing while also taking advantage of nationwide networks of waterways, railroads, and roads, using anything from bicycles and motor vehicles to camels, donkeys, and horses. The "biblioburro," for instance, began in the late 1990s in La Gloria, Colombia; run by public school teacher Luis Soriano, the program (still ongoing, as of 2021) features two donkeys who help distribute a collection of more than 4,800 materials to residents in rural areas.

SOURCES Lionel Casson, *Libraries in the Ancient World* (Yale University Press, 2001); "Moving Thinking," https://www.stedelijk.nl/en/exhibitions/moving-thinking-by-mariana-lanari; Diane Bashaw, "On the Road Again: A Look at Bookmobiles, Then and Now,"

Children & Libraries: The Journal of the Association for Library Service to Children (2010); Simon Romero, "Acclaimed Colombian Institution Has 4,800 Books and 10 Legs," *The New York Times* (October 19, 2008)

Book [43]

COUNTRY OF ORIGIN Parts of present-day Syria, Turkey, Iraq (originally Ancient Mesopotamia)
CREATOR(S) Unknown
EARLIEST KNOWN USE 2100–2000 BCE

BASIC INFRASTRUCTURE/MATERIALS Binding materials that could include rope, cord, plastic, wood, metal stable, glue, etc.; material for creating leaves that could include clay, stone, clay, leaves, papyrus, limestone, glass, ivory, wax, fabric, wood, paper, skin, hair, etc.; material for bound covers that could include wood, leather, or wood wrapped in leather, etc.; printing press, typewriter, computer; water/land/air transportation routes

RELATED Library [42], postal system [44], mundaneum [58], memex [60]

DESCRIPTION Some of the earliest cuneiform clay tablets could be considered a book, as the numbered tablets created a simple network of pages. However, once the codex emerged in ancient Rome as an object that was both bound *and* paginated, the book's networked qualities came to the fore. Umberto Eco states unequivocally that "books talk to each other" and these conversations between books (and writers) take place through complex webs of annotations, footnotes, citations, and intertextuality that exist both within and across books. Mikhail Bakhtin also coined the term "heteroglossia" in the 1930s to refer to the networks of different/competing discourses in a single text. Book networks are also often built or expanded by their travel through transportation networks such as the postal system [44] or even through hand-to-hand sharing as in the case of Samizdat (a form of Eastern bloc dissident activity that involved manually reproducing publications and then distributing them individually). As Robert Darnton points out, books are also part of larger circulation networks that include authors, publishers, printers, supplies, shippers, booksellers, readers, and binders.

EXPERIMENTS Particularly since the development of computer networks in the 1960s and 1970s, there have been many formal experiments with injecting networks into individual books themselves. For example, in addition to fiction by Jorge Luis Borges and Julio

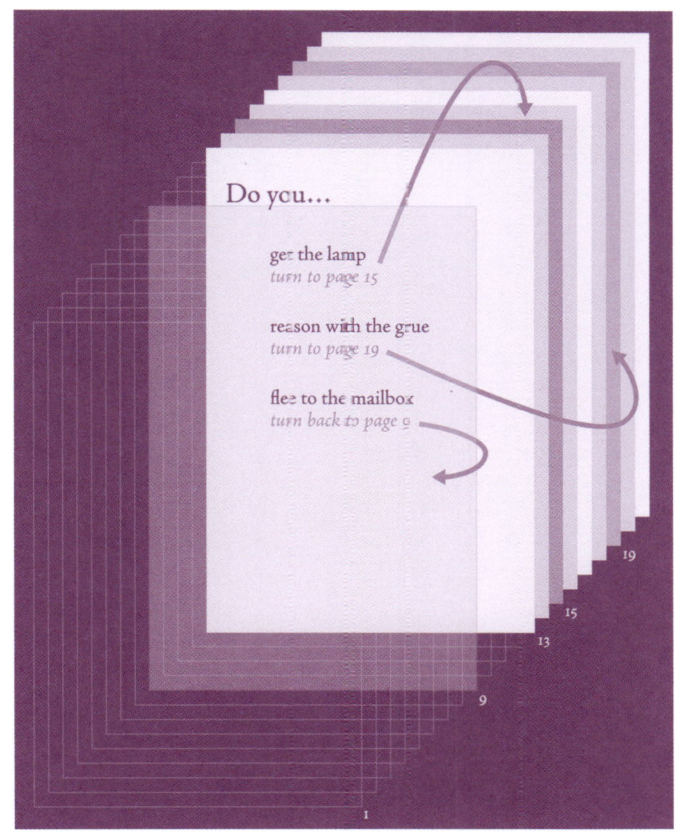

Choose Your Own Adventure books as illustrated by Christian Swinehart's "One Book, Many Readings."

Cortázar, the *Choose Your Own Adventure* series of books, which started in 1979, gave readers the ability to choose from among two or three options for the protagonist after several pages of reading. The books took on a rhizomatic structure that ultimately produced anywhere from seven to forty-four different endings for each book.

Offering a significantly more complex version of the book-as-network is Gilles Deleuze and Félix Guattari's *A Thousand Plateaus* (originally published in French in 1980 as *Mille Plateaux*), which was purportedly written so that the sections of the book (called rhizomes) could be read in any order, because every part connects to every other part. The complexity emerges in the process of reading since these connections have to be actively and conceptually produced; they are not part of any preexisting organization on behalf of the authors.

SOURCES Leila Avrin, *Scribes, Script and Books* (American Library Association, 1991); David Diringer, *The Book Before Printing* (Dover Publications, 1982); Kevin Smith, *The Structure of the Visual Book* (Keith A. Smith Books, 2003); Adrian Johns, *The Nature of the Book* (University of Chicago Press, 1998); Umberto Eco, "Borges and My Anxiety of Influence,"

[42] Small, informal, person-to-person network for information sharing made possible by a Chevrolet bookmobile in Charleston County (1946) parked next to the John's Island Community Center.

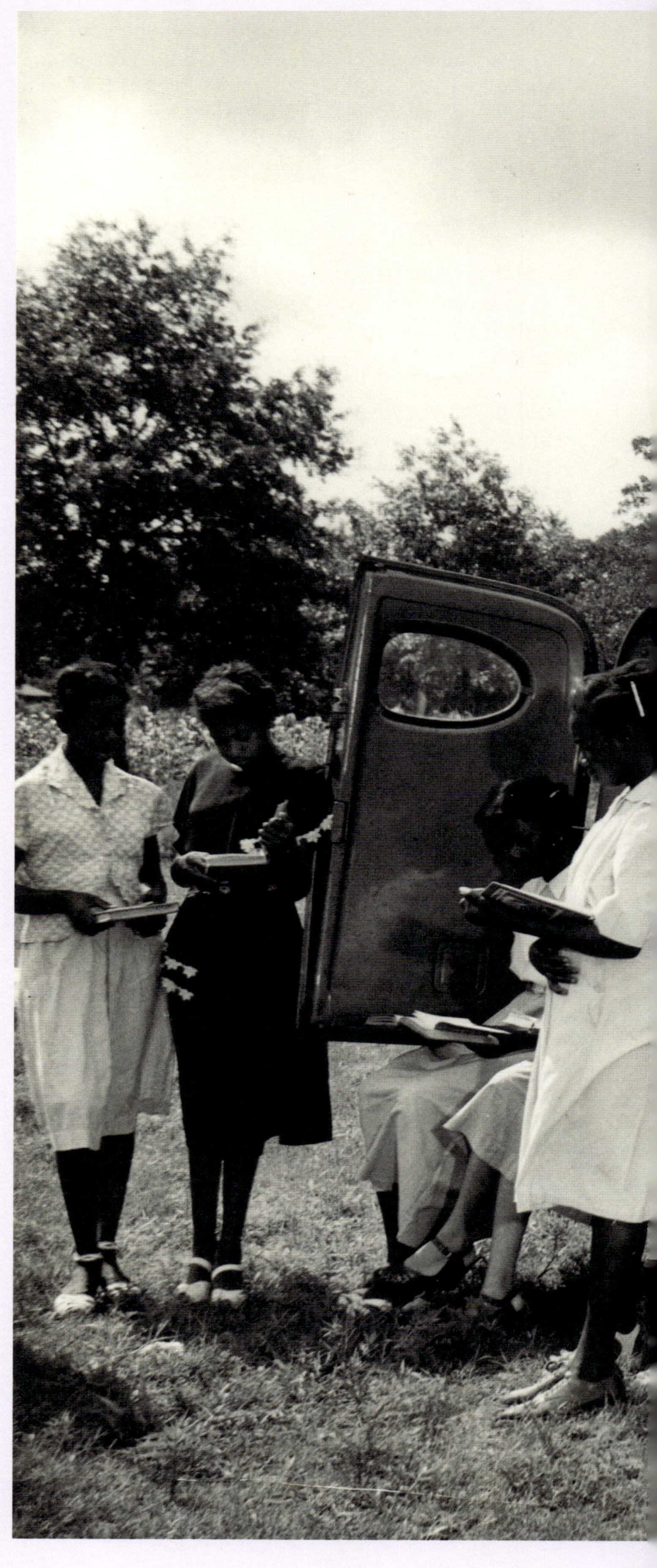

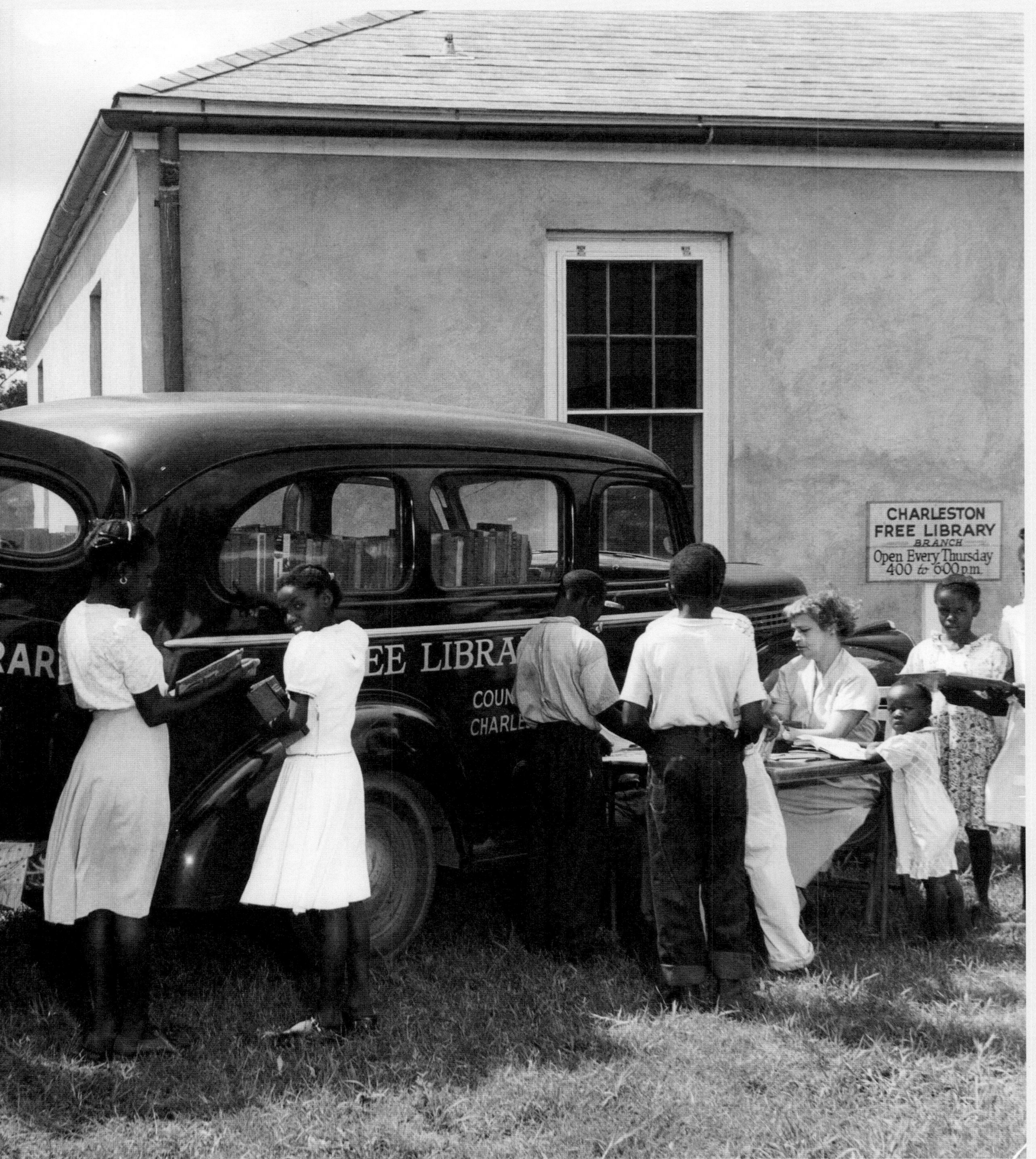
CHARLESTON
FREE LIBRARY
BRANCH
Open Every Thursday
4:00 to 6:00 p.m.

in *On Literature* (Harcourt, 2002); Mikhail Bakhtin, *The Dialogic Imagination* (University of Texas Press, 1981); Robert Darnton, *The Case for Books* (Public Affairs, 2009); Gilles Deleuze and Félix Guattari, *A Thousand Plateaus* (University of Minnesota Press, 1987)

Postal System [44]

COUNTRY OF ORIGIN Present-day Egypt (originally ancient Egypt)
CREATOR(S) Fifth and sixth dynasty pharaohs
EARLIEST KNOWN USE 1570 BCE (likely earlier)

BASIC INFRASTRUCTURE/MATERIALS Water/land/air transportation routes, stone or clay tablets, papyrus, limestone, paper, stations or offices

RELATED Book [43], library [42], pigeon post [44.1], projectile post [33.2], pony express [44.4], balloon mail [44.3], airgraph and v-mail [44.5], email letter [44.6], sneakernet [45]

DESCRIPTION The first system for delivering mail was a privately run courier service developed by ancient Egyptians for royalty; it consisted of horse-drawn chariots that would travel from the royal court to a series of relay stations and then on to the recipient. Citizens and state bureaucrats would also have to rely on individuals willing to deliver a letter for them while en route elsewhere, usually via horse-drawn chariot or carriage. It is worth noting, then, that the use of animals for mail delivery is part and parcel of the history of postal service; while pigeons, camels, dogs, cats, donkeys, and reindeer have all been used to facilitate postal delivery, the use of horses has the longest history and is the best-known example. While numerous state-run postal services were developed since the time of ancient Egypt, it wasn't until the nineteenth century that most countries had national and international postal delivery systems in place for the delivery of letters, postcards, and parcels. By the 1970s, mail delivery expanded to include electronic mail/messaging sent via any number of networks (including, for example, ARPANET), thereby introducing the now ubiquitous network for the exchange of digital communication and documents known as "email."

EXPERIMENTS The best-known artist experiments with the postal system began in 1962, with artist Ray Johnson sending letters with the instructions to "Please Add To and Return To Ray Johnson," thereby inaugurating the New York Correspondence School, which (along with other international art movements such as Fluxus) eventually evolved into the international movement known as mail art. Mail artists were, in a sense, hackers—in that they needed to know extensive details about packaging rules and they needed to be able to navigate/circumvent postal workers themselves, who had to be tricked into delivering everything from rubber stamps, envelopes, prints, lettering, banknotes, stickers, tickets, artist trading cards, badges, food packaging, diagrams, and maps. Still, it's also worth noting that the postal system, at least at this time, was an accessible and largely democratic network, and anyone with access to a mailbox and stamps could participate. In 1968, Fluxus artist Robert Filliou used the term "the Eternal Network" to refer to the networked communications of artists taking place via the postal system.

SOURCES David A. Warburton, "Letters, Letter Writing, Pharaonic Egypt," *Encyclopedia of Ancient History* (John Wiley & Sons, 2013); Edward Wente, *Letters from Ancient Egypt* (Scholars Press, 1990); Michael Crane and Mary Stofflet, *Correspondence Art* (Contemporary Arts Press, 1984); Natilee Harren, *Fluxus Forms* (University of Chicago Press, 2020); Chuck Welch, ed., *Eternal Network: A Mail Art Anthology* (University of Calgary Press, 1992)

Pigeon Post [44.1]

COUNTRY OF ORIGIN Ancient Rome
CREATOR(S) Unknown
EARLIEST KNOWN USE Approx. 50–40 BCE

BASIC INFRASTRUCTURE/MATERIALS Homing pigeon, small backpack or canister, paper or microfilm, metal leg band (optional), GPS tracker (optional)

RELATED Packet radio network [26], book [43], library [42], postal system [44], projectile post [33.1], pony express [44.4], balloon mail [44.3], airgraph and v-mail [44.5], email letter [44.6], sneakernet [45]

DESCRIPTION Pigeon post—also known as aerial courier, avian messenger, or bird post—is the use of homing pigeons to quickly transport very small letters, microfilm, tools for surveillance, trackers, and even narcotics over long distances. To achieve this, homing pigeons are first transported to their destination (via balloon, rail, train, automobile, or plane), at which point they return home. They have been bred to quickly navigate over very long distances, traveling back to their loft as far as 1,800 kilometers (1,100 miles) with a top speed of roughly 160 kilometers (100 miles) per hour. While it was likely not the first use of homing pigeons for transporting correspondence,

[44] Mail art by Ray Johnson, *Letters $10,000 Each* (ca. 1980).

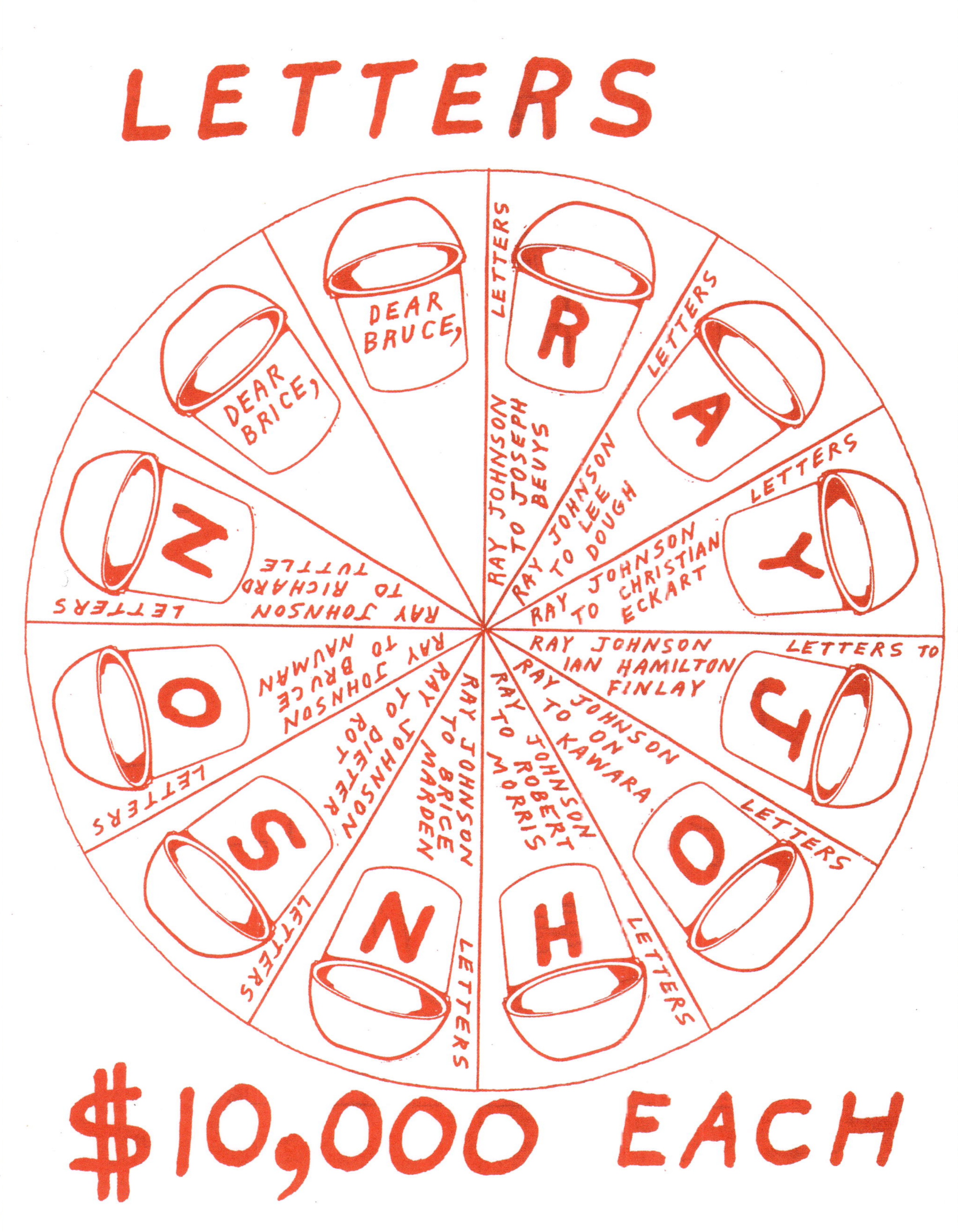

[44.1] Map of a pigeon post network from Moscow, Russia, in 1941 denoting communication lines between a central pigeon station, a nursery, and city pigeon stations. The network was used for communication about the front line of the war with Nazi Germany. The title of the map translates roughly to "Scheme #5, Defense Department's Pigeon-based Postal Service for the City of Moscow" (trans. Tatyana Huseynova).

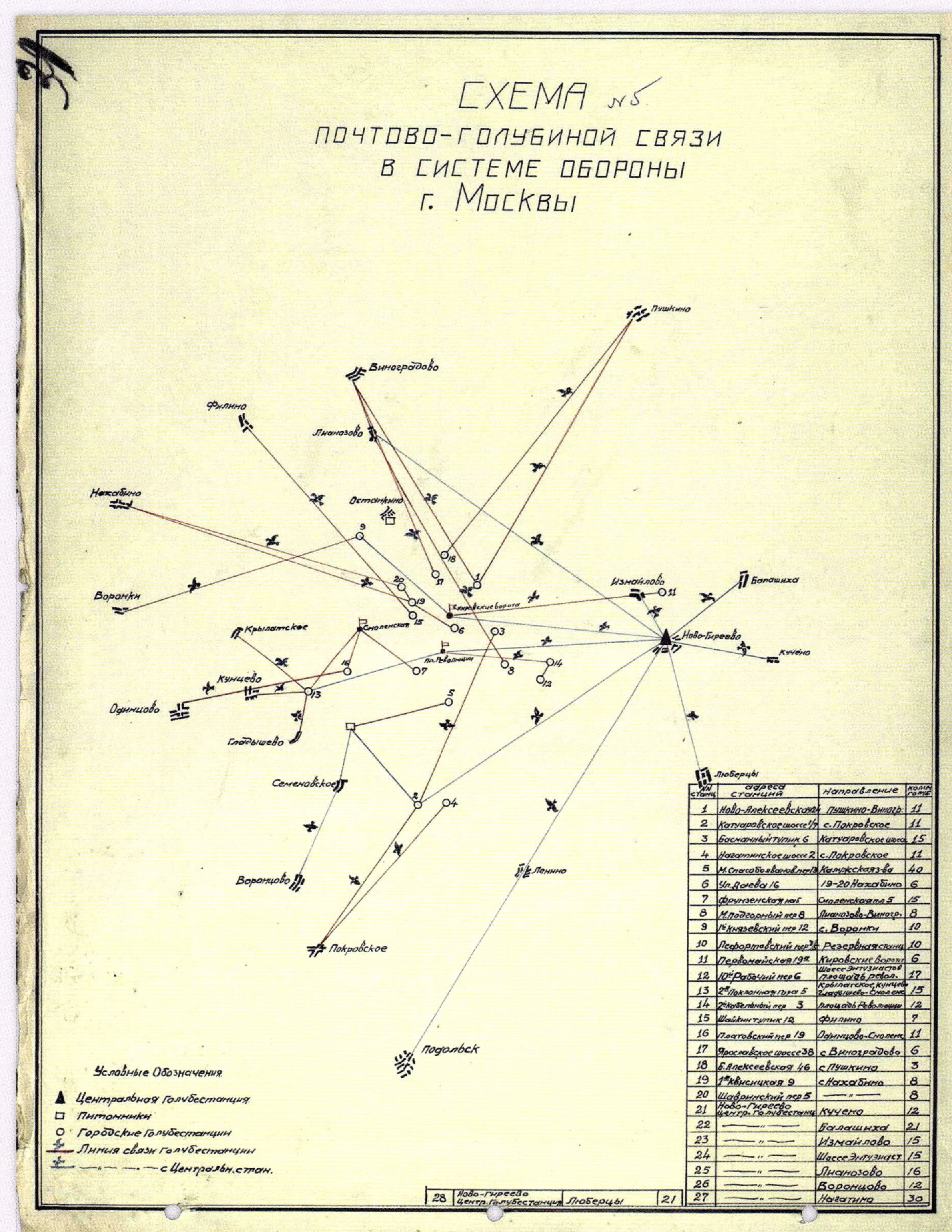

the earliest record of pigeon post appears in Roman author Sextus Julius Frontinus's *The Stratagems*, in which he records that pigeons were used to send messages between military consul Aulus Hirtius and Decimus Junius Brutus; given the years Hirtius and Brutus were active in politics and the military, I estimate the time period was likely somewhere between 50 to 40 BCE. Thereafter, pigeon post was used around the world for both commercial (stock-brokers, investors, merchants) and military purposes. Pigeons were particularly popular with the military after they played a pivotal role in transporting messages to Paris from the rest of Europe during the siege of the city in 1870–71. Messages were sent by placing as many as fifteen photo-micrography films in a small quill that was attached using waxed silk thread to one of the large feathers on the pigeon's tail. Thereafter, in spite of the very high mortality rate of these homing pigeons, they became a key component of military communication in countries such as the USA, Canada, the UK, Australia, India, Burma, and Russia; for example, the US Navy began using pigeons for communication and surveillance in 1896, followed in 1917 by the US Army Signal Corps's establishment of the pigeon service. While India was one of the last countries to discontinue its military use of pigeons, in 2008, countries continue to allege that neighboring states are using pigeons for communication or surveillance. Individuals also continue to use homing pigeons to send covert messages. There are even pigeon post businesses still in operation: one service that, as of 2014, was based in Buffalo, New York, asks: "Have a message or data card you need delivered? We have tailor-made backpacks that fit our individual pigeons." Further, the company declares they have "origami trained 'folders' that write tiny messages and fold them into shapes while groomers make sure the pigeons are ready to take flight. When the perfect pigeon is chosen to complement the right message he or she is let out the window." Finally, so-called narco pigeons are also commonly used to transport drugs into prisons around the world.

EXPERIMENTS Reminding us not only that there is more than one internet but also that internet protocols need not travel exclusively through wires and radiowaves, David Waitzman proposed "IP over Avian Carriers" (IPoAC) as an April Fools' Day joke on April 1, 1990. Submitted as RFC 1149 (a Request for Comments is a type of formal memo issued by the Internet Engineering Task Force for experimental specifications and proposals related to computer networking), the memo describes "an experimental method for the encapsulation of IP [internet protocol] datagrams in avian carriers." Perfectly appropriating the conventional discourse of computer networking, Waitzman

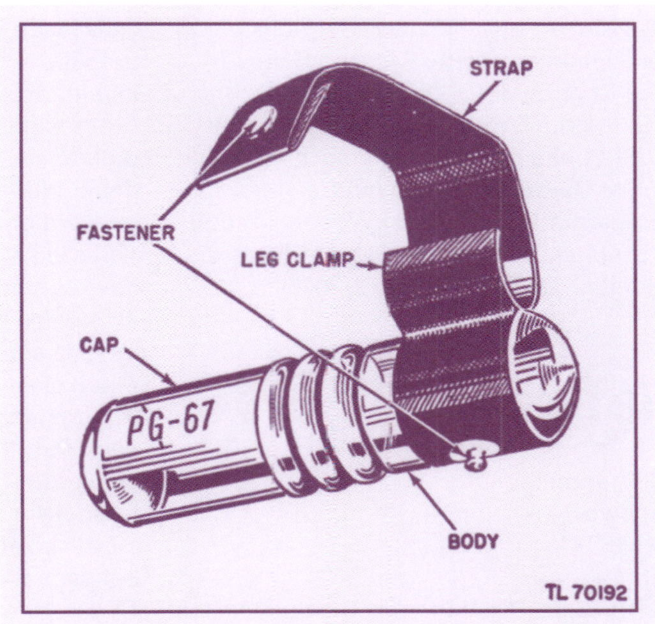

Schematic from a U.S. War Department Technical Manual (1945) of a standard message holder that would be attached to the leg of a homing pigeon.

continues: "Avian carriers can provide high delay, low throughput, and low altitude service. The connection topology is limited to a single point-to-point path for each carrier, used with standard carriers, but many carriers can be used without significant interference with each other, outside of early spring . . . The carriers have an intrinsic collision avoidance system, which increases availability. Unlike some network technologies, such as packet radio, communication is not limited to line-of-sight distance." In January 2002, the Bergen, Norway, Linux Users Group successfully demonstrated RFC 1149 by printing data on paper that was attached to pigeons' legs; once the pigeons arrived at their destination, the data was scanned and transferred to a computer. The transmission took one hour and forty-two minutes.

SOURCES Sextus Julius Frontinus, *The Stratagems, and the Aqueducts of Rome* (Harvard University Press, 1925); Andrew D. Blechman, *Pigeons: The Fascinating Saga of the World's Most Revered and Reviled Bird* (Grove Press, 2006); Francis D. Lazenby, "Greek and Roman Household Pets," *The Classical Journal* 44:5 (February 1949); Henry Teonge, *The Diary of Henry Teonge: Chaplain on Board H.M.'s Ships Assistance, Bristol and Royal Oak 1675–1679* (Routledge, 2005); "Rothschilds and Pigeon Post," Rothschild archive website; *American Air Mail Catalogue*, vol. 1 (American Air Mail Society, 1966); Thomas Appleton, *Usque ad mare: A History of the Canadian Coast Guard and Marine Services* (Department of Transport, 1968); "Fowl Play: Alleged Spy Pigeon Held in India," *The Express*

Tribune (May 28, 2010); "Carrier Pigeon Services," Danny's Carrier Pigeons website; Rhianna Schmunk, "Pigeon Wearing Crystal Meth 'Like a Backpack' Caught Inside B.C. Prison Yard," *CBC News* (January 6, 2023); David Waitzman, "A Standard for the Transmission of IP Datagrams on Avian Carriers," Request for Comments 1149, Network Working Group (April 1, 1990); Stephen Shankland, "Pigeon-Powered Internet Takes Flight," *CNET* (January 2, 2002)

Projectile Post [44.2]

COUNTRY OF ORIGIN Germany
CREATOR(S) Unknown
EARLIEST KNOWN USE 1475

BASIC INFRASTRUCTURE/MATERIALS Projectile launcher (a cannon, gun, rocket engine, etc.), ballistic projectile (cannon ball, bullet, rocket, missile, etc.), propellant (gun powder, fuel, etc.), internal parachute (optional)

RELATED Postal system [44], pigeon post [44.1], pony express [44.4], balloon mail [44.3], airgraph and v-mail [44.5], email letter [44.6], sneakernet [45]

DESCRIPTION "Projectile post" is not an established term but here refers to projectiles (bullets, cannonballs, rockets, and missiles) used for the long-distance transportation of communication using a weapon to propel an object into space. The earliest recorded instance of projectile post likely occurred in April 1475, when cannons and bullets were used during the Siege of Neuss (Germany) to transport letters between Neuss and Cologne. Since it was difficult to distinguish between a discharged bullet meant for opposing forces and a bullet bearing a message, optical signaling with fire and flags was used to indicate the successful transmission and reception of messages. Thereafter, the first recorded use of rockets to deliver mail took place in 1901 between the S.S. *Hauroto* (a passenger and cargo ship from New Zealand) and Niuafo'ou (an island part of the Kingdom of Tonga in the southern Pacific Ocean). Up to this point, residents of Niuafo'ou had used a "tin can mail service," whereby nearby ships would throw sealed tins of mail into the sea that were then retrieved by local swimmers. While rocket mail was intended to replace the tin can mail service, the latter proved to be a far more accurate and reliable means of mail delivery. The next use of rocket mail took place in 1932 in Austria, when rocket scientist Friedrich Schmiedl launched 102 pieces of mail between the towns of Schöckl and St. Radegund; this was quickly followed by rocket mail experiments in 1934 across the Scottish Hebridean islands, in 1934 from Saugur Island to shore in India, in 1936 across Greenwood Lake in the USA, and in 1936 across the USA-Mexico border. Few of these experiments were successful. In an effort to deliver mail even faster and more efficiently, and no doubt influenced by the Cold War race to land humans on the moon, the United States Postal Service participated in the first "missile mail" launch from the navy submarine U.S.S. *Barbero* to a naval auxiliary air station in Mayport, Florida, on June 8, 1959. The missile successfully carried 3,000 letters over a distance of a hundred miles in about twenty-two minutes. Postmaster General Arthur Summerfield famously declared, "Before man reaches the moon, mail will be delivered within hours from New York to California, to England, to India or to Australia by guided missiles." Likely the last delivery of rocket mail took place between the Mojave Air and Space Port to California City via the rocket-powered airplane known as XCOR EZ-Rocket on December 3, 2005; the rocket carried four pouches of mail across nearly ten miles of desert.

SOURCES "Archiv Fur Post und Telegraphie," *Beiheft Zum Amtsblatt Des Reichs-Postamts* 19 (October 1886); William Edgar Geil, *Ocean and Isle* (William T. Pater, 1902); "'Tin Can' Mail Service," *Waikato Times* 94:14730 (August 1921); R. Cargill Hall et al., *Essays on the History of Rocketry and Astronautics: Proceedings of the Third Through the Sixth History Symposia of the International Academy of Astronautics*, vol. II (1977); "The Experiment to Deliver Letters by Rocket in the Hebrides," *The Scotsman* (September 18, 2017); Rebecca Maksel, "When Missiles Delivered the Mail," *Smithsonian* (July 1, 2011); William I. Lengeman III, "Rocket Mail: Early Experiments Demonstrated Why This Idea Never Got Off the Ground," *Aviation History* (November 2007); "Missile Mail," USPS Postal History website; Cathy Hansen, "Remembering Dick Rutan's First Rocket Powered Air Mail Flight," *Aerotech News* (December 4, 2020)

Balloon Mail [44.3]

COUNTRY OF ORIGIN USA
CREATOR(S) John Wise
EARLIEST KNOWN USE 1859

BASIC INFRASTRUCTURE/MATERIALS Gondola or basket; one or more heat sources or burners; liquid propane, helium, or hydrogen; bag or envelope made from paper, fabric, or plastic; instrumentation (optional)

[44.2] Postcards used to commemorate the first missile-mail launch that took place off the coast of Florida in June 1959.

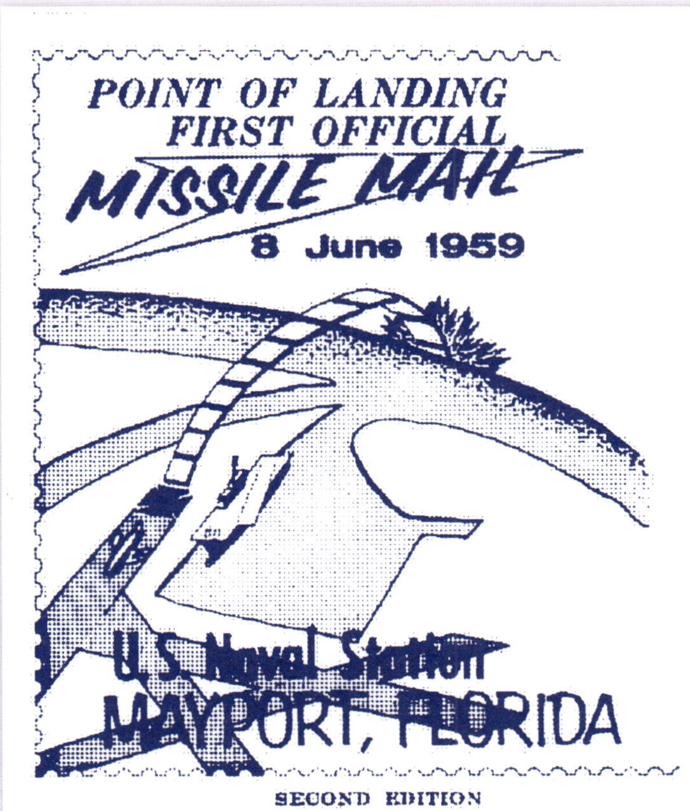

L. GORDON PIERSON
60 MAPLE AVE.
WEST ORANGE, N.J.

FIRST OFFICIAL MISSILE MAIL DELIVERED BY U. S. NAVY REGULUS I MISSILE AT THE U. S. NAVAL STATION, MAYPORT, FLORIDA, FROM SUBMARINE BARBERO IN INTERNATIONAL WATERS. DURATION OF FLIGHT ONLY TWENTY-TWO MINUTES.

SECOND EDITION

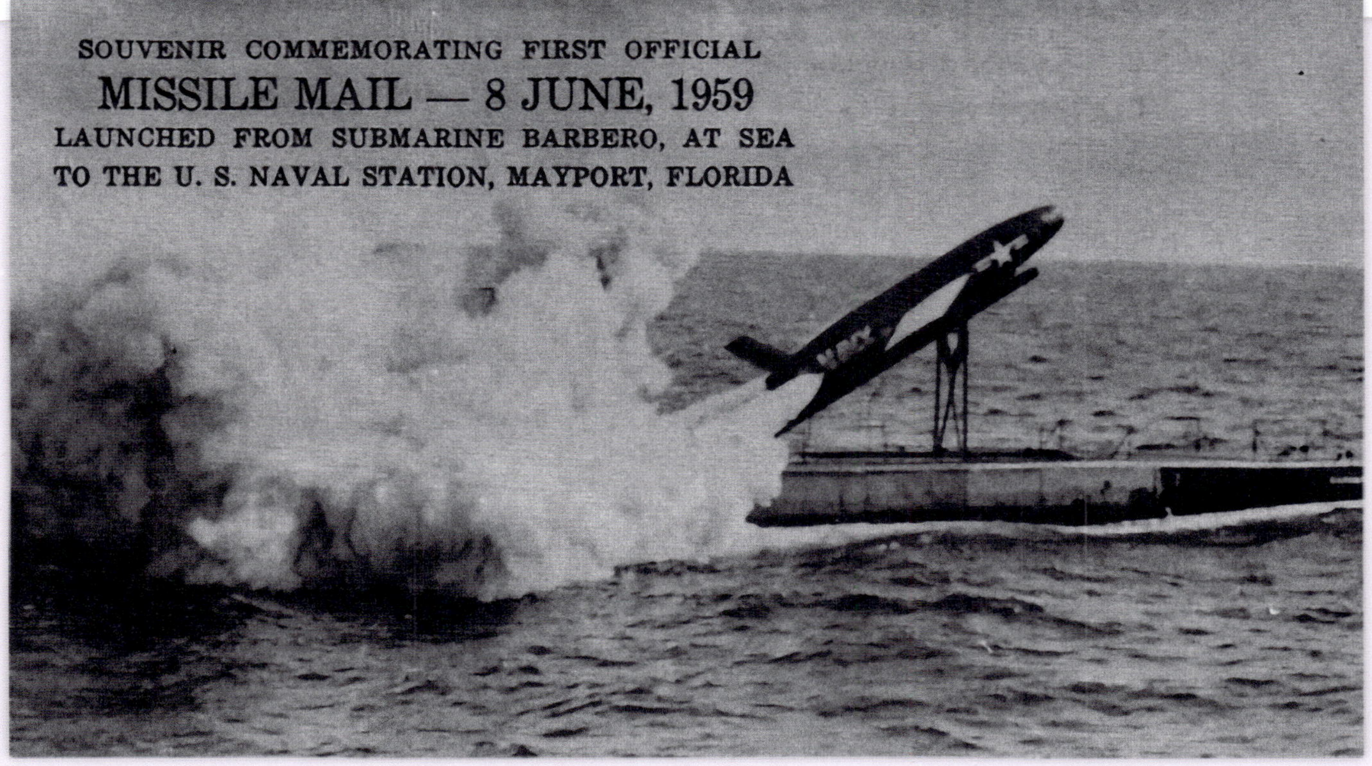

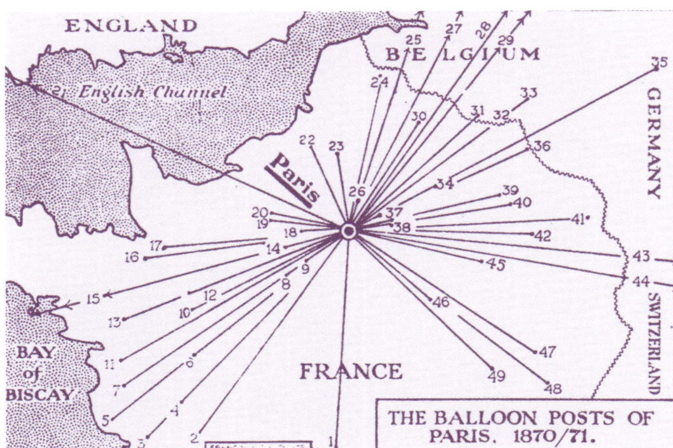

Balloon mail routes from Paris to Europe from 1870 to 1871, from *A Commercial & Historical Atlas of the World's Airways.*

RELATED Fire or smoke signals [3], postal system [44], pigeon post [44.1], airgraph and v-mail [44.5], sneakernet [45]

DESCRIPTION Balloon mail is the transportation of mail using a hot air balloon made out of paper, fabric, or plastic; filled with hydrogen, helium, or liquid propane; propelled by a heat source or a burner; and that may or may not have a pilot on board. While China developed airborne lanterns for signaling around 220 to 280 CE, balloons that could support the weight of a human pilot (who could also act as a steward for mail) were likely not invented until 1783 in France. In 1859, aeronaut and pilot John Wise attempted to deliver 123 letters from Lafayette, Indiana, to New York City. The flight made it as far as Crawfordsville, Indiana, at which point Wise was forced to land and have the letters delivered to their final destination via rail. However, the best-known and most thoroughly documented early use of hot air balloons for the delivery of mail was during the Franco-Prussian War and the siege of Paris that lasted from 1870–1871. Balloons or pigeons were the only means for mail to leave Paris during the time. According to the *American Air Mail Catalogue*, roughly fifty-five piloted balloons were released, carrying about 2.5 million letters, "238 passengers, 361 pigeons and 5 dogs." The longest flight from Paris to Norway took over fourteen hours and traveled a distance of 822 miles to carry five hundred pounds of mail. Balloons were also used in the UK in the early twentieth century as the earliest mail-carrying flights.

Hot air balloons have also been used throughout the twentieth and even the twenty-first centuries for disseminating information and propaganda to inaccessible areas. For example, in the mid-1950s, several hundred balloons capable of carrying 100,000 propaganda pamphlets were floated from West Germany to East Germany. According to the era's *Popular Mechanics* staff writers, "Nine pound packets of leaflets suspended by cord are attached to a [bicycle wheel] rim at regularly spaced points. An alarm clock causes two alarm powered razor blades to move slowly around the rim of the wheel and cut off the cords that were holding the packets at fixed intervals." More recently, it was reported in 2008 that waterproof leaflets were being sent from South Korea to North Korea via small helium-filled balloons.

SOURCES Tom Crouch, *Lighter Than Air: An Illustrated History of Balloons and Airships* (Johns Hopkins University Press, 2009); Owen Edwards, "Airmail Letter," *Smithsonian* 37:5 (August 2006); *American Air Mail Catalogue*, vol. 1 (American Air Mail Society, 1966); John Pringle, *Early British Balloons* (self-published, 1935); "Balloon Wheel Dumps Leaflets," *Popular Mechanics* (February 1956); John Sudworth, "Storm Looms Over N. Korea Balloons," *BBC News* (November 10, 2008)

Pony Express [44.4]

COUNTRY OF ORIGIN USA
CREATOR(S) Alexander Majors, William Waddell, and William Hepburn Russell
EARLIEST KNOWN USE 1860

BASIC INFRASTRUCTURE/MATERIALS Riders, horses, stations, land/water transportation routes

RELATED Radiotelegraphy [16.1], postal system [44], pigeon post [44.1], balloon mail [44.3], airgraph and v-mail [44.5], electrical telegraph [33], telephone [34], the clacks [64]

DESCRIPTION Lasting only eighteen months until the inauguration of the transcontinental telegraph, the Pony Express greatly reduced the length of time it took to send letters between Missouri and California. In the early months of its existence, the Pony Express involved 80 riders, 184 stations, 400 horses, and hundreds of station workers. However, prices were also initially 25,000 percent higher than the admittedly slower postal service. It took anywhere from one to ten days to send a letter via Pony Express. Despite its short lifespan, the Pony Express became a key part of American mythology around the settling of the West. It was even revived in 1952 as the "Electronic Pony Express" in Bell Telephone's advertising for its coast-to-coast microwave radio-relay [31] network, which happened to follow part of the Pony Express route.

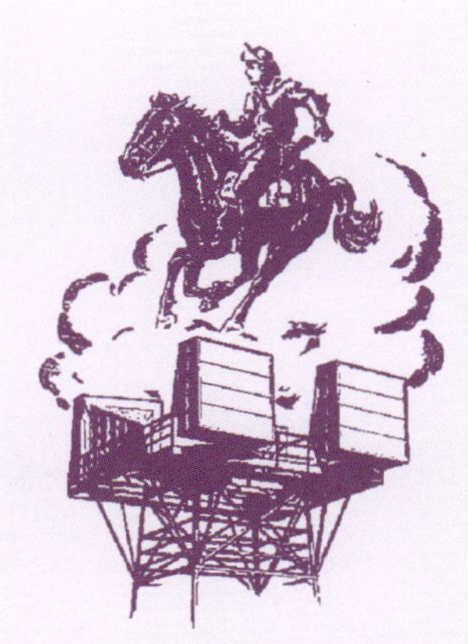

Electronic Pony Express

Telephone calls and television pictures now flash through the air along a radio-relay skyway from coast to coast. Part of the way the new route follows the trail of the Pony Express, old-time relay, whose riders galloped with urgent dispatches between Missouri and California, sometimes making a one-way trip in nine days. Today, riding microwave beams, long distance calls and television programs are relayed from tower to tower across the continent with the speed of light.

Distance has always been a challenge to Americans. This Almanac tells of some of the efforts at fast communication that have been made since earliest Colonial days. Also portrayed are some of the later steps leading toward radio-relay—the Bell System's most recent answer to the unending challenge.

The new radio-relay system is the seventh telephone highway placed in service across the United States since 1915. Supplementing wire and cable, radio-relay is helping to make America's vast communications network even stronger and more flexible. Tireless research and engineering made it possible.

And the research and engineering must go right on. As you read this, still further improvements and extensions of our service are in the making. For the telephone has become indispensable in the conduct of business, the administration of government, and in home and social life. It adds to national unity, aids national defense. Our responsibility—and our constant endeavor—is to make it always more dependable, useful and valuable to more and more people.

Promotional material in a 1952 issue of the *Bell Telephone Almanac* for their new coast-to-coast microwave radio-relay network.

EXPERIMENTS Prior to the creation of the Pony Express, the US government used camels for army transportation, trade, topographic surveys, and mail carriers from 1855 to 1863. More recently, in 2006, a class at MIT assessed the feasibility of whether a ball relay (with letters placed inside baseballs) could beat a horse relay such as the Pony Express. The answer: theoretically, yes. "Assuming an average ball carrier can throw a 40 mph ball 100 feet and catch-throw in 2 seconds, the ball relay is expected to be a blazing 20 mph, roughly 2× the Express's average speed of 10 mph. However, just one drop in 10 might allow the express to win."

SOURCES "Universal Service and the Postal Monopoly: A Brief History," USPS website (October 2008); "The Pony Express," Pony Express National Museum website (2012); Eva Jolene Boyd, *Noble Brutes* (Wordware, 1994); "Demise of the Pony Express: Idea Feasibility Testing," MIT Product Engineering Processes course website (September 2006); "Electronic Pony Express," *Bell Telephone Almanac* (1952)

Airgraph and V-Mail [44.5]

COUNTRY OF ORIGIN UK
CREATOR(S) British Post Office / Eastman Kodak Company
EARLIEST KNOWN USE 1941

BASIC INFRASTRUCTURE/MATERIALS Paper forms, Kodak Recordak automatic camera, microfilm, land/water/air transportation routes

RELATED Postal system [44], pigeon post [44.1], projectile post [33.2], airgraph and v-mail [44.5], email letter [44.6], telautograph [36]

DESCRIPTION In order to facilitate wartime mail delivery between Great Britain and Egypt, the British post office followed up on a 1932 proposal by the Eastman Kodak Company to use a machine called a Recordak to photograph documents onto microfilm, thereby greatly reducing the volume and weight of mail bags. Once the microfilm negatives arrived at their destination, they

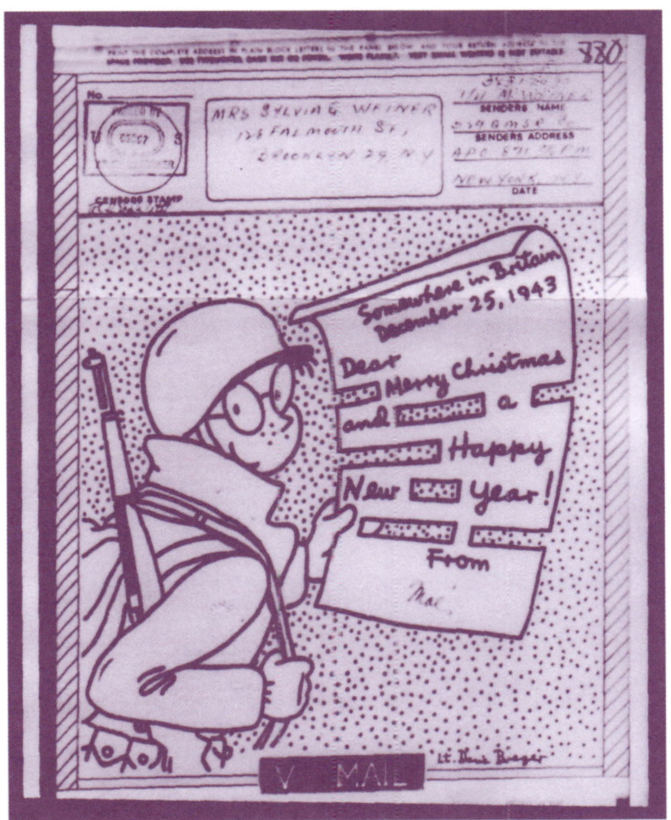

A V-Mail letter from December 25, 1943, satirizing the US Army's censorship practices.

[44.4] Illustrated map of the Pony Express route in 1860 by William Henry Jackson.

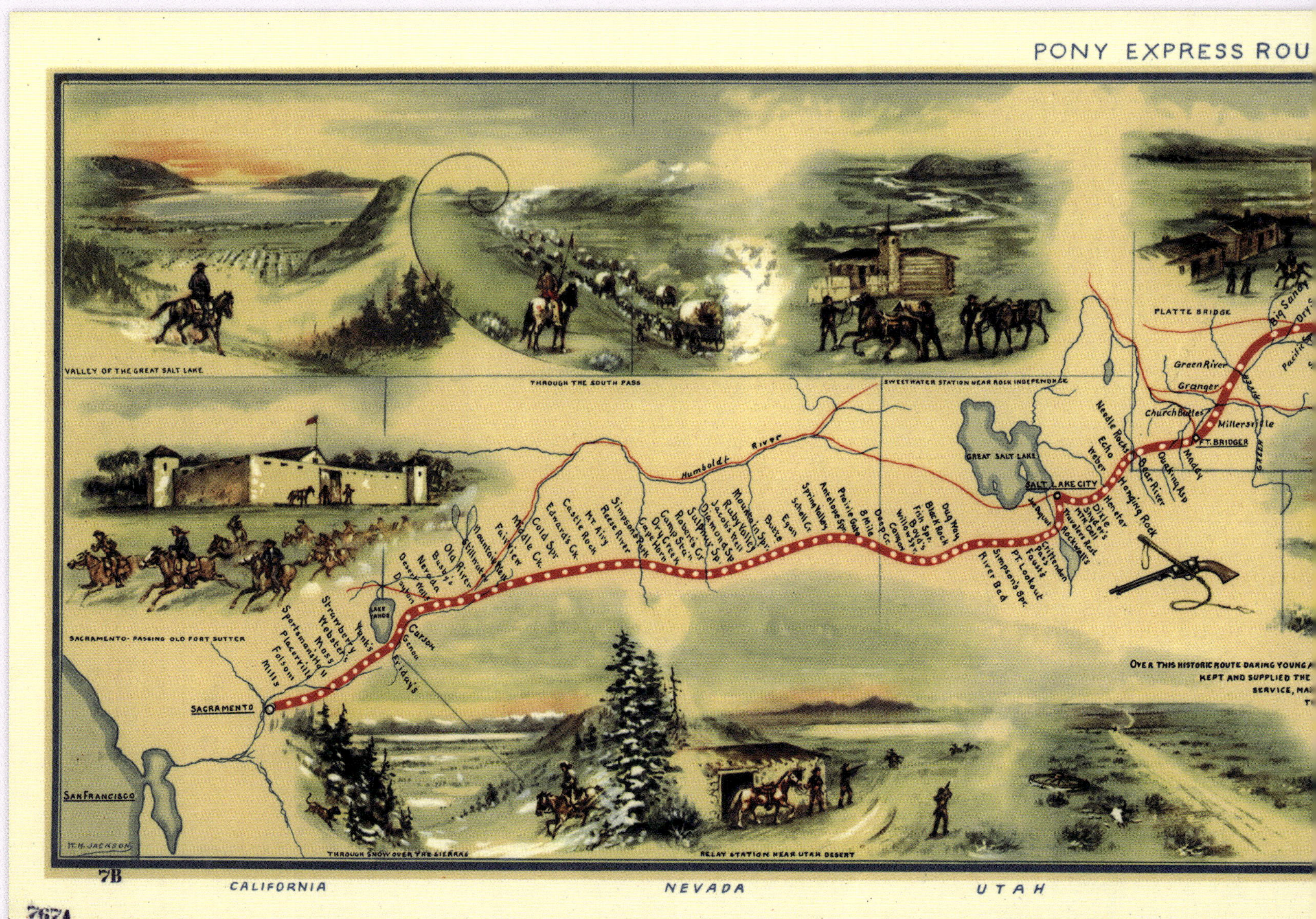

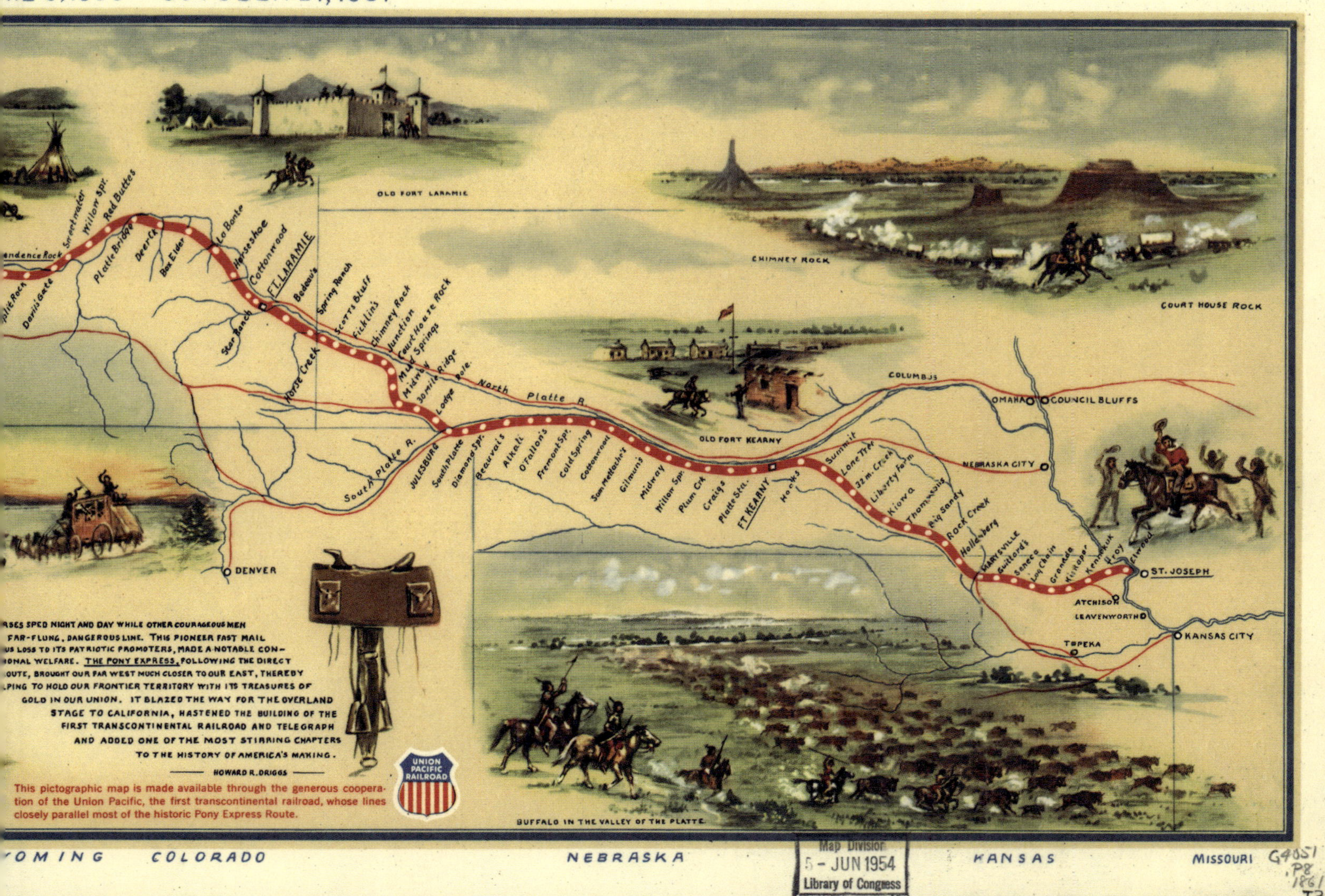
RIL 3, 1860 — OCTOBER 24, 1861
OLD FORT LARAMIE
CHIMNEY ROCK
COURT HOUSE ROCK
OLD FORT KEARNY
BUFFALO IN THE VALLEY OF THE PLATTE
FT. LARAMIE
FT. KEARNY
JULESBURG
DENVER
COLUMBUS
OMAHA
COUNCIL BLUFFS
NEBRASKA CITY
MARYSVILLE
ST. JOSEPH
ATCHISON
LEAVENWORTH
TOPEKA
KANSAS CITY
North Platte R.
South Platte R.
Red Buttes
Willow Spr.
Platte Bridge
Deer Cr.
Box Elder
La Bonte
Horseshoe
Cottonwood
Bedeau's
Spring Ranch
Scotts Bluff
Ficklin's
Chimney Rock
Junction
Court House Rock
Mud Springs
Midway
30 Mile Ridge
Lodge Pole
Star Ranch
Horse Creek
South Platte
Diamond Spr.
Beauvais
Alkali
O'Fallon's
Fremont Spr.
Cold Spring
Cottonwood
Gilmans
Midway
Willow Spr.
Plum Ck.
Craigs
Platte Sta.
Hooks
Summit
Lone Tree
32 m. Creek
Liberty Farm
Kiowa
Big Sandy
Rock Creek
Hollenberg
Guittard's
Seneca
Granada
Kennekuk
Troy
Elwood
…RSES SPED NIGHT AND DAY WHILE OTHER COURAGEOUS MEN
… FAR-FLUNG, DANGEROUS LINE. THIS PIONEER FAST MAIL
…US LOSS TO ITS PATRIOTIC PROMOTERS, MADE A NOTABLE CON-
…IONAL WELFARE. THE PONY EXPRESS, FOLLOWING THE DIRECT
…OUTE, BROUGHT OUR FAR WEST MUCH CLOSER TO OUR EAST, THEREBY
…LPING TO HOLD OUR FRONTIER TERRITORY WITH ITS TREASURES OF
GOLD IN OUR UNION. IT BLAZED THE WAY FOR THE OVERLAND
STAGE TO CALIFORNIA, HASTENED THE BUILDING OF THE
FIRST TRANSCONTINENTAL RAILROAD AND TELEGRAPH
AND ADDED ONE OF THE MOST STIRRING CHAPTERS
TO THE HISTORY OF AMERICA'S MAKING.
— HOWARD R. DRIGGS —
This pictographic map is made available through the generous cooperation of the Union Pacific, the first transcontinental railroad, whose lines closely parallel most of the historic Pony Express Route.
UNION PACIFIC RAILROAD
…YOMING
COLORADO
NEBRASKA
KANSAS
MISSOURI

were printed on photographic paper, carefully censored, and delivered as letters. The service, dubbed Airgraph, was inaugurated in 1941 between London and Cairo and was quickly adopted in 1942 by the USA as V-Mail (or Victory Mail). It took anywhere from two to five weeks to send a letter via Airgraph.

EXPERIMENTS Well aware of the government censorship of Airgraph letters and V-Mail, letter writers became clever at figuring ways to either circumvent or satirize censorship (by, for example, writing letters that appeared to already be censored).

SOURCES Camille Allaz, *The History of Air Cargo and Airmail from the 18th Century* (Gardners Books, 2005); "Microfilm Models-Precursors of V-Mail," Smithsonian National Postal Museum

Email Letter [44.6]

COUNTRY OF ORIGIN USA
CREATOR(S) Gene Johnson / United States Postal Service
EARLIEST KNOWN USE 1982

BASIC INFRASTRUCTURE/MATERIALS Paper, printer, envelopes, water/land/air transportation routes, stations or offices, computer, modem, telephone wire, electric power source

RELATED Radiotelegraphy [16.1], postal system [44], pigeon post [44.1], projectile post [33.2], pony express [44.4], balloon mail [44.3], airgraph and v-mail [44.5], telautograph [36], the clacks [64]

DESCRIPTION An email letter is a hybrid process of sending an electronic text-based message (now known as an email) to a mail delivery office at which point it is printed, enclosed in an envelope, stamped, and delivered as a letter. There is no established term for an email letter, but there are numerous examples around the world of services for transforming an email into a letter. Email letters were preceded by Western Union's mailgram service, which ran from 1970 until it was discontinued in 2006. Mailgrams were telegraphic messages received by United States Postal Service (USPS) offices via telephone [34], teletypewriter, or computer, and then placed in envelopes, stamped, and delivered via the postal service. In 1977, the Office of Technology Assessment (or OTA, an office of the US Congress that existed from 1974 to 1995) initiated a review of what was then called "electronic mail and message systems" to assess the USPS's future role in the delivery of electronic mail. A year later in 1978,

E-COM envelope, with prepaid postage, with printed letter inside.

E-COM (electronic computer originated mail) was proposed; it was approved in 1980 and began service two years later, ending in 1985. The service originally allowed senders to transmit up to two pages of text from their computers to one of twenty-five participating post offices across the USA The service initially cost twenty-six cents for the first page and five cents for the second along with a fifty-dollar annual fee. Notably, senders were required to send at least two hundred messages per transmission, meaning that the service was intended for marketing and promotion. Reportedly, most of E-COM's mail came from auto advertisers. One mailer declared, "This is your PERSONAL INVITATION to ATTEND the GREATEST AUTOMOTIVE INVENTORY REDUCTION SALE in the HISTORY OF MANASSAS. The Manassas Dealers involved MUST SELL 500 vehicles immediately!!"

MCI mail, one of the first email services in the world, which began in 1983, initially allowed MCI users to send electronic text-based messages to other users on the MCI network. It also gave users the option to send hard copies of their electronic messages via the USPS for the price of two dollars for up to three pages of text. Germany's Deutsche Post launched their own ePost service as late as 2010 and discontinued it in 2022.

SOURCES "Implications of Electronic Mail and Message Systems for the US Postal Service," National Technical Information Service Report #PB83-265017 (August 1982); Cecilia Hendrix, "6 Fascinating Things About Western Union's History," *Western Union Blog* (October 8, 2019); "E-COM, Electronic Computer Originated Mail," USPS Historian website (July 2008); Devin Leonard, "The Post Office Almost Delivered Your First E-mail," *Bloomberg News* (May 16, 2016); Stephen Manes, *The Complete MCI Handbook* (Bantam Books, 1988); Jürgen Olschimke, "Der neue alte E-Postbrief kommt," *Philately aktuell* 62:399 (September 2010)

Sneakernet [45]

COUNTRY OF ORIGIN USA
CREATOR(S) Unknown
EARLIEST KNOWN USE Roughly 1888

BASIC INFRASTRUCTURE/MATERIALS Portable storage media for electronic information that could include magnetic tape, floppy disks, optical disks, USB flash drives, computer hard drives, SD cards, or DVDs; digital or analog device for information recording and retrieval; water/land/air transportation routes; electric power source (optional)

RELATED Postal system [44], pigeon post [44.1], projectile post [33.2], pony express [44.4], balloon mail [44.3], airgraph and v-mail [44.5]

DESCRIPTION A sneakernet (also known as a floppy net, train net, pigeon net, tennis shoe net, air gap, and data mule) conventionally refers to the manual exchange of electronic information by way of portable, electronic storage media in lieu of sharing over a computer network. The exchange of information could take place hand-to-hand or by way of the postal system [44]. If the definition of a sneakernet is not limited to the era of computer networks, one could say that sneakernets began with sharing electronic information on any form of magnetic storage, such as that first proposed by the American engineer Oberlin Smith, who publicized his new method of wire recording using magnetic storage in 1888. Given their offline existence, sneakernets have and continue to be essential for those who value privacy and security, for those with inadequate internet access, and for those looking to efficiently share large amounts of data. Sneakernets thrive especially in countries whose governments are known for strict censorship and surveillance tactics, in countries where surveillance largely takes place at a corporate level, and countries whose citizens have limited internet access.

A fascinating example of sneakernets being used to aid political struggle was a secret communication network created as part of South Africa's liberation struggle from 1988 to 1991. According to Sophie Taupin, this network "allowed select exiled freedom fighters in and outside South Africa to communicate covertly with the senior leadership of the ANC exiled in Lusaka, Zambia, via London, UK . . . While the initial communication nexus was between Durban, London, and Lusaka, the communication infrastructure would later be operated from major cities in South Africa, namely, Cape Town, Durban, and Johannesburg as well as from Amsterdam, York (UK), Tall Cree (Alberta, Canada) and Harare." The process of delivering messages was extremely complicated, ingenious, and worth quoting at length:

> the South African Susan Tshabalala wrote a message on a laptop computer in a safe house in Durban and encrypted it with a home-made algorithm located on a floppy disk. Antoinette Voegelsang, a Dutch KLM flight attendant who regularly flew to Johannesburg, had secretly brought the disk into the country. Thanks to an acoustic coupler modem the written text was then converted into sound and recorded on a cassette tape. Then, Tshabalala went to a public phone, called the South African Tim Jenkin in London and played the encrypted sound message (a series of bip bip bip) by playing the tape recorder next to the public phone receiver. In London, the received message was recorded on an answering machine and decrypted by Jenkin using the same but reversed process. After reading the message, Jenkin decided where it was sent. Most of the time, it was transferred to Lucia Raadschelders, a Dutch woman located in Lusaka, using the same technical steps to inform the ANC leadership.

EXPERIMENTS Launched in 2016, Flash Drives for Freedom is an initiative run by the Human Rights Foundation that encourages people to donate memory devices or funds to purchase devices; devices are then "filled with content proven to inspire North Koreans to disbelieve Kim Jong-Un's propaganda . . . Contents include e-books, films, and an offline Korean Wikipedia"; the devices are then smuggled into North Korea by any means, ranging from balloons to cargo trucks and even person-to-person handoffs.

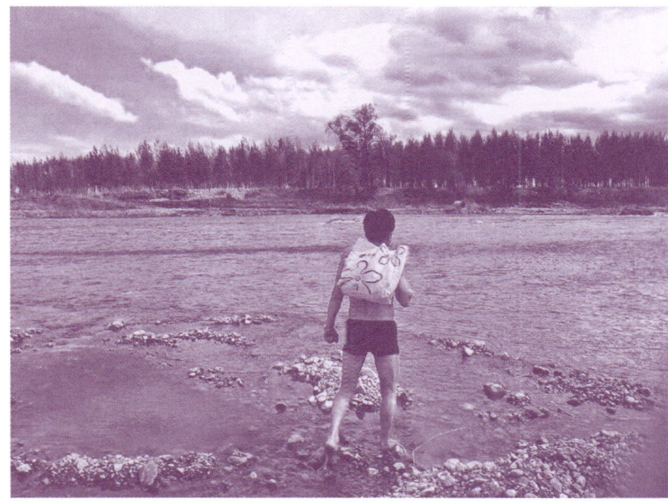

A North Korean smuggler carrying illegal USB drives on the Chinese side of the Tumen River, May 2013.

SOURCES Willy Ley, "For Your Information: The Galactic Giants," *Galaxy Science Fiction* (August 1965); A. C. Jaya, C. Safitri, and R. Mandala, "Sneakernet: A Technological Overview and Improvement," *2020 IEEE International Conference on Sustainable Engineering and Creative Computing* (2020); Kim Wall, "The Weekly Package: How Cubans Deliver Culture Without Internet," *Harper's Magazine* (July 2017); Eileen Guo, "Afghanistan's Real Internet Lives on Its Streets," *New York Magazine* (October 2018); Sophie Taubin, "Fugitive Infrastructure in the Fight Against South African Apartheid," *Third World Thematics: A TWQ Journal* (November 2022); Flash Drives for Freedom website

Radio Broadcast Network [46]

COUNTRY OF ORIGIN USA
CREATOR(S) American Telephone and Telegraph Company (AT&T)
EARLIEST KNOWN USE 1922

BASIC INFRASTRUCTURE/MATERIALS Power source, transmission lines (for example telephone or telegraph lines), transmitter, receiver, antennas (for example isotropic, dipole, monopole, array, etc.)

RELATED Radio broadcast [17], microwave radio-relay [31], communications satellite [32], television broadcast network [36]

DESCRIPTION Also known originally as "chain broadcasting," broadcast networks of radio stations, such

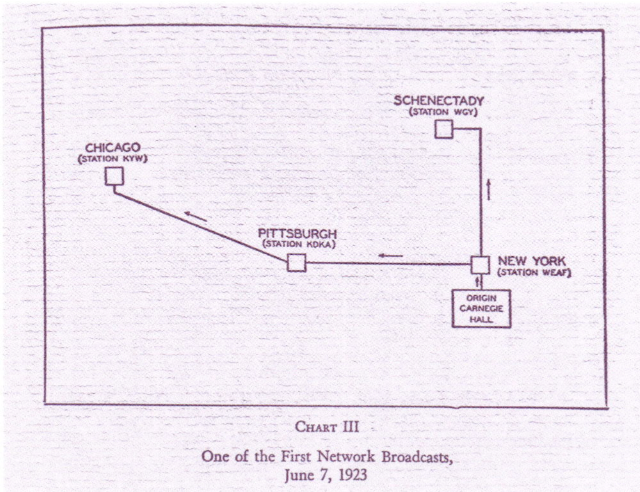

Chart delineating one of the first network broadcasts originating from Carnegie Hall, New York City, in 1923.

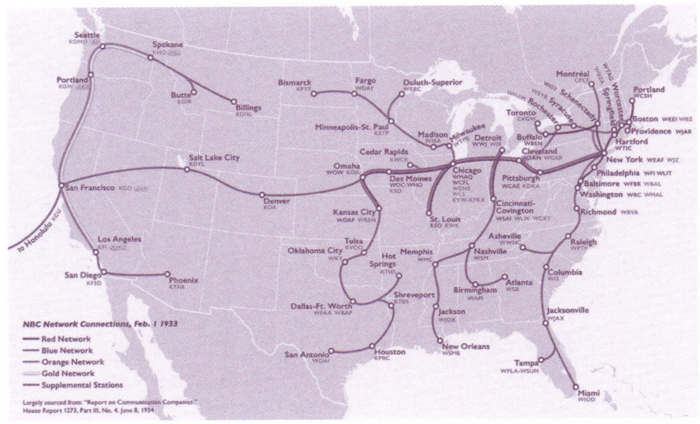

Artist's rendering of the five radio broadcast networks managed by the National Broadcasting Corporation (USA) in 1933.

as those pioneered by the American Telephone and Telegraph Company (AT&T) starting in 1922–1923, are networks of stations that share content, sometimes by way of syndication. In the Communications Act of 1934, "chain broadcasting" is defined as the "simultaneous broadcasting of an identical program by two or more connected stations." While each station broadcasts content using wireless radio, the stations were also linked via telephone [34] or telegraph lines. AT&T announced plans in 1922 for creating a radio network, but it took another year until they successfully demonstrated a network broadcast that linked WNAC in Boston with WEAF in New York City. The Radio Corporation of America (RCA) attempted to run a competing network out of WJZ in New Jersey; they broadcast their first network program in December 1923. One of the next most significant network broadcasts was of the annual meeting of the National Electric Light Association held at Carnegie Hall in New York City. The meeting and music from Carnegie Hall were broadcast across a chain of four stations (linking WEAF with radio stations in Schenectady, Pittsburgh, Pennsylvania, and Chicago) on June 7, 1923. This network then became "the WEAF chain," laying the groundwork for the creation of a series of broadcast networks operating across the USA Once RCA purchased WEAF from AT&T in 1926 and shortly thereafter created the National Broadcasting Corporation (NBC), the WEAF network became the Red Network. RCA also created the Blue Network out of the chain of stations connected to WJZ. By 1933, there were an additional three broadcast networks (orange, gold, and white) traversing the USA. Of course, numerous broadcast networks existed around the world at roughly the same time: for example, CNR Radio began in Canada in 1923, NHK began in Japan in 1925, and the BBC began in the UK in 1927. Within a year after the Communications Act of 1934, as Jonathan Sterne writes, "centralized, networked and mass-distributed radio broadcasting was enshrined in federal policy." Thus, the term "chain broadcasting" is no longer used, in part because the term "broadcast network" now means that broadcasting happens via numerous different infrastructures, including satellite and cable. These networks now exist in countries around the world, from Afghanistan to Zimbabwe, and they may be commercial or noncommercial, with the latter being either state or government sponsored.

SOURCES Michael J. Socolow, "A Wavelength for Every Network: Synchronous Broadcasting and National Radio in the United States, 1926–1932)," *Technology and Culture* 49:1 (January 2008); William Peck Banning, *Commercial Broadcasting Pioneer: The WEAF Experiment 1922–1926* (Harvard University Press, 1946); David Sarnoff, *Principles and Practices of Network Radio Broadcasting; Testimony of David Sarnoff, President, Radio Corporation of America, Chairman of the Board, National Broadcasting Company, Inc., Before the Federal Communications Commission, Washington, D.C., November 14, 1938 and May 17, 1939* (RCA Institutes Technical Press, 1939)

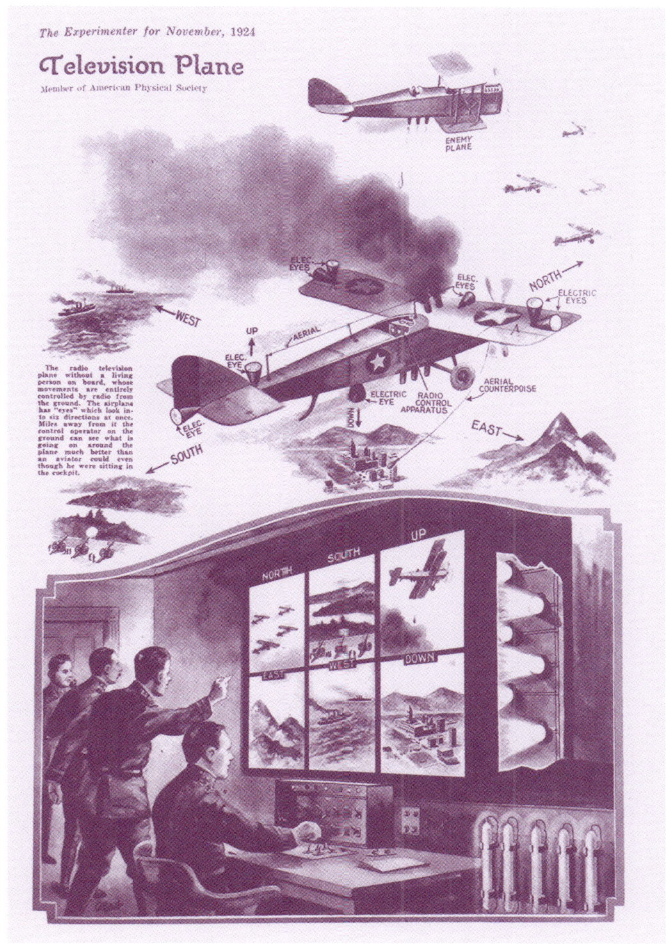

After witnessing an early demonstration of television by John Logie Baird in his laboratory, Hugo Gernsback predicted in November 1924 that radio-controlled "television planes" would immediately come into existence.

Broadcast Television [47]

COUNTRY OF ORIGIN Great Britain
CREATOR(S) John Logie Baird
EARLIEST KNOWN USE 1926

BASIC INFRASTRUCTURE/MATERIALS Power source, transmitters (including power supply, oscillator, amplifier, and modulator), receivers, antennas, television sets and/or monitors, transmission lines (coaxial cables and/or telephone/telegraph wires) and/or satellites, computers (optional), microphones (optional), earphones (optional)

RELATED Radio broadcast [17], microwave radio-relay [31], communications satellite [32], radio broadcast network [46], videophone [38], telefacsimile [37], radiofax [19]

DESCRIPTION Broadcast television—also known as terrestrial television, a television network, a television channel, and a television station—is the one-to-many broadcast of moving, sometimes live, images. Television broadcast networks are also referred to as broadcast television, television networks, television channels, and television stations. While most television broadcasts are part of larger national networks and, as such, they have used both wired and wireless transmission media since the 1930s, the history of television broadcasting in general is full of barely perceptible, largely undocumented shifts in how such broadcasting takes place. It is very difficult to treat the development of telegraphic or telephonic transmission of still images, the development of wired or wireless transmission of moving images, and the development of the television network as discrete histories. Notwithstanding the significant overlap between the three modes, Jonathan Sterne also points out that, unlike with radio, some of the first local television broadcasts were "represented by the networks as nodal events within a larger national network. Even before the means for networking television existed, the notion of networking was a key to the presentation of early broadcasts." Thus, it is also difficult if not impossible to determine the precise date at which most individual television stations became part of larger television networks. (And on occasion, it is even difficult to determine which individual television transmissions were real and which were hoaxes. In the

[47] Chris Burden's 1972 piece *TV Hijack*, in which he tried to subvert the norms of television broadcasting.

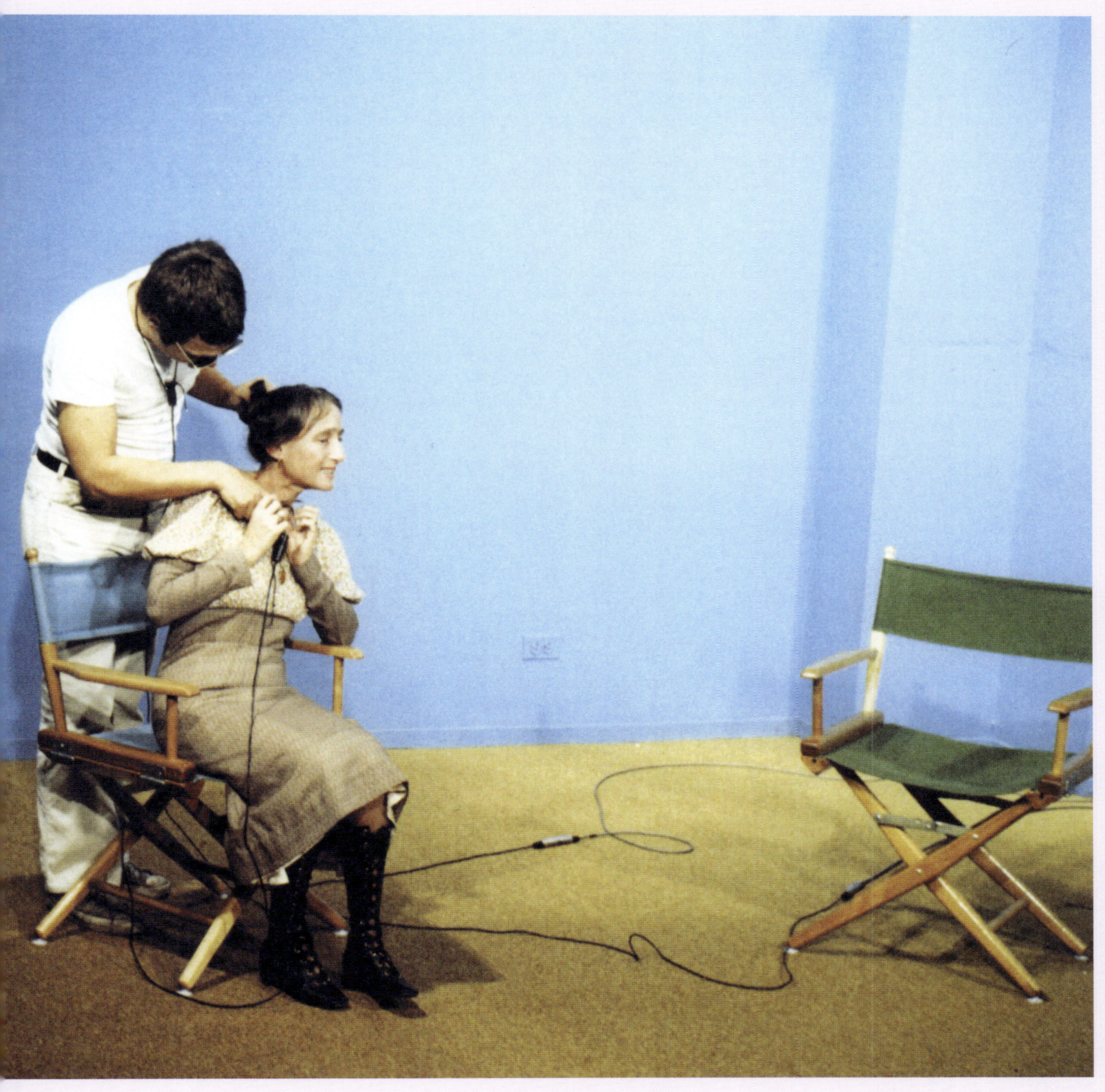

wake of successful transmissions of still images in the late nineteenth century, enthusiasm for the possibility of transmitting live images at a distance was so high in the early twentieth century that hoaxes started to emerge about, for example, a "seeing telephone" device called the "televue." As Doron Galili notes, one company even offered stocks for purchase in an advertisement that appeared in the *Los Angeles Herald* on February 17, 1907.)

We do know, however, that an April 1925 issue of *Nature* magazine reported that the month before, Scottish engineer John Logie Baird had given the first public demonstration of a wireless transmission of moving images of the silhouettes of ventriloquist dummies named James and Stooky Bill in a London department store. A few months later, in June of 1925, American engineer Charles Francis Jenkins also demonstrated the wireless transmission of moving images of silhouettes using a different technique (Jenkins's demonstration of his television to magazine publisher Hugo Gernsback several years earlier so inspired Gernsback that he speculated in a 1924 issue of *Radio News* that the "radio-controlled television plane will come into being immediately the minute the television problem is put on a practical basis"). In 1926, Baird demonstrated moving grayscale images, followed in 1927 by his successful transmission of both sound and images over a telephone line from London to Glasgow; in 1928, he managed to transmit over the Atlantic Ocean from London to New York. Contemporaneously, engineers in Japan such as Kenjiro Takanagi, as well as AT&T engineers, were also experimenting with transmitting moving images of people using telephone landlines combined with radio transmissions.

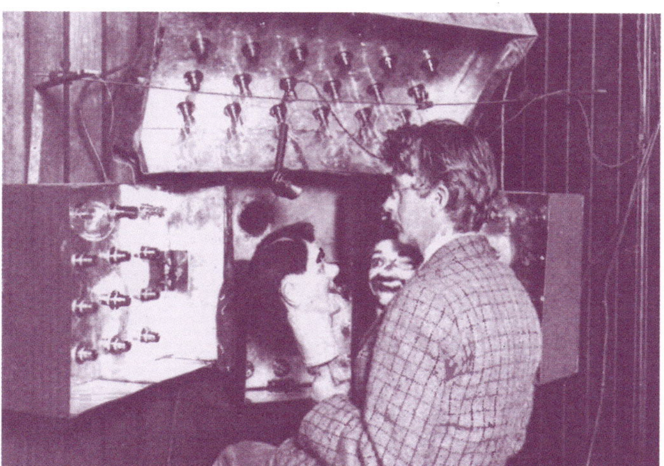

John Logie Baird demonstrating his television in March 1925 by using two ventriloquist's dummies named James and Stooky Bill.

What was likely the first entirely wireless television broadcast with the aim of financial profit (according to the International Telecommunication Union, this is in contrast to amateur radio [16] and amateur television [16.3], which lacks the geographical reach of a broadcast and also lacks a "pecuniary interest") that might also qualify as a television *network* broadcast took place in the USA on August 13, 1928. Radio station WRNY (owned by Hugo Gernsback's Experimenter Publishing Co. and based at the Roosevelt Hotel in New York City) broadcast images "by television over the radio." WRNY's television broadcasts used what's known as a forty-eight-line standard, whereby each element of the image is mechanically scanned once and in order for a total of forty-eight lines. The programs were simulcast by Gernsback's shortwave radio station W2XAL; however, the images and the sound could not be presented simultaneously. The two stations presented five minute programs every day on cooking, physical fitness, concerts, and upcoming events. These broadcasts took place a year after the 1927 Radio Act in the US, which created the Federal Radio Commission, the entity that allocated radio licenses and, later, those for television. Notably, Gernsback's broadcasts are not usually viewed as an early instance of a television network, whose process of sharing content was originally known as "chain broadcasting"; still, these simulcast broadcasts do adhere to the definition of chain broadcasting given in the Communications Act of 1934, which specifies the "simultaneous broadcasting of an identical program by two or more connected stations." Within a year of these broadcasts, television broadcasts were then introduced in the UK, followed by Germany, Australia, and the Netherlands. Outside of the USA, most of the TV broadcasts that became part of networks (such as the British Broadcasting Corporation or BBC) were public broadcasters—funded by citizens and built, managed, and regulated by government agencies. Throughout the 1930s, broadcasters worked to develop the best techniques for television networking, including moving from mechanical to electronic televisions, and the use of coaxial cable, microwave radio-relays [31], airplane relays, and teletranscription. After a pause in the technology's development during World War II, the majority of TV broadcasts relied on some combination of wired and wireless transmission media (later including communication satellites), and the broadcasts were available in most countries around the world by the 1960s. Today, nearly all television broadcasts are digital and take place over coaxial cable, satellite, and/or internet.

EXPERIMENTS While nearly all early television broadcast networks were experimental by definition, one noteworthy experiment was a public viewing event

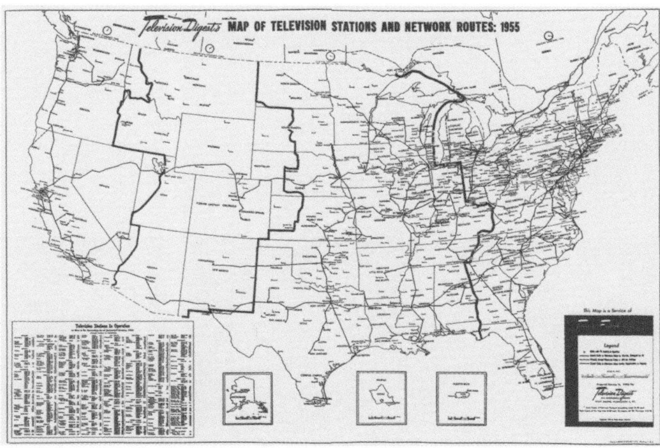

Map of television stations and network routes using coaxial cable and/or microwave relays along with planned routes across the US (1955).

of musicians and comedians displayed on six television sets that took place in fascist Italy in July 1939, in Rome and then in Milan. The broadcasts were the result of years of work by the fascist regime's radio broadcasting office, Ente Italiano per le Audizioni Radiofoniche (EIAR). As Danielle Simon writes, "More than simply another way to entertain EIAR's growing population of subscribers, these experimental broadcasts served as evidence of Italian Fascism's standing on the world stage. The Magneti Marelli equipment used for the transmissions, developed in consultation with engineers from the Radio Corporation of America (RCA), was regularly cited as proof of Italy's rapid technological development, and thus of the nation's hard-won progress." The broadcasts (which went so far as to include messages written on background walls proclaiming that "Mussolini is always right") ended less than two years later, but nonetheless represented one of the fascist regime's concerted attempts to remind viewers that "the televised content, the devices, and even the building in which they were consumed [were] products of Italian Fascist policy and ingenuity."

In stark contrast is the fact that, beginning in the 1960s, broadcasters themselves were invested in transforming television into a more interactive medium, thereby trying to undo its association with fascism and large-scale control of the masses. As Tilman Baumgärtel points out, television broadcasters have tried to read out letters from viewers, air live telephone calls, incorporate electronic or televoting, and also use touchtone signals to introduce greater interactivity. In 1964, the German TV show *Der Goldene Schuss* (or *The Golden Shot*) was centered around a crossbow mounted on a tripod with a TV camera stationed behind it, simulating the point of view of the shooter. Viewers could "shoot" at a target on the TV screen by using verbal commands issued over the telephone, telling the blindfolded camera person where to aim. However, as discussed in greater detail in the videophone [38] entry, it was not until Van Gogh TV's *Piazza Virtuale* in 1992 that, writes Baumgärtel, "the telephone became an input medium for programme sequences on the monitor." This principle was used shortly afterwards on TV2 Denmark's 1994 interactive TV show *Hugo*, about a Scandinavian troll that viewers could remotely control with a telephone touchpad; the show's custom computer hardware converted the signals generated by the tone dials into directions for the character.

According to Ieuan Franklin, there are also countless examples of artists intervening in the regime of television, either by producing experimental programs for television networks or by intervening in the very functioning of broadcasting itself. For example, in 1972 American artist Chris Burden performed his piece *TV Hijack* during a live television interview (also recorded by his own camera crew) during which he held the interviewer Phyllis Lutjeans at knifepoint, insisting he would kill her if the broadcast ended. Burden and Lutjeans were already friends, and she had insisted on the interview in spite of her knowledge of Burden's history of using weapons in his performances; nonetheless, as Mike Boehm succinctly puts it, "In an era when airliner hijackings were regularly in the news, [Burden] hijacked the show." From 1973 to 1977, Burden also bought airtime on broadcast television stations in New York and Los Angeles during which, instead of airing a commercial, he would, for example, recite the names of artists.

SOURCES Jonathan Sterne, "Television Under Construction: American Television and the Program of Distribution, 1926–62," *Media, Culture & Society* 21:4; Doron Galili, *Seeing by Electricity: The Emergence of Television, 1878–1939* (Duke University Press, 2020); "Current Topics and Events," *Nature* 115:2892 (April 1925); "Radio Shows Far Away Objects in Motion," *The New York Times* (June 14, 1925); Orrin Dunlap Jr., "The Televisor," *Popular Radio* 10:7 (November 1926); Hugo Gernsback, "A Radio-Controlled Television Plane," *The Experimenter* 1:4 (November 1924); H. de A. Donisthorpe, "British Listeners Hear Peculiar Sounds as New Machine Broadcasts Pictures of Moving Objects on the 200-Meter Wave," *The New York Times* (October 17, 1926); "Kenjiro Takayanagi, Electrical Engineer, 91," *The New York Times* (July 25, 1990); Robert Hertzberg, "Successful Television Programs Broadcast by RadioNews Station WRNY," *Radio News* (November 1929); "Compilation of the Communications Act of 1934 and Related Provisions of Law: Including Communications Act of 1934,

[47] Viewers of the 1964 German broadcast television show *Der Goldene Schuss* (*The Golden Shot*) could attempt to shoot at a target on-screen by issuing commands via telephone to a blindfolded cameraman.

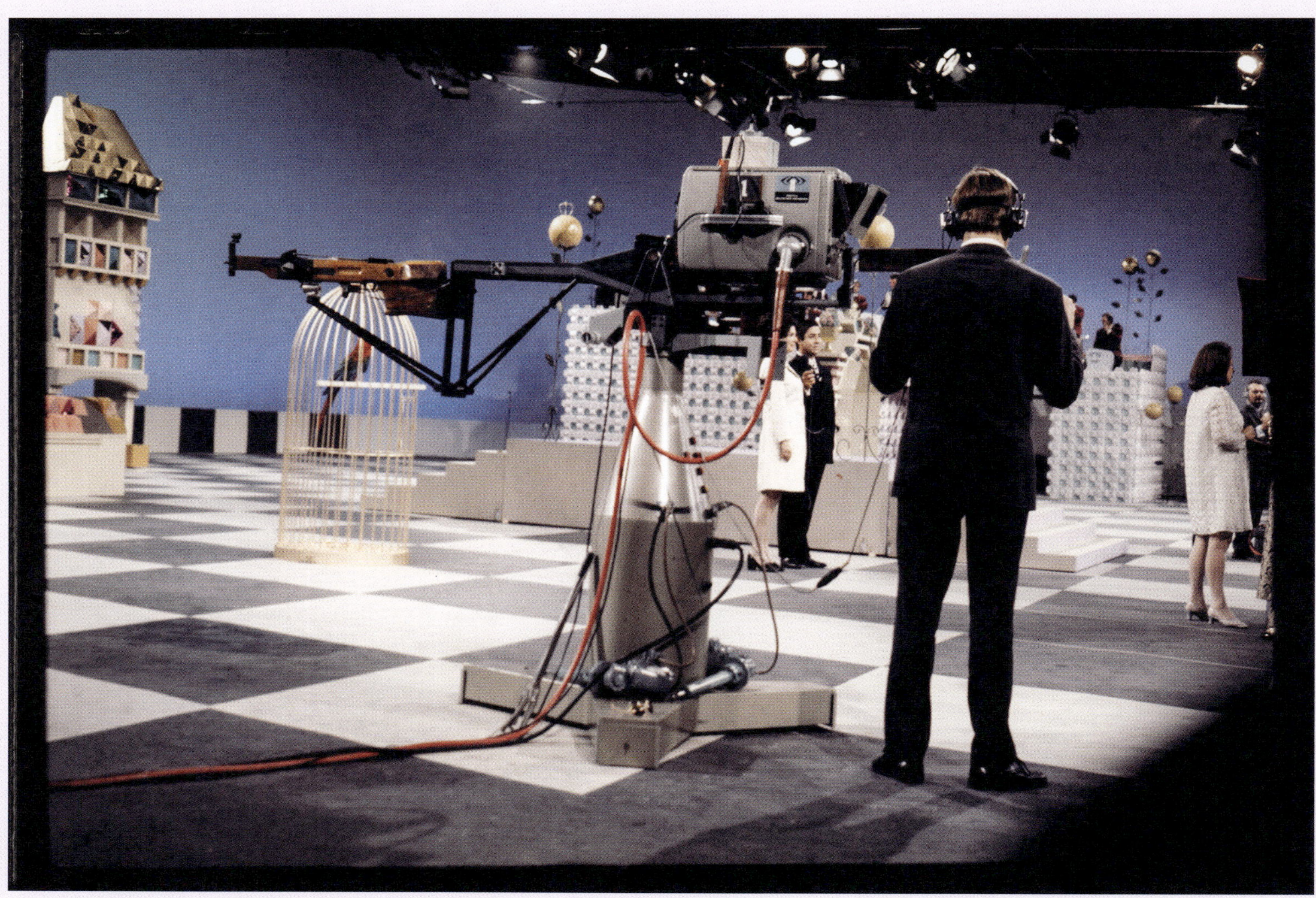

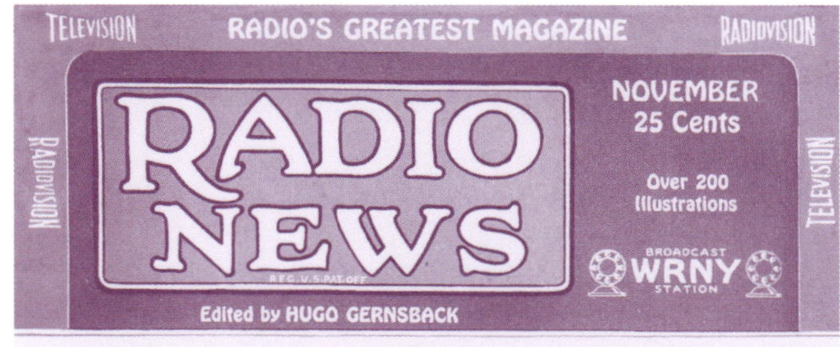

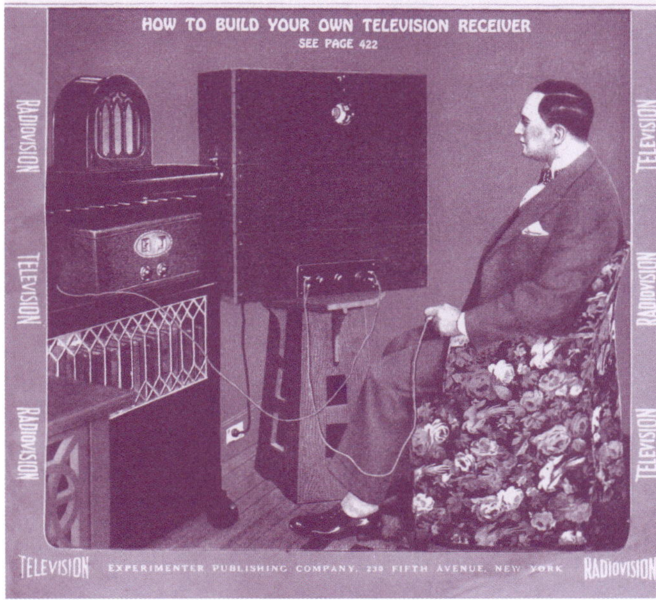

Hugo Gernsback viewing WRNY television from his New York City apartment in August 1928. The image appeared on the cover of *Radio News*, November 1928.

Communications Satellite Act of 1962, Selected Provisions from the United States Code" (United States General Post Office, 1995); Joseph H. Udelson, *The Great Television Race: A History of the American Television Industry* (University of Alabama Press, 1982); R.W. Burns, *Television: An International History of the Formative Years* (Institution of Electrical Engineers, 1998); Susan J. Douglas, *Inventing American Broadcasting, 1899–1922* (Johns Hopkins University Press, 1989); Erik Barnouw, *A History of Broadcasting in the United States, Volume 1: A Tower of Babel* (Oxford University Press, 1966); Anne-Katrin Weber, *Television Before TV* (Amsterdam University Press, 2022); Danielle Simon, "From Radio to Radiovisione: Italian Radio's Television Experiments, 1939–1940," *Representations* 151:1 (Summer 2020); Tilman Baumgärtel, *Van Gogh TV's "Piazza Virtuale": The Invention of Social Media at Documenta IX in 1992* (Transcript Verlag, 2022); Von Hartmut Goege, "Als 'Der Goldene Schuss' zum letzten Mal in ZDF lief," Deutschlandfunk website (February 7, 2020); "Hugo erövrar världen," *Datormagazin* (September 1994); Chris Burden, "*TV Hijack*, February 9, 1972," Metropolitan Museum of Art website; Mike Boehm, "Chris Burden's Youthful Edge Grabbed One TV Host by the Throat," *Los Angeles Times* (May 12, 2015)

Cable Television [48]

COUNTRY OF ORIGIN USA
CREATOR(S) Unknown
EARLIEST KNOWN USE 1947

BASIC INFRASTRUCTURE/MATERIALS Radio tower, antennas, control station, coaxial cables and/or fiber optic cables, television set, utility poles (optional), cable converter box (optional),

RELATED Microbroadcast [27], radio broadcast [17], broadcast television [47], pirate television [25], NABU network [48.1], telephonoscope [55], mundaneum [58]

DESCRIPTION Cable television (cable TV), or community antenna television (CATV), transmits television programs by way of antennas, amplifiers, and coaxial cables that carry radio frequency signals or by way of fiber optic cables. Cable TV emerged as a way for areas at the far reaches of over-the-air television signals to access the medium. As Patrick Parsons makes clear, "CATV began and proliferated just at the edge of the receivable signal. The fringe was one of those problems that illustrated the evolutionary diversity of technology by inviting a host of technical solutions, CATV being only one. Even before the nearest TV station was scheduled to go on the air, people at the fringe, and beyond, were erecting multistory antennas and creating imaginative home-made receivers in an effort to catch the weak and distant signal."

While the German Third Reich used coaxial cable to televise the Olympics in 1936 from Berlin to Leipzig, followed by the use of cable to transmit from London to Birmingham and then from New York to Philadelphia later in the same year, these transmissions were experimental, short-lived, and involved dozens of small video cameras, a very large video camera called a "television camera," and several different types of video camera tubes. By contrast, cable television was not so much an invention as a technology that emerged from entrepreneurial adaptions of techniques borrowed from radio, namely the use of antennas, amplifiers, and boosters to better receive television broadcast signals from nearby urban centers. While numerous people across the US have been cited as "inventors" or initiators of cable television from 1947 to 1949, reportedly the first person to erect a tower and antenna for this purpose was Ed "Peanuts"

Trusky, who began displaying television programming in his local shop in Mahanoy City, Pennsylvania, in 1947. Writes Patrick Parsons: "Three doors down, the clerics at the neighborhood church were sufficiently enchanted with the idea that they convinced Peanuts to extend the antenna line into the rectory. A friend of Peanuts, a block in the other direction, reportedly did the same. According to one version of the story, one of the many clients at the poolroom, John Walsonovich, got his idea for a CATV business in the TV glow at Trusky's poolroom. Bringing television 'down the mountain' was an idea that caught on quickly." While the early decades of cable television were defined by its ability to boost broadcast television [47] signals to local communities, after several rulings and interventions by the FCC to protect the broadcast television industry throughout the 1960s, it evolved into a practice dedicated to "narrowcasting," or the dissemination of largely local programming by, to, and for local audiences.

By 1970, when Ralph Lee Smith published a lengthy account of cable television in *The Nation*, cable television had expanded to 2,500 different systems across the US and about a total of 4.5 million subscribers. By this time, the setup for cable TV was also, necessarily, more complex than it had been in 1947: "A tower is constructed on a hill or other spot selected for good reception. The antenna system on the tower is carefully engineered, usually with a separate antenna for each channel that is to be received. In some instances, distant signals come to the tower by a relay system of one or more microwave transmitters. At the foot of the tower is a small control station, called the 'headend,' where the signals are brought up to maximum strength and clarity. Here, also, some of the signals may be rechanneled . . . Amplifiers placed at distances of 1,500 to 2,000 feet along the trunk line into town keep the signals strong. From the main cable, feeder lines, 'tapoffs,' and 'housedrops' carry the signals to individual streets and the subscribers' homes." The result was that suddenly more channels were available to more people for far less money than what the commercial broadcast television "oligarchy" charged. What's more, since cable companies profited from subscriber fees rather than from the broadcast of particular programs, they were also free to pursue diversified programming.

The year 1972 brought about significant changes in cable TV, as the FCC mandated that the hundred most profitable cable operators provide access channels—either noncommercial public access channels, educational access channels, government access channels, or leased access channels that charged a use fee. Public access channels were a particular boon to media activists and artists keen on exploring television and video as medium for subversion, intervention, education, and experimentation, and even as a new medium for distribution outside of conventional museums and galleries. Writing on the Alternate Media Center that opened the year before at New York University, John J. O'Connor declared in *The New York Times* in 1972: "The content can be miserable. The technical quality is often atrocious. And, as established television executives insistently note, nobody is watching anyway. Yet the experiments with public access on cable television continue to be among the more significant in contemporary communications. On specific channels set aside by a cable company, groups or individuals are afforded, without charge, an opportunity to present themselves directly, undiluted by the direction or inhibitions of media professionals. The only restrictions on content at present relate to laws on libel and profanity." O'Connor ends, "Eventually, it seems, television's monologue may have to make room for cablevision's dialogue."

While the 1980s brought about numerous FCC rulings aimed at deregulating the cable industry in the US and

HERB SCHILLER
READS
The New York Times

FOR IMMEDIATE RELEASE — NOVEMBER 10, 1981

CONTACT:

DeeDee Halleck, Producer
COMMUNICATIONS UPDATE
165 West 91
New York, NY 10024
212 362-52877

HERB SCHILLER
READS
THE TIMES
Manhattan Cable Public Access
Wednesdays at 10 PM
Channel D

For the next four Wednesdays, a special LIVE series of Communications Update will feature media critic Herbert Schiller. Professor Schiller is in the Communications Department at the University of California at San Diego, and is the author of Who Knows? Information in the Age of the Fortune 500, Ablex, Norwood, NJ. Included in the discussions will be an analysis of the New York Times Board of Directors and their various corporate interlocks. Of particular interest is the way in which The Times covers the information industry and international communications issues. Brian Winston, TV Critic for The Soho News, says, "Schiller is exactly the sort of person who should be heard outside the confines of mass communication sociology." The show is at 10 PM on Wednesdays on Channel D, and is re-broadcast of Fridays at 2:30 PM. The show is produced with the assistance of: Esti Marpet, Media Bus, Pennee Bender, Mary Feaster, Marty Lucas, Richard Linette, Vickie Gholson, Valerie Van Isler, Diana Agosto and Daniel Brooks.

Paper Tiger TV co-founder DeeDee Halleck announces their first show, *Communications Update*, in November 1981, which featured communications professor Herbert Schiller analyzing *The New York Times*.

[48] Sherry Miller Hocking, pictured here in 1972, ran workshops for the Experimental Television Center.

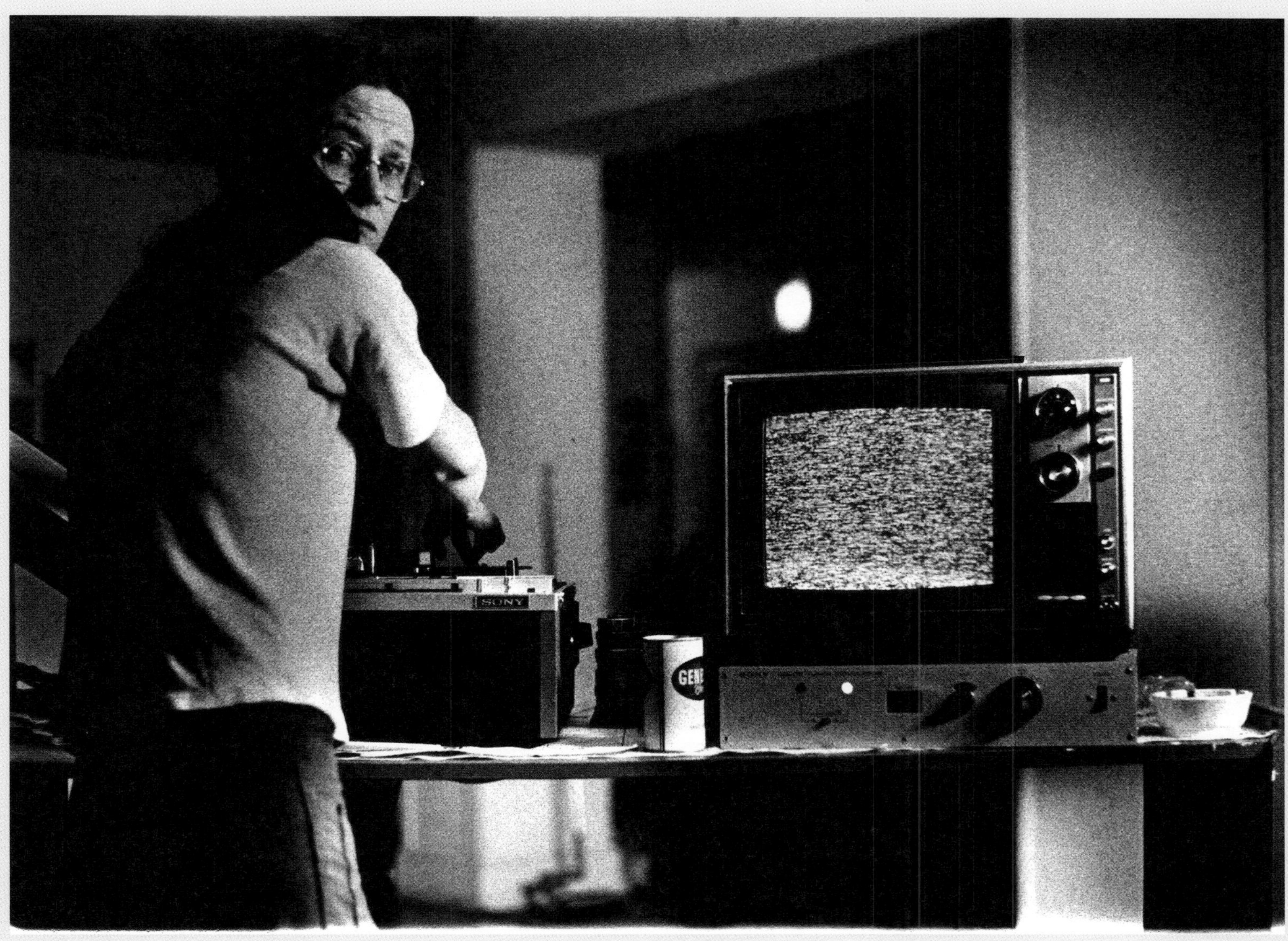

[48] Artist Jaime Davidovich produced *The Live! Show*, which aired on Manhattan cable TV and included appearances by him as “Dr. Videovich.”

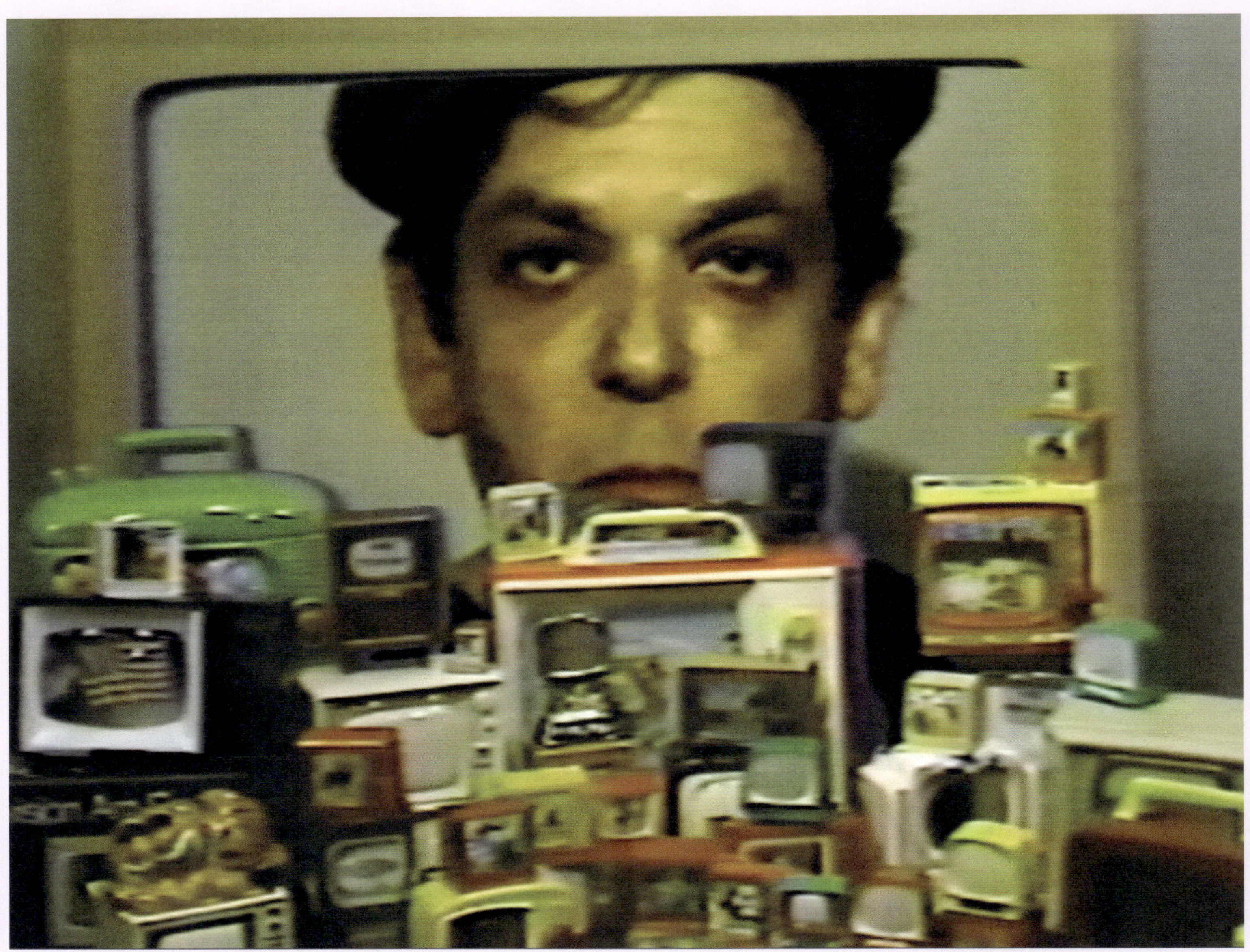

eliminating the obligation for cable operators to carry local programming, there was also a simultaneous increase in its use. In 1988, media activist and founder of Paper Tiger TV DeeDee Hallack described the range of groups using cable TV: "It is true that neo-Nazis have taken advantage of the cable access policies in many cities and towns. The program *Race and Reason* is a despicable mixture of florid racism, homophobia, and anti-semitism, broadcast weekly on dozens of public access stations in major cities . . . [journalists] forget to . . . mention the other groups who regularly use cable in many cities: the League of Women Voters, the Boy Scouts, the high school news programs, the welfare rights organizations, the Gay Men's Health Collective, and so forth."

Public access television still exists in many countries around the world, including Canada, Australia, Germany, Norway, and Sweden. As with many other networks, cable television has also almost entirely been converted to digital and, as many viewers now opt for internet-based streaming services instead, cable television may be in danger of becoming obsolete.

EXPERIMENTS Artists began to be more aware of the potential of cable television for community use and experimentation starting in the late 1960s and early 1970s, combined with the emergence of affordable video equipment (partly evidenced by the fact that the underground *Radical Software* magazine published a number of pieces on the potential of cable TV in their second issue in 1970). However, in the US, the 1972 provision by the Federal Communications Commission for cable television stations to provide "access channels," especially public access channels that should be set aside for noncommercial, community use, suddenly opened the door for a wide and wild range of experiments with local television. The Experimental Television Center, started in 1969 in Binghamton, New York, began airing a community cable show called *Access* in 1972 in which they featured tapes that simply explored the possibilities of video. Writes Ann-Sargent Wooster on the Experimental Television Center website, "Before MTV, before Industrial Light and Magic, a radical group of people believed that television was an art medium. They felt free to play with the television signal to make funky, sophisticated, chaotic, poetic, raw, cutting edge, disruptive, politically savvy, artistically elegant tapes that were the antithesis of the broadcast television of then and most of now."

In 1976, Argentinian artist Jaime Davidovich and others affiliated with New York–based art venues such as the Kitchen, Global Village, and Anthology Film Archives created Cable SoHo (soon renamed the Artists' Television Network, or ATN) in an attempt to get video/TV art aired on Manhattan Cable TV. ATN produced a show inspired by the Cabaret Voltaire (the nightclub frequented by Dadaists in 1916) called *The Live! Show* that aired from 1979 to 1984. It featured performances, videos, interviews, and skits; Davidovich would frequently appear as the character Dr. Videovich, "specialist in curing television addiction," as William Grimes put it in *The New York Times*.

Paper Tiger TV (PTTV) also emerged in New York City in 1981 as an initiative by activists, academics, and artists seeking to critique mainstream broadcast TV networks as well as major news outlets. PTTV began as a series called *Communications Update*, a show that opened with the words "It's eight-thirty. Do you know where your brains are?" and which initially featured Professor Herbert Schiller, who would analyze *The New York Times*. As co-founder DeeDee Halleck put it, PTTV was part of the public access movement, which "took root at a moment of maximum disillusionment with network television." It was also intended to be an alternative to cable TV monopolies—"to allow individual citizens the right to electronic expression." Their aim was to demystify television technology by deliberately embracing a DIY aesthetic that included hand-held title cards. As Halleck described it in 1983, "If there is a specific look to the series, it is handmade, a comfortable nontechnocratic look that says friendly and low budget. The seams show." While the initial broadcasts took place over local cable TV, they ended up being part of a network of broadcasts as, according to Jesse Drew, "the Paper Tiger collective took on the task of dubbing and hand-delivering or mailing copies of Paper Tiger programming to other supporters in other cable systems for playback." In 1986, PTTV launched Deep Dish TV—the first independent satellite network in the US Halleck writes that the group would rent several hours of satellite time a week during which local cable programmers would "downlink" and tape programs and then air them on local channels.

SOURCES "Early Electronic Television: The 1936 Berlin Olympics," EarlyTelevision.org; Patrick Parsons, *Blue Skies: A History of Cable Television* (Temple University Press, 2008); Ralph Lee Smith, "The Wired Nation," *The Nation* (May 18, 1970); *The Electromagnetic Spectrum*, Radical Software 1:2 (1970); John J. O'Connor, "Public Access Experiments on Cable TV Advancing," *The New York Times* (June 6, 1972); "Experimental Television Center Archives," ETC website; William Grimes, "Jaime Davidovich, Artist Whose Videos Bypassed the 'Gatekeepers of Culture,' Dies at 79," *The New York Times* (August 30, 2016); Julia Fiore, "The Avant-Garde

[48.1] Undated image from internal corporate promotional material of a family enjoying the NABU Network.

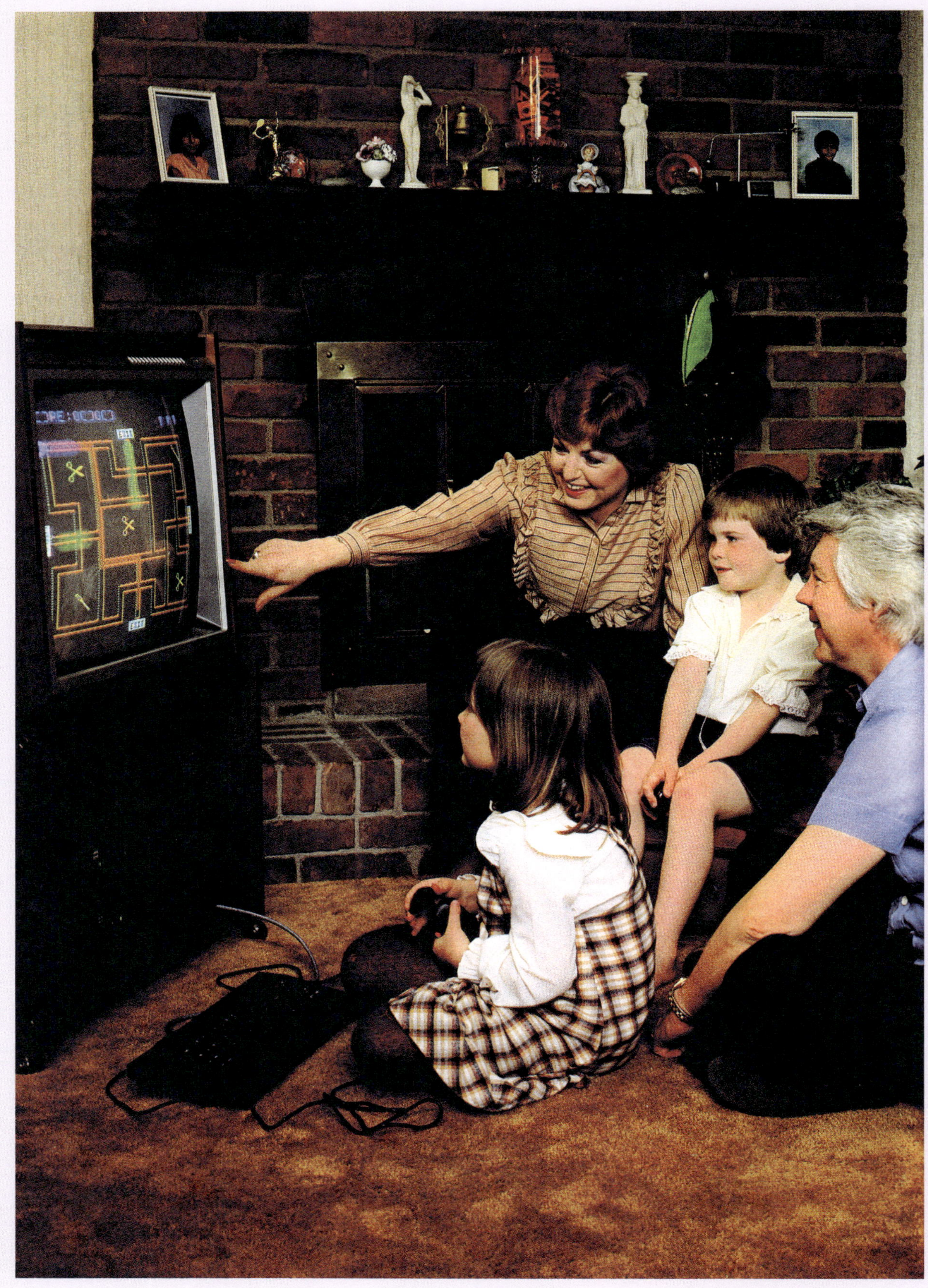

ArtThat Was Made for TV," Artsy.net (January 23, 2019); Bill Olson, "The History of Public Access Television," Billolsonvideo.com (March 8, 2015); DeeDee Halleck, "Paper Tiger Television: Smashing the Myths of the Information Industry Every Week on Public Access Cable," *Media, Culture and Society* 6 (1984); DeeDee Hallack, *Hand-Held Visions: The Impossible Possibilities of Community Media* (Fordham University Press, 2002); Jesse Drew, "From the Gulf War to the Battle of Seattle: Building an International Alternative Media Network," *At a Distance: Precursors to Art and Activism on the Internet*, Annmarie Chandler and Norie Neumark, eds. (MIT Press, 2005); Ieuan Franklin, "TV Interventions: Artists, Activists, and Alternative Media," in *Contemporary Radical Film Culture* (Routledge, 2020)

NABU Network [48.1]

COUNTRY OF ORIGIN Canada
CREATOR(S) John Kelly
EARLIEST KNOWN USE 1982

BASIC INFRASTRUCTURE/MATERIALS NABU personal computer, mini computer or "head end," combiner, adaptor, television set, coaxial cable, printer (optional), disk drive(s) (optional)

RELATED Cable television [48], videotex [52]

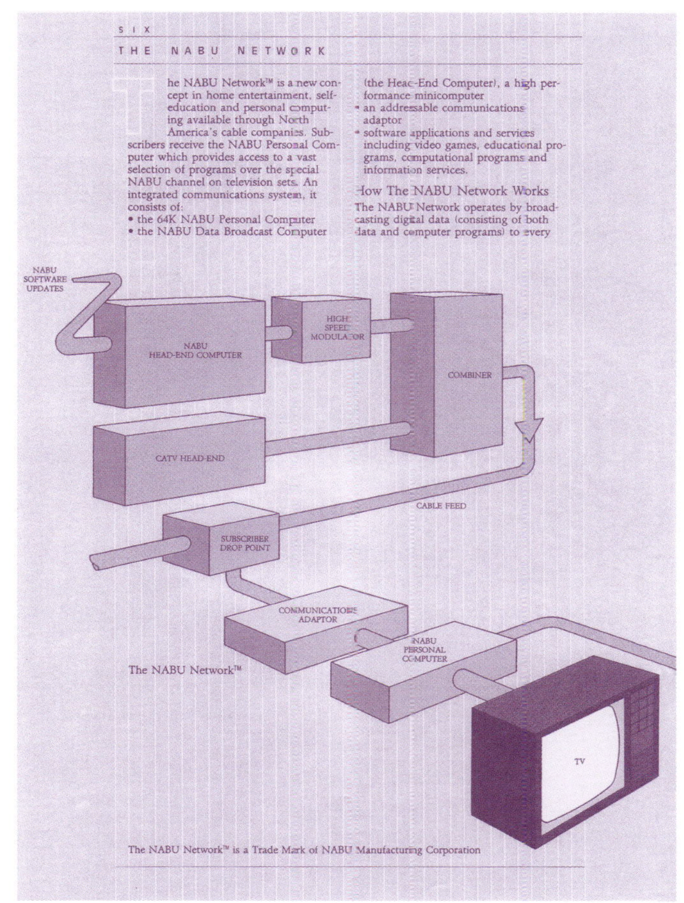

SIX

THE NABU NETWORK

The NABU Network™ is a new concept in home entertainment, self-education and personal computing available through North America's cable companies. Subscribers receive the NABU Personal Computer which provides access to a vast selection of programs over the special NABU channel on television sets. An integrated communications system, it consists of:
• the 64K NABU Personal Computer
• the NABU Data Broadcast Computer (the Head-End Computer), a high performance minicomputer
• an addressable communications adaptor
• software applications and services including video games, educational programs, computational programs and information services.

How The NABU Network Works

The NABU Network operates by broadcasting digital data (consisting of both data and computer programs) to every

The NABU Network™ is a Trade Mark of NABU Manufacturing Corporation

Depiction from a 1982 NABU Manufacturing Corporation pamphlet of how the NABU Network transmitted via cable TV and a NABU "head-end" computer.

DESCRIPTION The NABU Network (or the Natural Access to Bi-Directional Utilities Network) was the only instance of a two-way network that operated over cable television [48]. Users could rent or purchase a NABU Network PC, which used their home cable TV subscription to access NABU's servers and, for a monthly fee, provided access to software and information. The computer could also function on its own, as it had a 80K capacity along with a sound generator and a graphics processor. The NABU servers were minicomputers called "head ends" whose digital output had to be modulated with an RF modulator before it could be transmitted over coaxial cable; the transmission then moved through what NABU called a "combiner," a piece of equipment that merged NABU programs with the other information being transmitted over cable; the transmission subsequently moved through the "adaptor," which was the interface between the cable and the user's NABU PC, thereby making content appear on the user's TV set.

NABU Network service was first offered to select subscribers in Ottawa (Canada) in the spring of 1982. One of the appeals of NABU was that its transmission speed was, according to one of its corporate pamphlets, over 21,000 times faster than the transmission rate over telephone network [34]. In fall 1983, it was announced in the trade publication *Cable Marketing* that NABU would extend its Ottawa service, which had access to 85,000 cable subscribers, to Richmond, Virginia (USA), and to Vancouver (Canada), which had access to 250,000 cable subscribers. The extension was only one-way, with plans for two-way distribution; it is unclear whether two-way service was ever available to these extensions. In February 1984, it was announced in *Cablevision* that NABU was also extending to 5,000 cable subscribers in Alexandria, Virginia: "The Alexandria site was chosen, officials say, because of its proximity to Washington with the chance to show off the virtues of the network to officials in Congress, the NCTA [the National Cable & Telecommunications Association], the FCC and other important parties." In December 1985, the publication *The Hard Copy* reported that NABU made an agreement with the Japanese firm ASCII that involved the installation of fifty NABU systems in select Japanese homes.

In August 1986, NABU sent letters to its subscribers informing them that "Due to a low subscriber base and the ongoing problem of acquiring interesting and valuable software, we regret to inform you that THE NABU NETWORK will not be continuing its service beyond August 1986 . . . We hope you have benefited from your exposure to THE NABU NETWORK and that you will continue to pursue the benefits of computer technology."

According to a price list dated September 1984, it cost $19.95 CDN a month to rent the hardware and software with the option to add on entertainment, education, or home management software for anywhere from $4.95 to $7.95 a month, and another disk drive for $15.95 a month (adjusted for 2024, one would pay anywhere from $44 USD to over $100 USD a month).

Ex-FCC Chairman Tom Wheeler, who was president of NABU for its final year of service, wrote an editorial on net neutrality for *Wired* magazine in 2015 in which he used NABU as an example of what happens when networks are closed. For Wheeler, even though "NABU was delivering service at the then-blazing speed of 1.5 megabits per second," which was hundreds of times faster than their competitor America Online, NABU folded because they "had to depend on cable television operators granting access to their systems," whereas America Online "had access to an unlimited number of customers nationwide who only had to attach a modem to their phone line" to receive service. Simply put: "The phone network was open whereas the cable networks were closed." However, Scott Wallsten responded to Wheeler's editorial by pointing out that NABU failed not so much because it relied on a closed network, but because it could not provide enough meaningful resources to compel people to subscribe and also because cable infrastructure was not equipped to handle more than minimal two-way traffic.

EXPERIMENTS Beginning in December 2022, ex-NABU engineer Leo Blinkowski and DJ Sures, whose family members helped found NABU, launched the NABU Preservation Group (also known as RetroNET) as part of their now successful attempt to develop a NABU Internet Adapter software. As they write on their website, "This software emulates the NABU Network Adapter, which would have been the hardware used to connect NABU Personal Computers to the NABU Network. The NABU Internet Adapter provides entertainment channels, such as the original NABU Network from 1984 and new homebrew software, demos and utilities."

SOURCES "NABU PC Technical Specifications," Spec. 50-90020490, Nabunetwork.com; "Nabu Comes On Line," *Cable Marketing* 3:9 (October 1983); "Heading South for the Market," *Cablevision* (February 6, 1984); NABU letter to customers, dated February 6, 1984, York University Computer Museum NABU archives; "Price List and Subscription Rates," York University Computer Museum NABU archives (NABU Network, September 1984); *The Hard Copy* 1:1 (undated), York University Computer Museum NABU archives; Tom Wheeler, "This Is How We Will Ensure Net Neutrality," *Wired* (February 2015); Scott Wallsten, "The NABU Network: A Great Lesson, but Not About Openness," The Technology Policy Institute website (February 5, 2015); "About RetroNet," Nabu.ca

Cellular Network [49]

COUNTRY OF ORIGIN USA
CREATOR(S) Douglas H. Ring and W. Rae Young / Bell Telephone Laboratories
EARLIEST KNOWN USE 1947

BASIC INFRASTRUCTURE/MATERIALS Power source, mobile transceiver, stationary transceiver, antennas, microphone, microchip, communications satellite (optional), transmission lines (optional; telephone wires or fiber optic cable)

RELATED Pager [21], Wi-Fi [29], Bluetooth [30], microwave radio-relay [31], communications satellite [32], telephone [34]

DESCRIPTION A cellular network transmits wireless radio signals primarily for suitably equipped mobile transceivers such as cell phones by dividing a geographical area into a grid of interlocking, regular shapes (often hexagonal) called "cells." Each cell includes at least one transceiver traditionally affixed to a cell tower that is responsible for sending and receiving transmissions in the form of voice, data, or other forms of content to/from a neighboring cell. Any given area can be divided and subdivided into smaller and smaller cells. The result of this structure is that radio spectrum is shared more efficiently, the network has a greater overall capacity than that of a single, large transmitter or satellite, and, most important, service is maintained even if the user moves between cells. Mobile or cellular phone (as they are referred to in North America) networks are the most common form of cellular network.

The cellular network structure was first developed in 1947 by Douglas H. Ring and W. Rae Young while they

[48.1] Images of the NABU Network, including the bootup, category, and NABU FILER menus, accessed via emulation by Znigniew Stachniak in 2005.

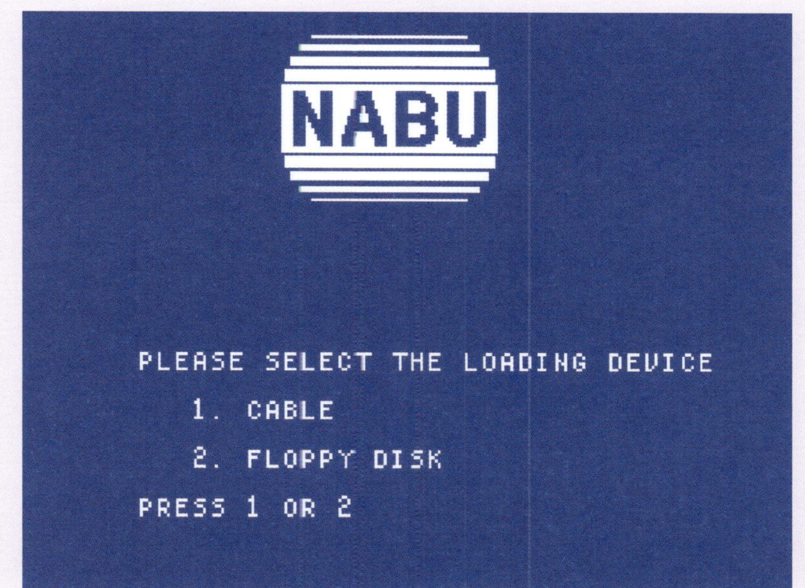

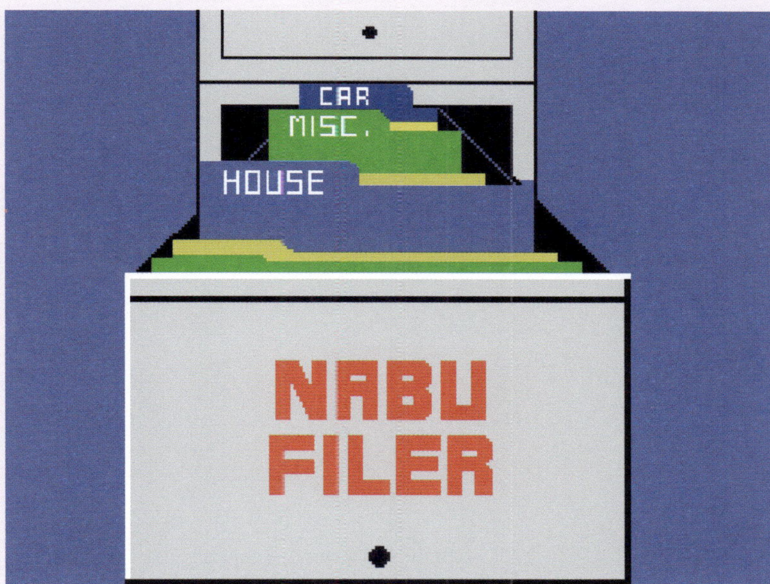

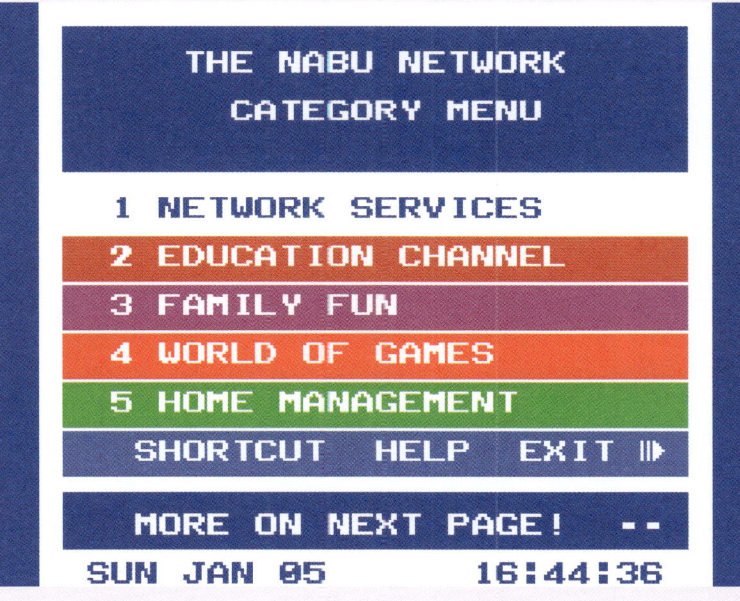

[49] Illustration of Bell Labs' 1947 concept of the cellular network in which "electronic switching equipment automatically hands off a call from one to another of the radio transmitters serving those cells without interrupting the call."

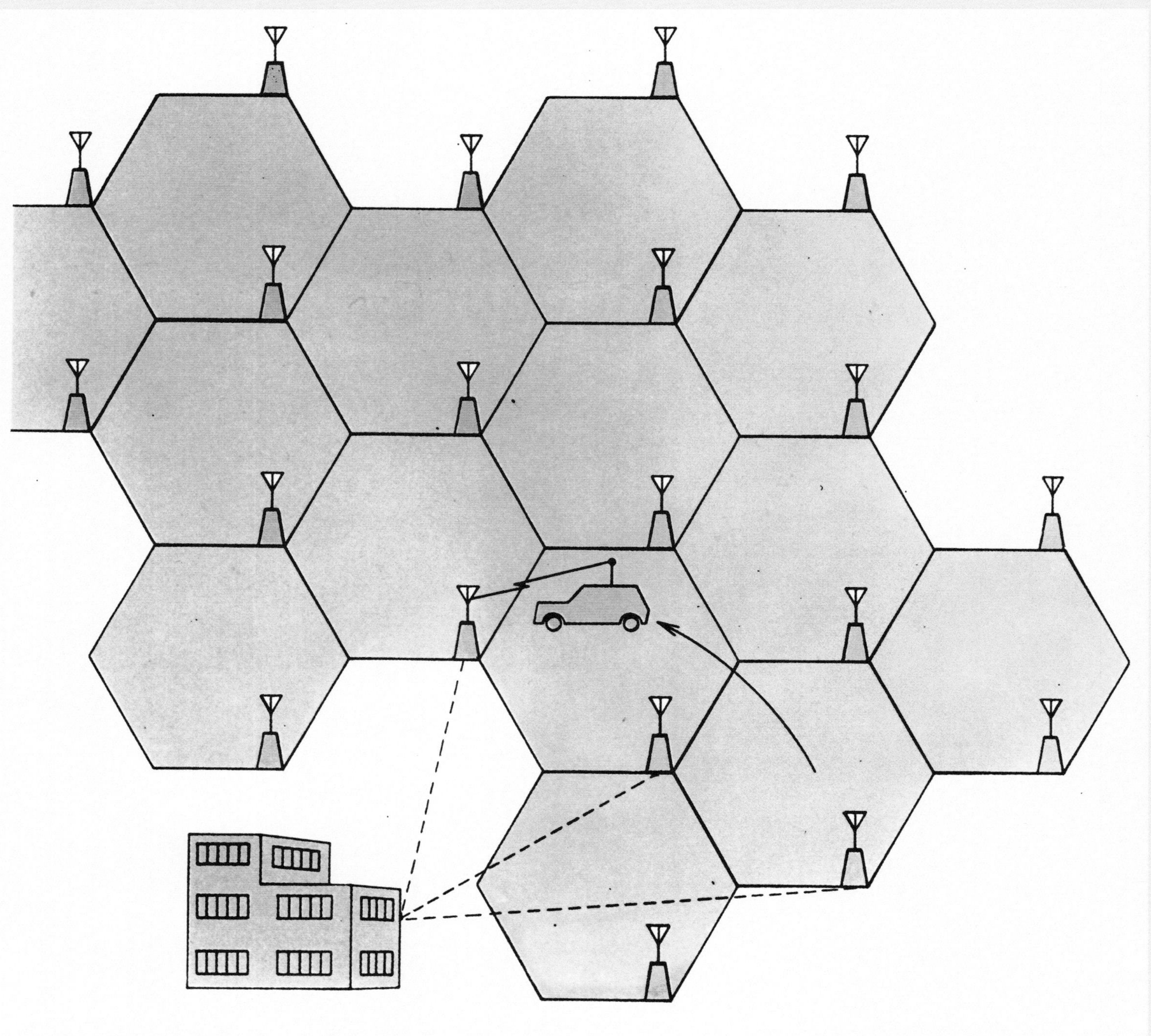

they were working for Bell Telephone Laboratories; their goal was to create a more efficient method than the existing mobile telephone service by creating a cellular structure for radio transmissions could "provide service to any equipped vehicle at any point in the whole country." According to Ring, "land transmitter bands should be used in blocks with one block assigned to each primary area, and the various blocks packed as closely as possible and adjacent to each other. This minimizes the possibility of interference between land transmitters and car transmitters since only one guard band between land and car frequencies is required." Thus, the key feature of a cellular network is that each cell uses certain assigned frequencies that cannot be used by neighboring cells; the frequencies *can* be reused by other cells in the network provided they are far enough away from each other. However, it was not until some decades later that the frequencies and technology for cellular networks became available. On April 3, 1973, Martin Cooper (the so-called father of the handheld cell phone) and John Mitchell of Motorola demonstrated the first known public two-way call to Joel Engel of AT&T using cell phones. Thereafter, the first generation of a commercial cellular network (1G) was an analog, voice-only network launched in Tokyo, Japan, in 1979 by Nippon Telegraph and Telephone and subsequently launched in the USA in 1983. 2G, the first digital cellular network, was launched in Finland in 1991, leading to the largely worldwide launch of cellular. 2G also marked the beginning of the inclusion of one-to-many broadcasting (an integration of pager [21] technology) in cellular standards; the capability is known as cell broadcast or short message service-cell broadcast, and is used for national public alerting systems such as the Wireless Emergency Alerts sent out by the Federal Emergency Management Agency in the USA. 2G also introduced short message service (SMS, or texting), which eventually paved the way for the transmission of other data services over cellular. As of 5G, most traffic moves from the mobile transceiver (whether cell phone, tablet, or laptop), to the cell tower/base station, through the radio access network (responsible for linking devices to other, larger networks), to the backhaul network (which used to be composed largely of copper wired networks or satellite but now primarily uses fiber optic cable or microwave radio-relay [31]), and to the core network (which used to be circuit based, but is now packet switched).

EXPERIMENTS In 2015, in the wake of discoveries that state and local law enforcement agencies in the US were using fake cell towers to track and intercept cellular network transmissions, artist and engineer Julian Oliver launched his project *Stealth Cell Tower*, which uses objects such as a modified Hewlett Packard Laserjet 1320 printer along with a Raspberry Pi microprocessor as a way to install "rogue cellular infrastructure." As Oliver describes it, "'*Stealth Cell Tower*' surreptitiously catches phones and sends them SMSs written to appear as though they are from someone who knows the recipient. It does this without needing to know any phone numbers. With each response to these messages, a transcript is printed revealing the captured message sent, alongside the victim's unique IMSI number and other identifying information. Every now and again the printer also randomly calls phones in the environment and on answering, Stevie Wonder's 1984 classic hit 'I Just Called to Say I Love You' is heard."

SOURCES Gerard Goggin, *Cell Phone Culture: Mobile Technology in Everyday Life* (Routledge, 2006); D. H. Ring, "Mobile Telephony: Wide Area Coverage," *Technical Memoranda* (Bell Telephone Laboratories, December 11, 1947); Nathan J. Muller, *Wireless A to Z* (McGraw-Hill, 2002); Christopher Cox, *An Introduction to 5G* (Wiley, 2020); "Technical Realization of Cell Broadcast Service," Specification #23.041, 3rd Generation Partnership Project (3GPP) website; Larry Greenemeier, "What Is the Big Secret Surrounding Stingray Surveillance?," *Scientific American* (June 25, 2015); Julian Oliver, "Stealth Cell Tower: Rogue Cellular Infrastructure Disguised as Office Printer," Julian Oliver website (2015)

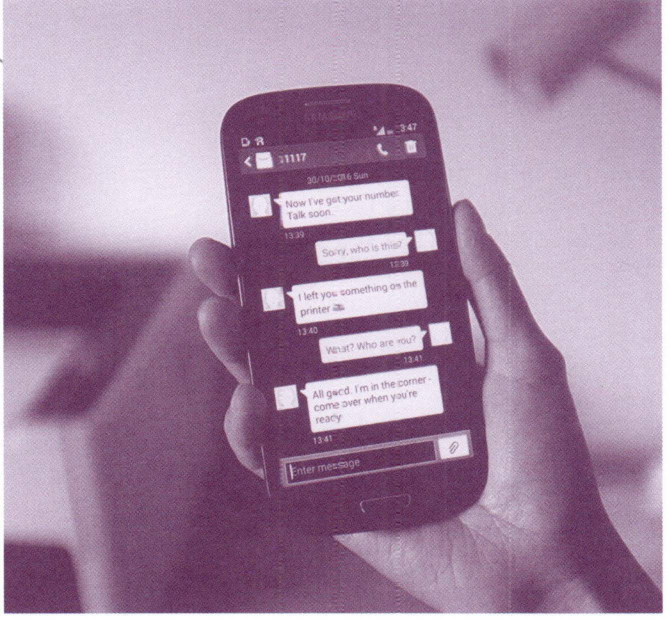

Text message exchange with Julian Oliver's *Stealth Cell Tower* (2015).

Time-Sharing Network [50]

COUNTRY OF ORIGIN USA
CREATOR(S) John McCarthy / MIT Computation Center
EARLIEST KNOWN USE 1961

BASIC INFRASTRUCTURE/MATERIALS Electrical wires, coaxial cables (optional), power source, telephone circuits, mainframe computer or minicomputer or microcomputer, terminal(s)

RELATED Telephone [34], packet radio network [26], microwave radio-relay [31], cable television [48], Project Xanadu [62]

DESCRIPTION Time-sharing networks were the first networks linking digital computers, usually in a star formation, whereby two or more remote terminals connect to a central mainframe computer. The first time-sharing networks were circuit switched (utilizing the public switched telephone network [34]) but later they were also packet switched and also on occasion incorporated coaxial cable connections and microwave radio-relay [31]. While "time-sharing" was originally proposed by British computer scientist Christopher Strachey in 1959 to mean "multi-programming" on a single, large computer (or, simply, the ability of a computer to run two or more programs at the same time), by the mid-1960s "time-sharing" had come to refer to numerous people accessing a single mainframe computer via physically distanced terminals or consoles. One can imagine how revolutionary this network must have seemed in the 1960s, as it turned the one-to-many model of the radio and television broadcast networks that were so dominant at the time into a many-to-one and even many-to-many model of networking. Anticipating the explosion of networks and networking in the 1980s that followed the arrival of the affordable personal computer, in 1966 *Scientific American* described time-sharing as a technique not so much for sharing access to computer programs but rather one that "allows communication *among users* . . . The time-sharing computer system can unite a group of investigators in a cooperative search for the solution to a common problem, or it can serve as a community pool of knowledge and skill on which anyone can draw according to his needs. Projecting the concept on a large scale, one can conceive of such a facility as an extraordinarily powerful library serving an entire community—in short, an intellectual public utility" (emphasis my own).

Despite how revolutionary these networks were, the history of the first time-sharing networks largely revolves around high-level, seemingly sedate, and arguably insular early computer science projects that were undertaken at large research institutions. Thus, the first time-sharing network was the Compatible Time-Sharing System (CTSS), which was developed at MIT and demonstrated in 1961 on an IBM 709 computer. A second version of the CTSS was launched in 1963 and, notably for the time, included off-campus users from California, South America, Scotland, and England. After the PDP-1 computer was released in 1959 and became known as the first computer designed with user interaction in mind, another early time-sharing network emerged in 1962: the Bolt, Beranek, and Newman (BBN) network, which allowed five different users at teletype terminals in Washington, D.C., and Cambridge, MA, to access BBN's PDP-1 computer in Boston. The hope was that the network would, according to J. McCarthy et al. in their 1963 report, "increase the effectiveness of the PDP-1 computer . . . by allowing each of the five users, each at his own typewriter, to interact with the computer just as if he had a computer all to himself." After this point, time-sharing networks began to emerge mostly across North America and Europe—the most noteworthy being the PLATO II time-sharing network (1962), the Dartmouth Time Sharing System (1963), Tymshare (1964, which turned into Tymnet in 1970), Compuserve (1969, which turned into an online information service in 1979), and also IPSANET (1973).

All of the aforementioned networks were essential to the development and public acceptance of telecommunications networks. PLATO, however, was particularly important for the role it played in laying the foundation for online communities (a concept that was practically unthinkable prior to the early 1960s).

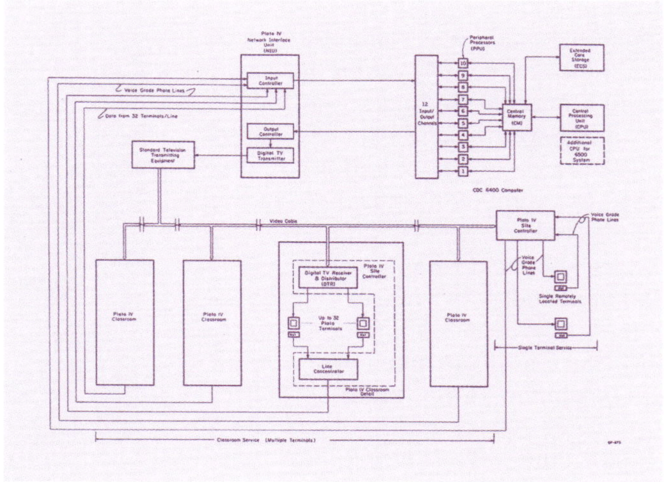

Diagram illustrating the PLATO IV (launched in 1972) networking connections between numerous classrooms, remote terminals, and a CDC 6400 mainframe computer using coaxial cable and leased telephone lines.

PLATO (Programmed Logic for Automatic Teaching Operations) was created at the University of Illinois in 1960, under the leadership of Donald Blitzer, as an experiment in computer-mediated education. As Joy Lisi Rankin narrates in *A People's History of Computing in the United States*, while the first version only allowed one user at a time access to the keyboard and television set connected to a nearby ILLIAC I mainframe computer, with the introduction of PLATO II in 1961 the system had become a time-sharing network, as two users could simultaneously use two different terminals. By 1969, PLATO III could host twenty different terminals, including a terminal at a high school in nearby Springfield, Illinois, about ninety miles away. The PLATO IV system that was developed between 1972 and 1975 inaugurated significant innovations in online communication that had not existed before at this scale and level of accessibility. Running on a CDC 6400 mainframe computer and utilizing internal coaxial cable connections, leased lines from the telephone network [34], and occasionally microwave radio-relay [31], Jack Stifle described the typical PLATO IV system in 1972 as including "approximately 2000 remotely located computer terminals, each requiring an input bandwidth of approximately 1200 bps . . . connected to one centrally located computer. Most of these terminals will be located in classroom installations of up to 32 terminals each. Some of these classroom installations will be located within a few miles of the computer site, while others will be [within] a few hundred miles . . . a limited number of individual remotely located terminals must also be serviced." Now with a sizeable number of users, PLATO IV's introduction of two applications in particular for online communication became even more important: Notes and Talkomatic. David Woolley created Notes in 1973, initially as a way for users to report system bugs; he writes, "I came up with a design that allowed up to 63 responses per note and displayed each response by itself on a separate screen. Responses were chained together in sequence after a note, so that each note could become the starting point of an ongoing conversation. This is what would today be called a linear discussion forum, and PLATO Notes was apparently the first of its kind." Woolley later added a top-level menu that allowed users to "choose among three notes files: System Announcements, Help Notes, and General Notes." Every note had a twenty-line limit so that it appeared in its entirety on a single screen. A few months later, Doug Brown created a program called Talkomatic, which supported chat among PLATO users. Woolley recounts that "The real magic of Talkomatic was that it transmitted characters instantly as they were typed, instead of waiting for a complete line of text. The screen was divided into several horizontal windows,

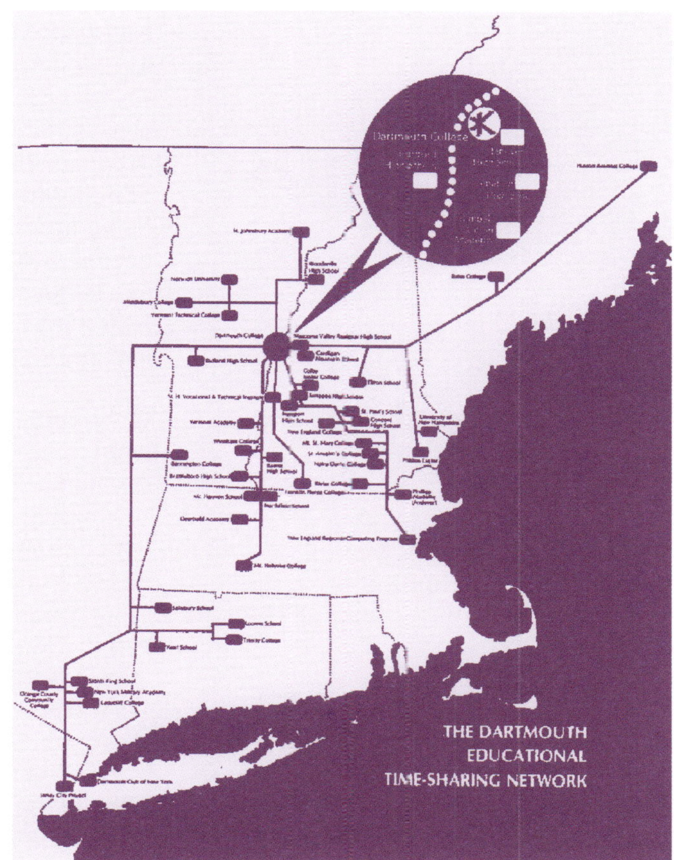

Undated map of the Dartmouth Time Sharing System's connections between in-state and out-of-state colleges and high schools.

with one participant in each. This allowed all the participants to type at once without their messages becoming a confusing jumble. Seeing messages appear literally as they were typed made the conversation feel much more alive than in line-by-line chat programs. I worked with Doug to expand Talkomatic to support multiple channels and add other features. Each channel supported up to five active participants and any number of monitors who could watch but couldn't type anything." Finally, PLATO was also remarkable for its time because, while packet-switched networks such as ARPANET struggled to connect incompatible machines, PLATO solved the problem of incompatibility by ensuring all its terminals were interchangeable.

The Dartmouth Time Sharing System (DTSS) was launched a few years after PLATO in 1964 as a time-sharing network initially connecting twenty teletype terminals across Dartmouth College along with a terminal at a nearby high school to a GE-225 mainframe computer, soon replaced by a GE-235 computer. Similar to PLATO, numerous different iterations of DTSS were launched over the years, resulting in, for example, the creation of the Kiewit Network, which

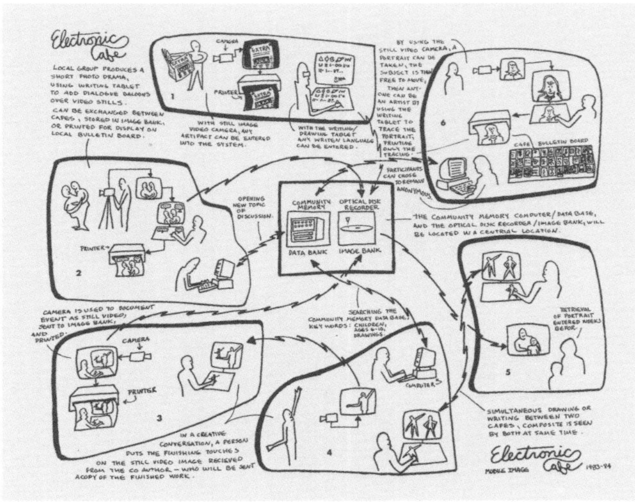

Brochure for Kit Galloway and Sherrie Rabinowitz's *Electronic Café* (1983–1984), detailing the equipment set-up at each node, which revolves around the Community Memory database.

provided multiple high schools and colleges (as far away as Montreal, Canada) access to DTSS. While it was launched a few years after CTSS, BBN, and PLATO, it is nonetheless considered the first large-scale time-sharing network that was also, for the first time, widely available to thousands of users. DTSS was also part of Dartmouth's cutting-edge commitment to providing computer literacy skills to all of its students. Remarkably, DTSS co-creator John Kemeny reported in 1972 that 90 percent of Dartmouth students had acquired computer experience because of the widespread use and availability of DTSS.

So many time-sharing networks emerged from the mid-1960s to the early 1970s that two-hundred-page reports comparing the virtues of different systems were being published in 1973. By this time, the setup for time-sharing networks had also become fairly standardized, and remained so until the networks were phased out by what would soon become affordable personal computers and their own networking capabilities. As the authors of the *Auerback Guide to Time Sharing* wrote, "Terminals, computer hardware and software form the nucleus of the time-sharing system. To gain access, users at remote locations employ various types of terminal devices. In conversational time sharing, about 90 percent of the terminals used today are teletypewriters . . . Communications facilities connect the remote terminals to the process center. The transmission line is usually a normal telephone circuit operating at 1200 baud . . . Many time sharing companies use the public dial-switched telephone network, where the user is charged on a per-call basis. An alternative is the leased line, which gives the user exclusive use of a communications line and at a flat rate."

EXPERIMENTS Community Memory (CM) started in 1973 as a small network of teletype terminals set up across Berkeley and San Francisco and which connected to an SDS 940/XDS 940 time-sharing computer. It also represented one of the most significant re-imaginings of time-sharing networks, transforming them from being networks used almost exclusively in corporate or research settings to a network that was publicly available, easily accessible—requiring almost no prior experience with computing—and whose content was entirely driven by its users.

CM emerged not only in the context of the cutting-edge research computing taking place nearby at the University of California, Berkeley; Stanford University; and Xerox Parc but also, more important, in the context of the Free Speech Movement that began on the campus of UC Berkeley in 1964–1965 in reaction to campus rules forbidding most kinds of political activity on campus, including advocacy and fundraising for the Civil Rights Movement. On December 3, 1964, student activist Mario Savio gave a speech at the beginning of a sit-in during which he made the famous declaration: "There is a time when the operation of the machine becomes so odious, makes you so sick at heart, that you can't take part; you can't even passively take part, and you've got to put your bodies upon the gears and upon the wheels, upon the levers, upon all the apparatus, and you've got to make it stop. And you've got to indicate to the people who run it, to the people who own it, that unless you're free, the machine will be prevented from working at all!" Of course, the machine to which Savio referred was more figurative than literal (referring to the machinery of imperialism, colonialism, racism, etc.), but his message influenced members of the counterculture enough that by 1973, Lee Felsenstein, Efrem Lipkin, Ken Colstad, Jude Hilhon, Mark Szpakowski, and members of the nonprofit Resource One were working not to reject the technology of their time but rather create nonhierarchical, community-oriented technological alternatives.

Resource One also played a significant role in the creation of CM. Founded in 1970, the nonprofit was headed by Pam Hart, who was dedicated to making computer services available to members of the counterculture. In a 1972 issue of *Rolling Stone*, Stewart Brand interviewed Hart, who described how she and others convinced the Trans-America Corporation to donate a refurbished SDS 940 mainframe computer. In a Resource One newsletter from 1974, eight months after the first CM terminal became operational, an anonymous writer explained that their mission was to explore "the fundamental tension between person and machine: Can this tool of a militarized society be made directly useful to people? How? . . . Are we risking dependence on an overgrown, high-level technology? What should we do with all the technological tools we've acquired?" CM was, therefore, an attempt to answer these questions about the role of computers and advanced technology, framed within the context of countercultural beliefs about communality and anti-authoritarianism. (Mark Szpakowski points out that CM was also influenced by public infrastructures such as flea markets and the free healthcare services offered by the Haight-Ashbury Free Clinics, along with the Chilean socialist government Project Cybersyn. This aspect of CM is discussed at greater length in telex [39].)

In terms of infrastructure, the Free Speech Movement's telephone room was particularly influential in how it was structured for open, nonhierarchical public information sharing, and it became the prototype, according to Felsenstein, "of not only the next few years of switchboards—which were these special interest info referral places, a phone and a file card box maybe and a bulletin board—but also the prototype of the social internet." CM was also modeled after the structure of a Library [42]. Felsenstein recalls that during the development phase of CM, a librarian commented to members of Resource One that they seemed to be trying to create "a library with no books on the shelf. They suggested that we put some 'books on the shelf' and then see where we could go with that. Lipkin . . . took this as a challenge to put one or more terminals in public locations to see what information might accumulate on the system."

Felsenstein describes the first iteration of CM, which was set up in a nonprofit community record and music store, Leopold's Records in Berkeley, and then at other locations around Berkeley and San Francisco: "Before the bulletin board was a table and on it sat a squat box made of corrugated cardboard with a window on its sloping front made of clear vinyl plastic. Two holes in the front face revealed vinyl sheets with asterisk-shaped cutouts, like flexible cat doors. On the right side of the box was bolted a small cooling fan that purred . . . Inside the box was a teleprinter—a Teletype Model 33 ASR that had gone through three years' service as a commercial time-sharing computer terminal. Urethane foam glued inside of the cardboard muffled the whirr of the teleprinter's motor and the 'chunk-chunk-chunk' of its print head." Users were then invited to put their hands through two holes in the front-facing plastic to use the keyboard: "Pushing the return key caused a prompt to be printed on the paper visible through the window; 'Type a word you're interested in and push the green button.' Doing so usually (but not always) resulted in the printer typing out the first lines of a number of bulletin-board items that had earlier been entered by other users who had indexed it using that word among others." The information was transmitted to and from the mainframe computer that was being stored in a warehouse in San Francisco. Sitting under the table was a three-hundred-baud modem, and "in the modem box sat the handset of a Bell model 500 telephone connected to a 'foreign board' phone line in Oakland with local dialing rates to San Francisco. Each morning the terminal's attendant would place a single call to the computer's number and place the handset in its cradle. Direct electrical connection to the phone network was still forbidden and the subject of a lawsuit. The data were coupled acoustically to and from the handset, and the free call lasted all day."

Given the relative inaccessibility of time-sharing networks in the early 1970s, CM's powerful appeal came from the fact that it was, as Michael Rossman wrote in 1975, "inescapably political. Its politics are concerned with people's power—their power with respect to the information useful to them, their power with respect to the technology of information . . . The system democratizes information, coming and going." In an archival document titled "Proposal for a Community Memory

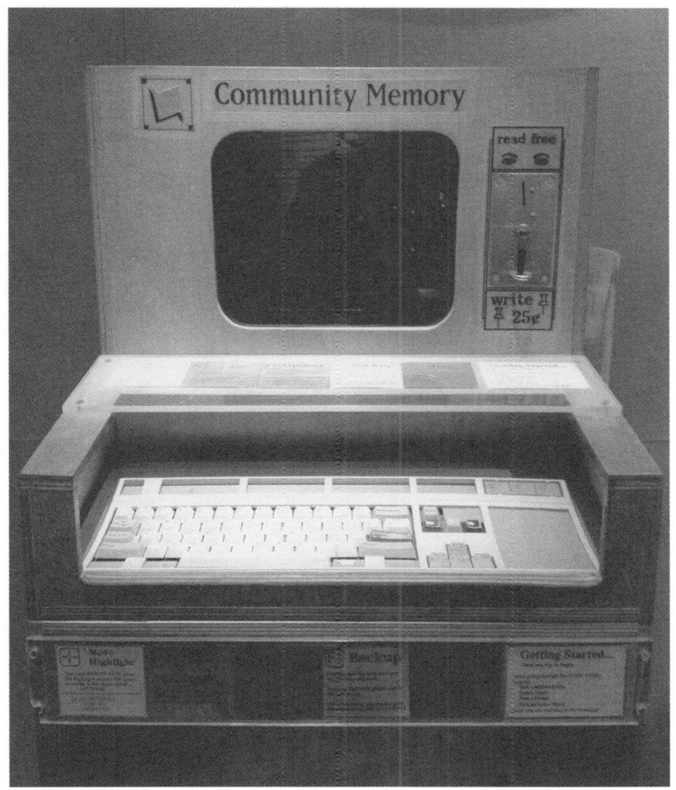

A coin-operated, third, and final generation Community Memory terminal now housed in the Computer History Museum; it was free to read messages and twenty-five cents to post.

[50] The first Community Memory terminal setup at Leopold's Records in Berkeley, California, in 1974.

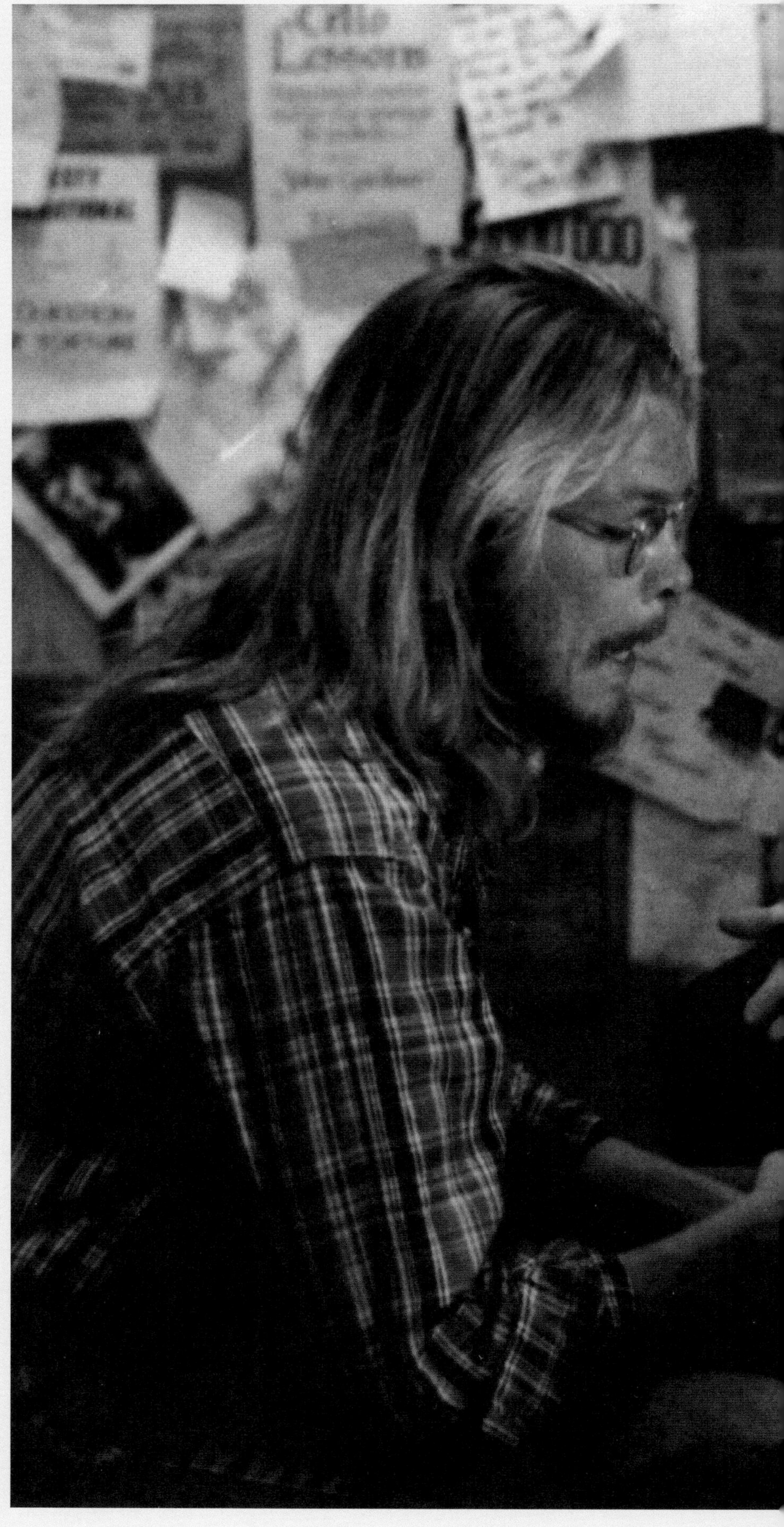

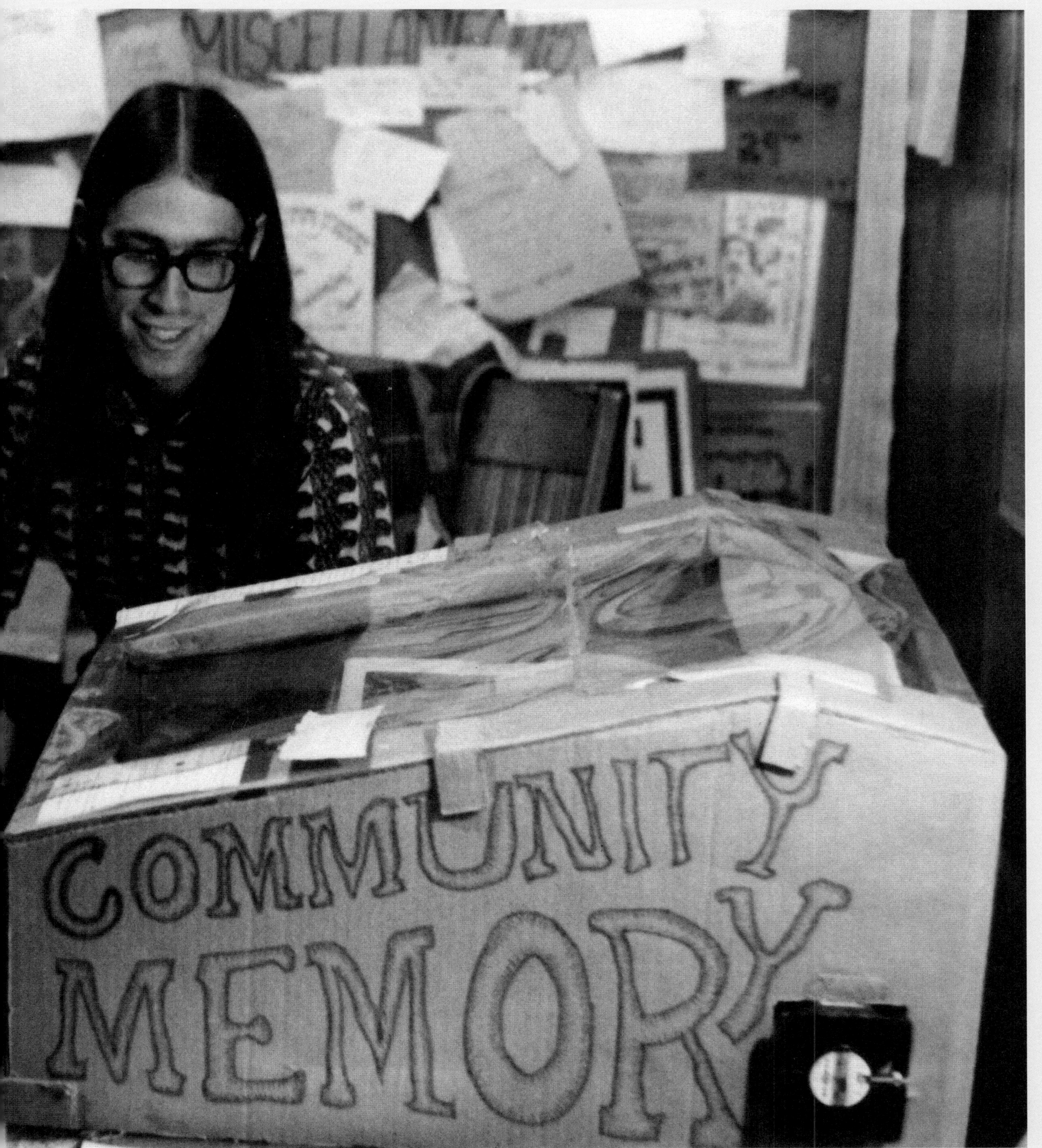
COMMUNITY
MEMORY

Pilot Project," the writer recounts that the first iteration of CM was in use up to 70 percent of the time the terminals were available, with an average of thirty people using each terminal.

This first iteration ended in January 1975, and a nonprofit called Community Memory Project was formed several years later. This group launched a second iteration of CM in 1984 that lasted until 1988. This version of CM was used by Kit Galloway and Sherrie Rabinowitz's ambitious project *Electronic Café* (*EC*, also discussed in the videophone [38] entry). The five *EC* nodes located across Los Angeles were similarly aimed at inducing members of the public to familiarize themselves with the capabilities of telecommunications equipment that would soon change the shape of the twentieth century. In their proposed budget for *EC*, Galloway and Rabinowitz estimated that of their total $185,451 budget (about $574,000 in 2024), $91,445 would be spent on audio/video/computer equipment and $7,323 on telecommunications. Of the computer equipment, *EC* terminals used Community Memory's custom-built database software, which allowed visitors to enter comments, responses, observations, or listings through the terminal keyboard. Every entry was labeled with keywords chosen by the users, so that the keywords would turn into extended discussions on particular topics. The final iteration of CM was launched in 1989 as ten coin-operated IBM PCs connected via leased lines to an IBM 386 XT computer that acted as a server and ran custom conferencing software; CM finally came to an end in 1992. It's worth noting that, in 2016, Lee Felsenstein stated that "*Electronic Café* came closest to realizing our vision for Community Memory—a network of neighborhood information exchange points enabled with multiple media and open to all comers on an ongoing basis. It can still be done!"

SOURCES Christopher Strachey, "Time Sharing in Large, Fast Computers," UNESCO Information Processing Conference (June 15, 1959); J.A.N. Lee, "Claims to the Term 'Time-Sharing,'" *IEEE Annals of the History of Computing* 14:1 (1992); R.M. Fano and F.J. Corbató, "Time Sharing Computers," *Scientific American* (September 1966); J. McCarthy et al., "A Time-sharing Debugging System for a Small Computer" (ACM Press, 1963); David Walden and Tom Van Vleck, eds., *The Compatible Time Sharing System (1961–1973) Fiftieth Anniversary Commemorative Overview* (IEEE Computer Society, 2011); *Auerbach Guide to Time Sharing* (Auerbach Publishers, 1973); Joh Kemeny, *Man and the Computer* (Charles Scribner's Sons, 1972); David R. Woolley, "PLATO: The Emergence of Online Community," in *Social Media Archaeology and Poetics*, ed. July Malloy (MIT Press, 2016); Joy Lisi Rankin, *A People's History of Computing in the United States* (Harvard University Press, 2018); Jack Stifle, "The PLATO IV ARCHITECTURE," CERL Report X-20 (rev. May 1972); Lester F. Eastwood, Jr. and Richard J. Ballard, "The Plato IV CAI System: Where is it Now? Where Can it Go?" *Journal of Educational Technology Systems* 3:4 (1975); "Episode 4: Community Memory and the Computing Counterculture," Artists and Hackers podcast (March 18, 2021); Robert Cohen and Reginald Zelnik, eds., *The Free Speech Movement: Reflections on Berkeley in the 1960s* (University of California Press, 2002); "Mario Savio's speech before the FSM sit-in," Free Speech Movement archives website; Stewart Brand, "SPACEWAR: Fanatic Life and Symbolic Death Among the Computer Bums," *Rolling Stone* (December 7, 1972); "WHO WE ARE," *Resource One Newsletter* 2 (April 1974); Mark Szpakowski, Bluesky post to Lori Emerson (February 23, 2024); Lee Felsenstein, "Community Memory: The First Public-Access Social Media System," in *Social Media Archaeology and Poetics*, ed. July Malloy (MIT Press, 2016); Ken Colstad and Efrem Lipkin, "Community Memory: A Public Information Network," *Computers and Society* 6:4 (December 31, 1975); Michael Rossman, "Implications of a Community Memory," *Computers and Society* 6:4 (December 31, 1975); "Proposal for a Community Memory Project," Community Memory archives, Computer History Museum (undated); "Electronic Cafe Budget Outline," Community Memory archives, Computer History Museum (undated)

Teletext [51]

COUNTRY OF ORIGIN UK
CREATOR(S) British Broadcasting Corporation
EARLIEST KNOWN USE 1973

BASIC INFRASTRUCTURE/MATERIALS Electric power source, television set or CRT monitor, remote control or keyboard, teletext decoder box, coaxial cable (optional), antenna (optional), personal computer (optional)

RELATED Radio broadcast [17], packet radio network [26], communications satellite [32], television broadcast network [36], videotex [52]

DESCRIPTION Teletext is a one-way broadcast service that uses the vertical blanking interval of a standard TV signal. The vertical blanking interval is the time it takes for an electron beam to move on a raster display from the bottom of one page back up to the top of the next page over broadcast television [47]; this interval is normally unused during a television broadcast.

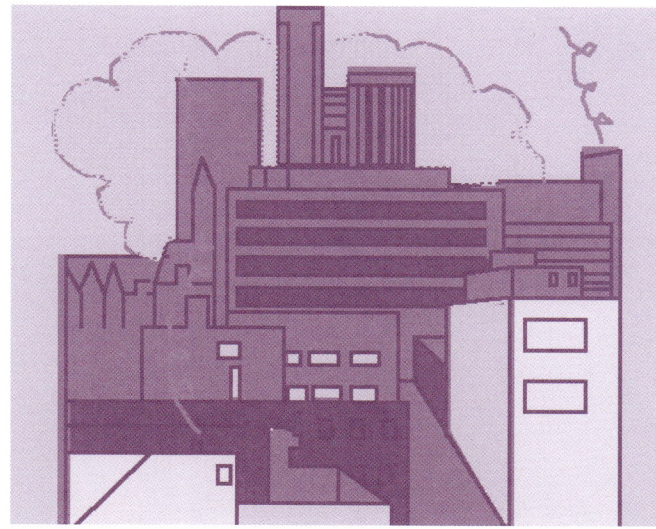

A page from Elaine Cohen's *City*, distributed in 1984 on Canada's teletext service, Telidon; recovered and restored by digital preservationist John Durno.

Unlike videotex [52], which was originally designed to be a two-way service over telephone lines [34], teletext was originally designed to work over coaxial cable, a VHF television signal, satellite, or, in the case of Telidon (Canada's teletext service), even over an X.25 packet-switched network. The roots of teletext lie in RCA engineer William D. Houghton's experimental system called Homefax, which he described as "a new concept in home information services." Homefax began in 1967 and lasted until the early 1970s and was, as Leonard R. Graziplene explains it, the first system to attempt to use "a spare line in the field blanking interval of the television signal to transmit a slow-scan image from a camera, which was then read as hard copy at the destination site. A whole frame of information took about 10 seconds to deliver." Thereafter, the BBC announced the first teletext system in October 1972, called Ceefax, which differed significantly from Homefax in that rather than printing information as a facsimile, it appeared on a television screen. Ceefax also digitally encoded text, which later made it possible for teletext to work with personal computers. The service began experimental transmissions in March 1973. Over the next year, the BBC and other organizations created a unified standard, and in April 1974, Ceefax was officially launched for the public; it offered a range of linked pages on news, weather, and TV schedules, which users could navigate using a remote control. In the early years of service, viewers needed to purchase a teletext decoder that was attached to their television sets; by 1977, televisions were manufactured with built-in decoders. In 1981, the BBC introduced a teletext system for its microcomputers that included Telesoftware—a datacasting service that distributed computer software over the vertical blanking interval and allowed computer users to download these programs on disks or cassette tapes.

Ceefax prompted the launch of many different teletext systems around the world, including Antiope (France, 1976), ORACLE (UK, 1978), Telidon (Canada, 1978), Electra (USA, 1982), Televideo (Italy, 1984), JTES (Japan, 1983), and Aertel (Ireland, 1986). Of all the aforementioned services, Canada's Telidon was the most significant because it worked over broadcast television, radio signal, satellite, or even packet-switched network. It also featured superior graphics because of its coding scheme, called the Picture Description Instructions. In addition to services available across the province of Ontario in cities, elementary and secondary schools, universities, colleges, and libraries, Telidon was also adopted by New York University's Alternate Media Center starting in 1981 and the University of Alaska's distance learning programs in 1983. Because of its attractive graphics, relatively wide availability, and heavy promotion by the Canadian government, Telidon was used by a number of new media artists in the early 1980s.

There are many live teletext feeds available through the internet from countries ranging from Armenia to Turkey. One of the last government-supported teletext services, Sweden's SVT Text, is still functioning as of 2024, after forty-five years of service.

EXPERIMENTS Pierre Moretti created the first Telidon artwork in 1980—a short film titled *Graphic Variations on Telidon*—for the National Film Board of Canada. Of particular historical importance is the work

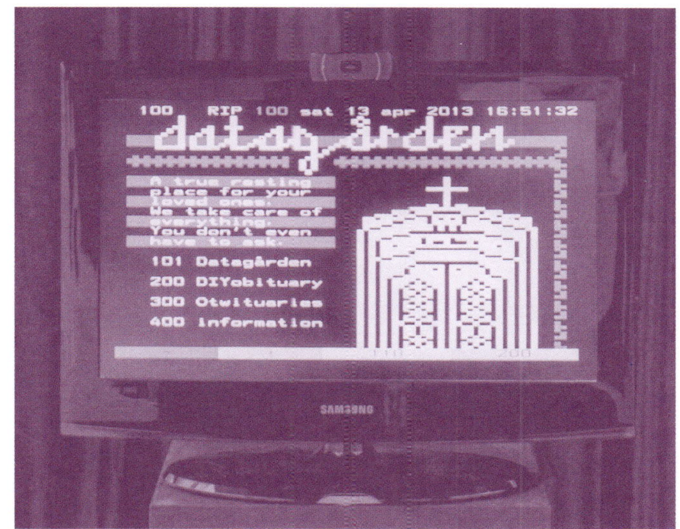

The opening page from a 2013 exhibition in Bonniers Konsthall (Stockholm, Sweden) of GOTO80, Raquel Meyers, and Possan's *Datagården*.

[51] Advertisement for the first teletext system, the BBC's Ceefax.

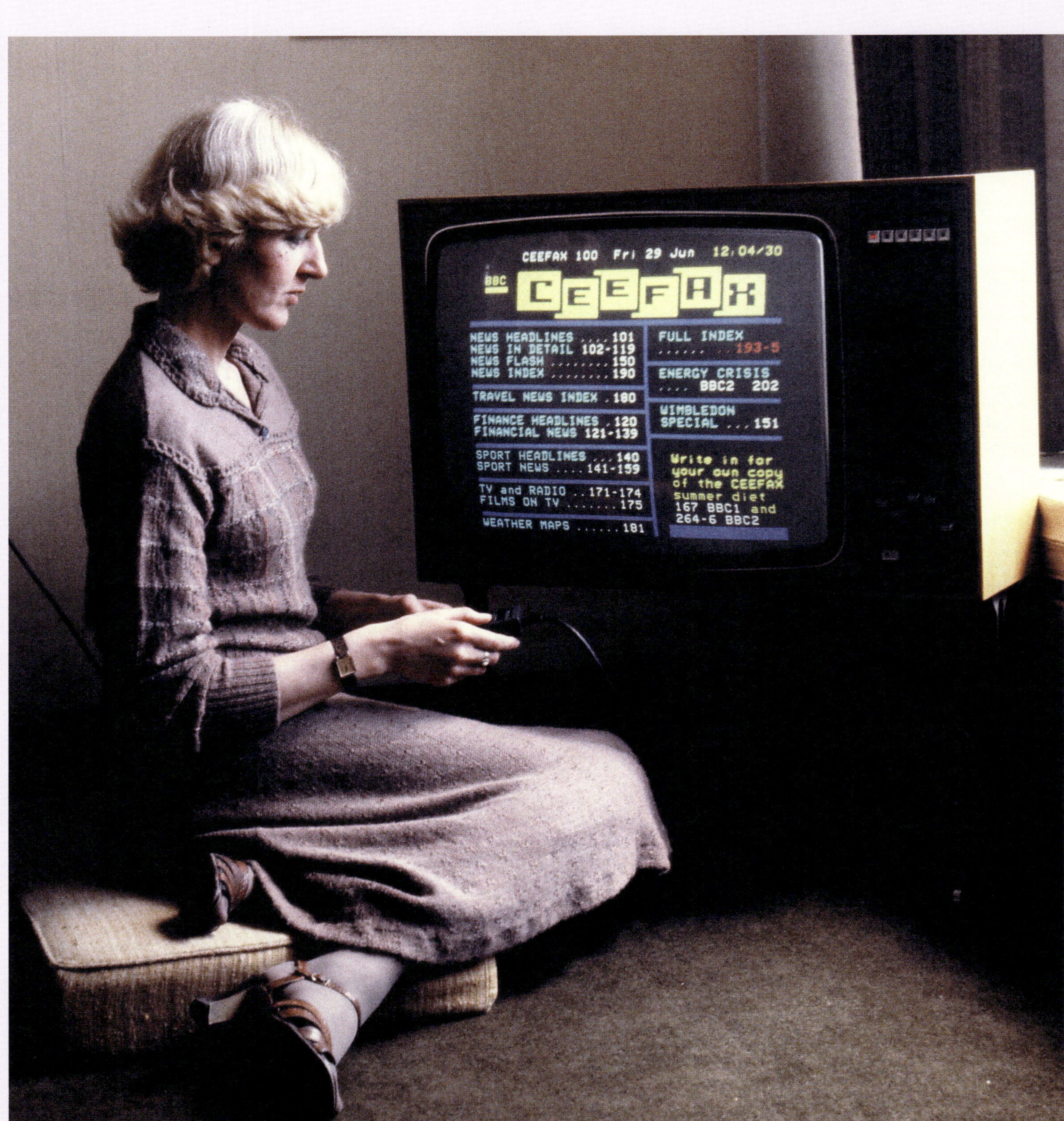

spawnedby Toronto Community Videotex (now known as Inter/Access), a nonprofit organization founded by Bill Perry in 1983 that grew out of his project Telidon at Trinity Square Video, which provided artists access to an IPS II workstation that had been lent to him by Bell Canada in 1981. Thereafter, Telidon art was featured at the 1983 São Paulo Biennale and Expo '86. More recently, there have been teletext festivals nearly every year since 2012, which feature artists from across the world (but especially from Europe) who are still actively creating art using teletext. For example, GOTO80, Raquel Meyers, and Possan created *Datagården* for the festival Art Hack Day in Stockholm, Sweden, in March 2013. The piece is a "teletext cemetery with obituaries of people's online lives"; one page is "a speculation about how automatic obituaries might look like in the future, by scraping together all the data trails we leave behind," while another uses a webcam "to show your face in teletext in realtime." The online exhibition catalog states that "Datagården works on any television, using Text-tv and a normal remote control. Text-tv is one of the largest mass media of Sweden, with two million viewers everyday . . . Datagården is most likely the first installation in the world to use a custom teletext signal. It features video input, video feedback and Twitter-feeds along with custom-made graphics/text/software and C64-music." Since 2013, Meyers in particular has continued to be a prolific creator and performer of teletext art.

SOURCES John Durno, personal email correspondence (October 2023); W.D. Houghton, "Homefax: A Consumer Information System," *RCA Engineer* 16:5 (February/March 1971); Leonard R. Graziplene, *Teletext: Its Promise and Demise* (Lehigh University Press, 2000); "BBC Engineering: Information Sheet," 4008:5 (October 1975); "Broadcast Teletext Specification" (BBC, Independent Broadcasting Authority, and British Radio Equipment Manufacturers' Association, 1976); *BBC Micro Teletext System User Guide* (Acorn Computers, 1981); H.G. Brown, C.D. O'Brien, W. Sawchuk, and J.R. Storey, "A General Description of Telidon: A Canadian Proposal for Videotex Systems," *CRC Technical Note* 697-E (December 1978); "What Is Telidon?," Inter/Access website; Pierre Moretti, "Graphic Variations on Telidon," National Film Board of Canada (1980); Jordan Pearson, "The Original Net Artists," *Vice* (July 21, 2015); "The Teletext Museum—Live Feeds," teletext.mb21.co.uk; "Datagården," Goto80.com; "Art Hack Day: Larger Than Life," arthackday.net; Raquelmeyers.com; Miranda Bryant, "Teletext Lives On in Sweden Thanks to Nostalgia and Trusted Content," *The Guardian* (January 11, 2024)

Videotex [52]

COUNTRY OF ORIGIN UK
CREATOR(S) Samuel Fedida / British Telecom Research Laboratories
EARLIEST KNOWN USE 1974

BASIC INFRASTRUCTURE/MATERIALS Electrical wires, telephone circuits, keyboard or keypad, television set or terminal, decoder

RELATED Telephone [34], videophone [38], cable television [48], teletext [51]

DESCRIPTION Videotex refers to a wide range of two-way communication systems that use a television set or terminal and telephone network [34], sometimes in conjunction with a packet-switched network, and modem for sending/receiving information. The British Telecom Research Laboratories first developed videotex and, in 1974, launched their experimental system, named Viewdata, for requesting and retrieving data from a central database. Similar to the one-way teletext [51] systems being developed at the same time, viewdata could only display 40 × 24 characters at one time. British Telecom then launched Prestel in 1979 as the world's first videotex system for the general public to access information from travel agencies, banks, newspapers, and other information services. By 1983, users could select information with either a numeric keypad or a keyboard; the information itself was presented as menus on pages, and data was sent to users over telephone lines at about 1,200 bits per second, while users could send data at about 75 bits

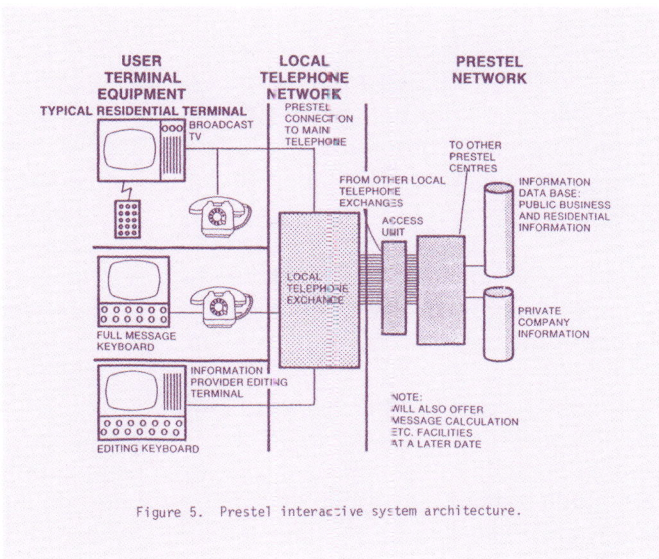

Illustration of the Prestel architecture from a 1980 report prepared for the US Department of Commerce.

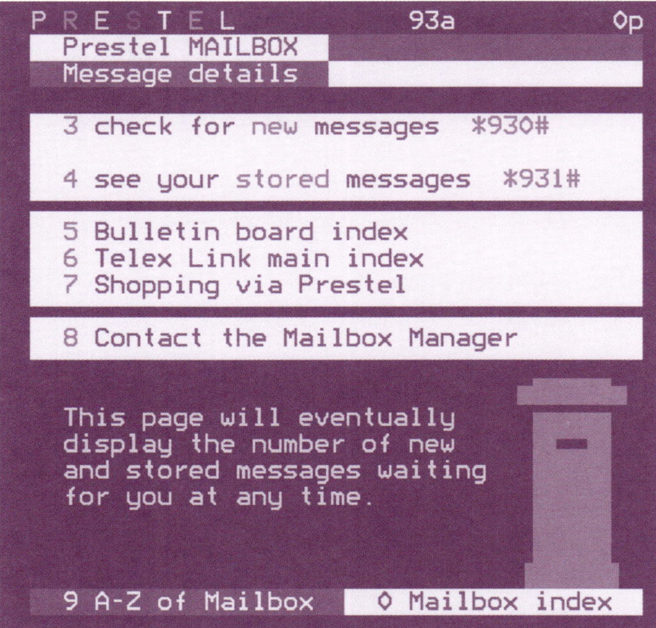

Screenshot of the MAILBOX application available on Prestel.

per second. John Carey and Martin Elton point out that communication was limited to the transmission of "a simple pre-formatted message along the lines of: 'I am running late. Expect me at [fill in time].'" With a complex and expensive pricing system, Prestel was slow to catch on. Darby Miller reported in a special issue of *Byte* magazine dedicated to videotex developments around the world that while Prestel may have been the world's largest videotex provider, only 24,000 people had subscribed as of 1983. While Prestel never gained traction among UK consumers and was phased out in 1994, its system was taken up by Sweden, the Netherlands, Finland, and West Germany, and influenced the creation of over a dozen different iterations of videotex in countries around the world, including Australia and New Zealand. Japan also launched its own Captain videotex network in 1984 that transmitted more than 3,500 different kanji characters as pre-rendered images rather than as alphanumeric characters.

But one of the best known and most successful implementations of videotex was France's Minitel network, launched in 1983. While the network was in fact called Télétel and only used terminals called Minitel, the latter name is now synonymous with the network itself. From 1983 to 1987, the government freely provided Minitel terminals to all French telephone subscribers; the terminals connected to Télétel's regional switching stations that would in turn use France's packet-switched network Transpac to transmit data. Minitel was initially used as an online phone directory (intended to replace the then ubiquitous paper-based tomes), as a way to access online goods and services, and as an online entertainment provider. The entertainment ranged from video games to online sex chat forums, known as "le Minitel rose," or "pink Minitel." Jeff Nagy delves deeper into Minitel's "pink messaging services," finding that they "specialized in pay-by-the-minute erotic chat—and users flocked to them to trade mediated come-ons anonymously with each other, or, unbeknownst to them, with professional *animateurs*, chat hosts employed to keep their clients logged on as long as possible." And since the service could rarely find enough women to take on these roles, most *animateurs* were "generally young men who shaped their online personas to meet the specific sexual proclivities of their mostly male interlocutors, gender bending for pay." Also unusual for the time, rather than requiring users to subscribe to a monthly service, Minitel users were charged by connection time, thereby making the "pink messaging" a financial boon to the French communications administration known as Postes, Télégraphes et Téléphones.

Julien Mailland and Kevin Driscoll describe how one might initially use a Minitel:

> a typical first foray into the world of Minitel was to query the electronic phone book. Calls to the phone book were free of charge for the first three minutes, which provided a low-risk opportunity for experimentation. With the Minitel 1 properly installed and powered on, one initiated a connection by lifting the handset, dialing "11," and waiting for an answer. As the phone rang, the call was being routed to a regional office where a computerized switch . . . provided a gateway between the standard telephone network and the electronic phone book's data network. On successful routing, a high-pitched tone . . . would sound in the earpiece. By pressing the button marked Connexion/Fin on the keyboard, the Minitel would enter "remote" mode, join the call, and handshake with the remote service. At this point, one returned the telephone handset to its cradle and watched as the menu of the electronic phone book appeared on the screen at 1200 bits per second. At the conclusion of the session, pressing the Connexion/Fin button caused the Minitel to "hang up" and return to local mode.

Mailland and Driscoll also note that by 1989, "there were approximately five million Minitel terminals in use, leading James Gillies and Robert Cailliau of the European Organization for Nuclear Research (CERN) to describe France as 'the world's most 'wired' country' in their published account of the period." Minitel systems were implemented all around the world

[52] Poster from roughly 1987–1988 advertising a Minitel online sex chat forum.

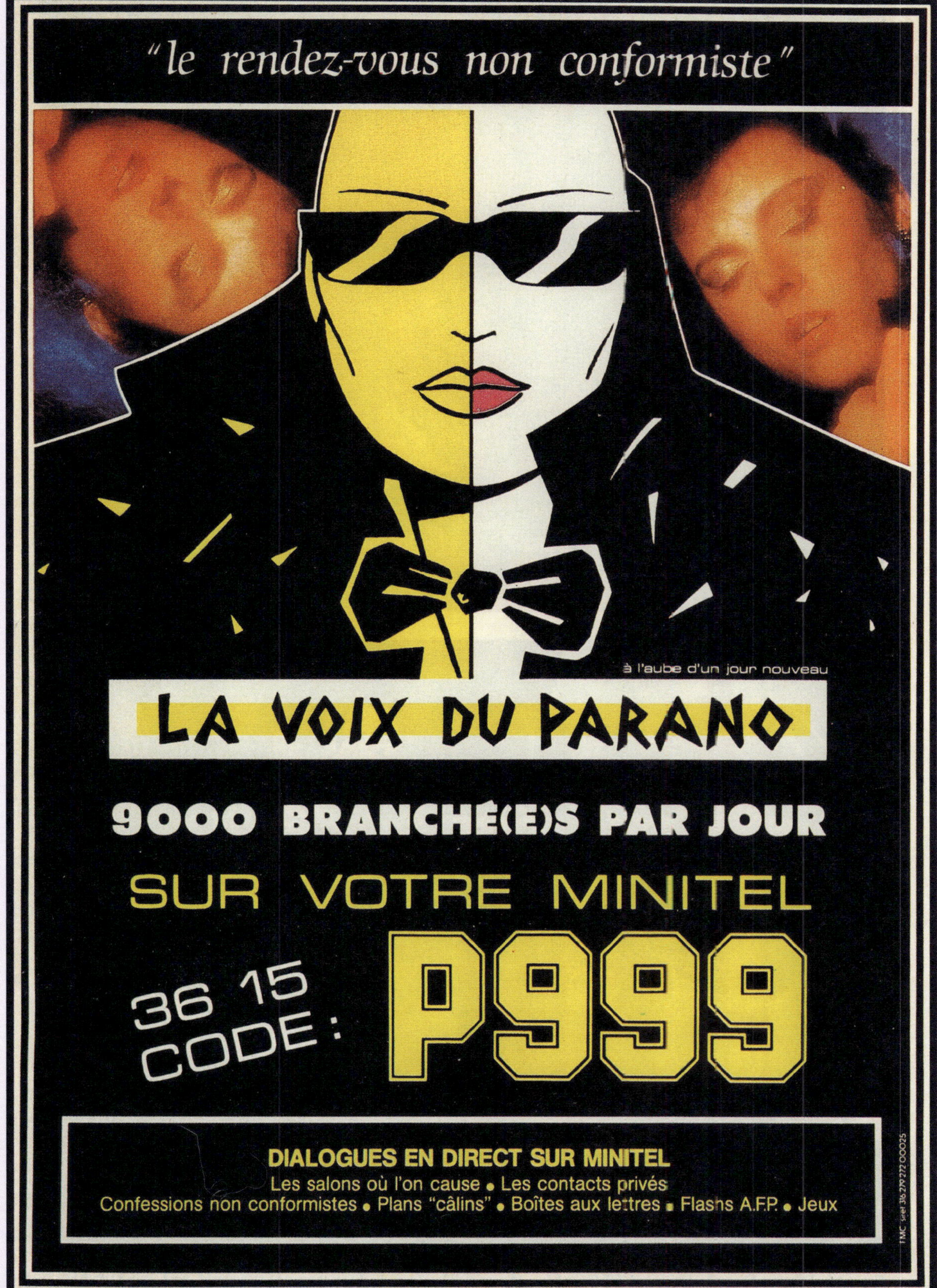

[52] Eduardo Kac's Minitel work *Reabracadabra*, originally from 1985 and restored in 2016 with the help of a French research lab called PAMAL.

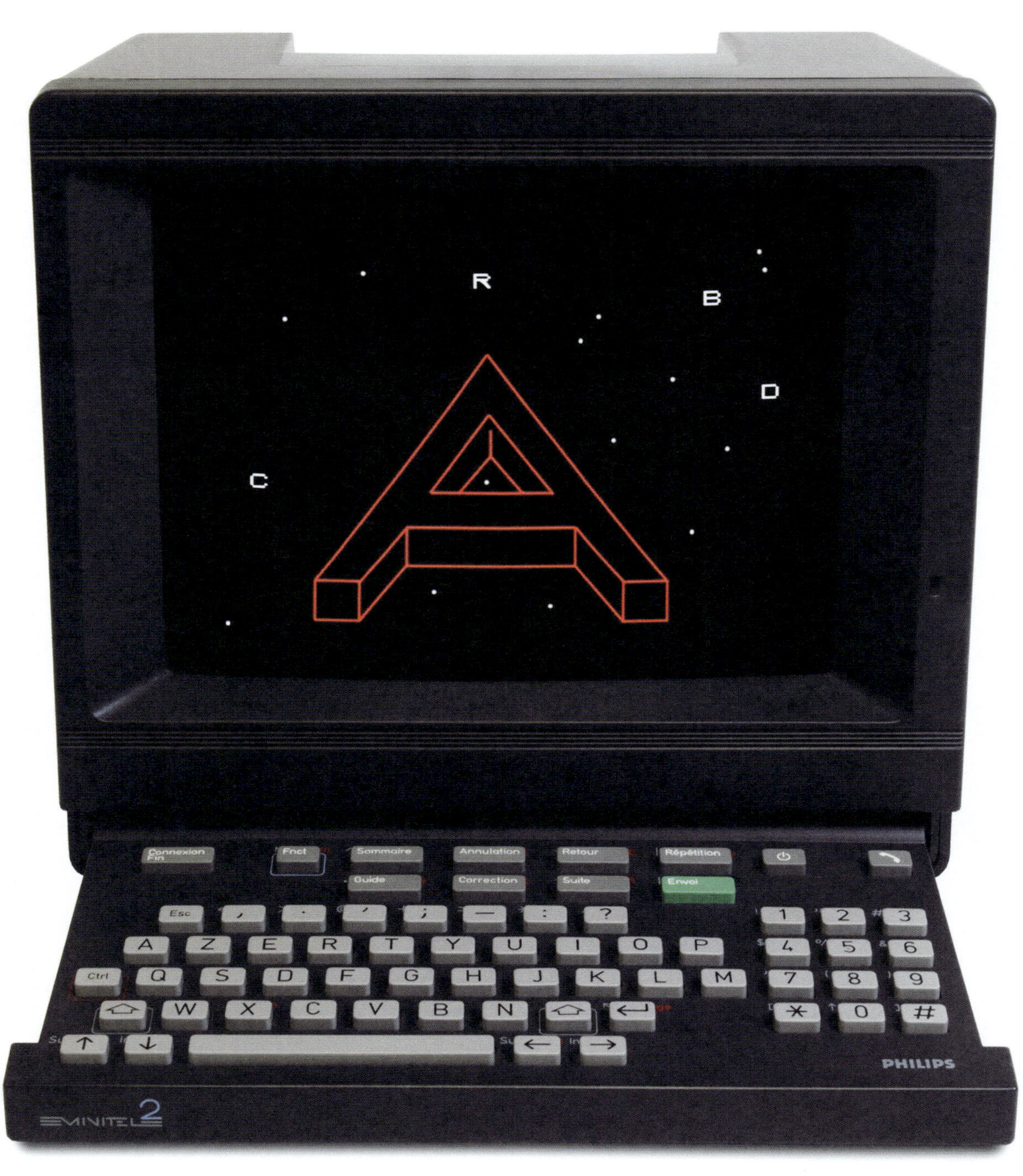

by countries such as Belgium, Brazil, Greece, Spain, Ireland, Canada, and the United States. Minitel was officially discontinued in France in 2012.

EXPERIMENTS In the early 1980s, Brazil purchased a Minitel system from France, renamed it Videotexto, and installed the network in large urban centers such as São Paulo and Rio de Janeiro. Soon after, artists Mario Ramiro and Jose W. Garcia created Clones: A Simultaneous Radio, Television and Videotex Network. This was a short-lived, experimental, intermedia network from 1983 that attempted to integrate three different communications media through the simultaneous transmission of representations of a horizontal red bar. Lasting only four minutes, Ramiro describes the work as follows: "On nine videotex terminals connected to nine different phone lines, viewers saw a horizontal red bar displacing itself upon the plane surface of the two-dimensional grid of the system. The image of the same horizontal bar received from a live broadcast in the city moved in perspective away from the surface of television monitor screens, while slowly going out of focus and losing its rigid geometric shape. An acoustic representation of this object, based on the sounds produced by an iron bar falling and hitting the ground, was broadcast on the radio." Transmissions took an average of four minutes to travel from sender to receiver.

A few years later, from 1985 to 1986, multimedia artist Eduardo Kac also began experimenting with creating artworks for Videotexto, including a series of animated poems that were exhibited in a local videotex art gallery run by the telephone company Companhia Telefônica de São Paulo; Kac used videotex screens to display the poems as looping videos. One work in particular, *Reabracadabra* from 1985, demonstrated Kac's ties to earlier Brazilian concrete and conceptual poetry. As Anders Carlsson describes it, "When users of Brazilian videotex accessed *Reabracadabra*, they would see diagonal lines appear on their screen, which were drawn to form triangular shapes that eventually grew into the letter A in 3D. The consonants of the poem's title, R, C and D, orbited around the A, just like particles around a nucleus or planets around the sun."

Minitel also made possible Olivier Auber's 1987 creation of the still ongoing social media game (or continuously evolving global work of art) *Poietic Generator* (*PG*). As Auber describes it, *PG* is "a social network game which may be envisioned as a 100% human 'Game of Life,' that is to say a cellular automata where every single cell is manipulated by a single human being. It allows everybody . . . all together regardless of his/her language, culture and educational background, to participate in real time (with a PC or mobile device) in the process of self-organization at work in the continuous emergence of a global picture." More directly put, anyone may contribute a 20 × 20 pixel image that is added to the (therefore continuously changing) global image constituting *PG*. In a recent interview with Judy Malloy, Auber makes it clear that his choice to use Minitel was more pragmatic than aesthetic: "To realize the *PG* in 1987, it would have been easier to use PCs or Macintoshes rather than Minitels. Nevertheless, I chose the Minitel, even though I had no experience with it, only because it was at the time available to all French users. It should be noted that for technical reasons the PG never went on the mainstream Minitel market. Indeed, the screen of the Minitel had too low a resolution to display the global image of the PG. The Minitel was primarily used as a user interface (a very inconvenient one, no mouse, keys only)." In contrast to the Brazilian artists experimenting with Videotexto at the same time, for Auber "there was very little place for works of art in the so-called Minitel culture. It was seen as a cash machine by the French Telcos and broadcasters." While PG was first implemented on Minitel in France, it then migrated to the World Wide Web in 1997, where it remains today. No doubt partly because of its longevity, PG has gone on to be one of the most influential, even canonical, works of early net art.

SOURCES L.R. Bloom et al., *Videotex Systems and Services*, NTIA-Report-80-50 (US Department of Commerce, October 1980); J. Tydeman et al., *Teletext and Videotex in the United States* (McGraw-Hill, 1982); Graham Hudson, "Prestel: The Basis of an Evolving Videotex System," *Byte* 8:7 (July 1983); Darby Miller, "Videotex: Science Fiction or Reality?," *Byte* 8:7 (July 1983); John Carey and Martin Elton, "The Other Path to the Web: The Forgotten Role of Videotex and Other Early Online Services," *New Media & Society* 11:1&2 (2009); Julien Mailland and Kevin Driscoll, *Minitel: Welcome to the Internet* (MIT Press, 2017); Jeff Nagy, "Pink Chat: Networked Sex Work Before the Internet," *Technology and Culture* 62:1 (2021); Mario Ramiro, "Between Form and Force: Connecting Architectonic, Telematic and Thermal Spaces," *Leonardo* 31:4 (1998); Eduardo Kac, "Minitel Works," Ekac.org; Anders Carlsson, "When Net Art Outlives the Net: Eduardo Kac's Poetry for Videotexto," Rhizome.org (November 3, 2016); Olivier Auber, *Poietic Generator*, poietic-generator.net; Judy Malloy, "Interview with Olivier Auber," Narrabase.net (undated)

[52] Exhibition of Mario Ramiro's and Jose W. Garcia's "Clones" in the Museum of Image and Sound in São Paulo, Brazil, 1983.

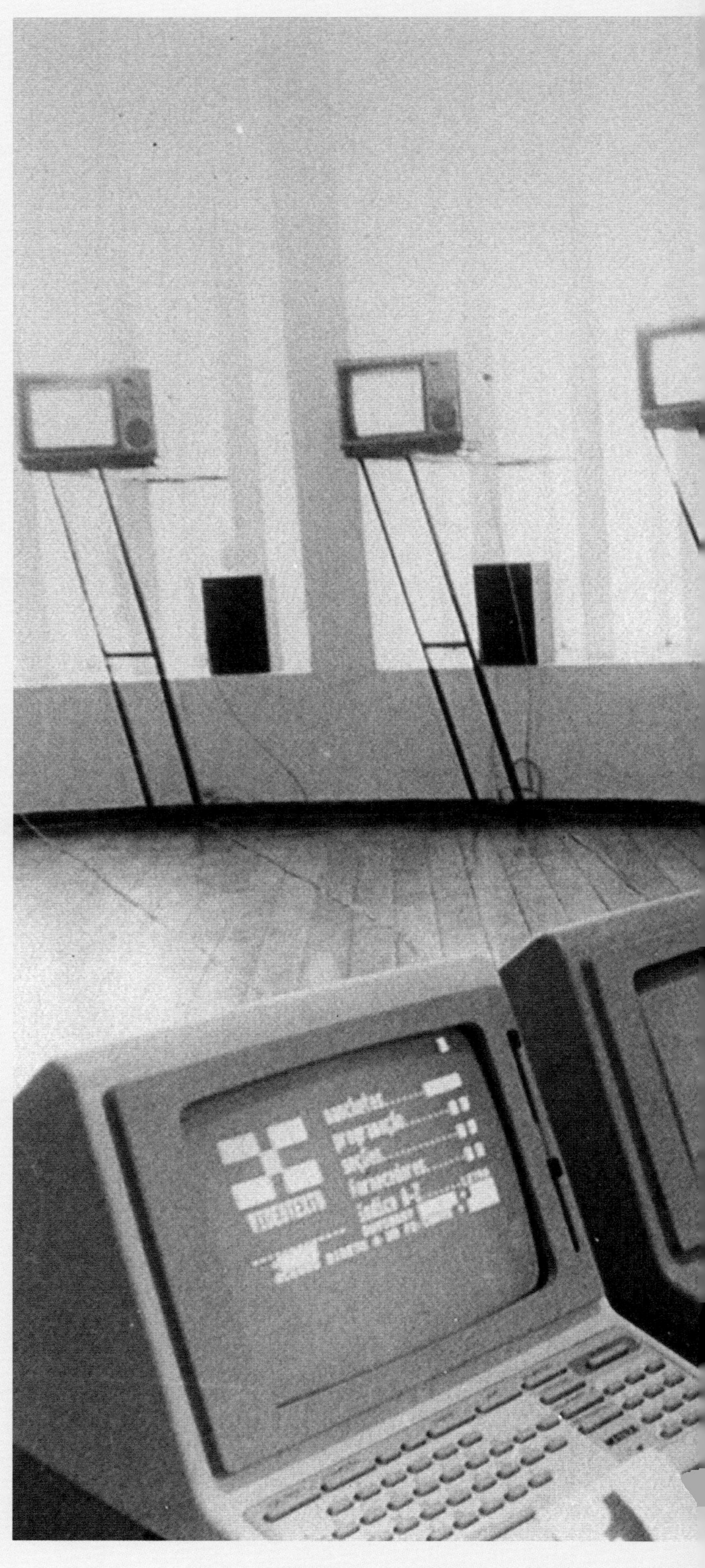

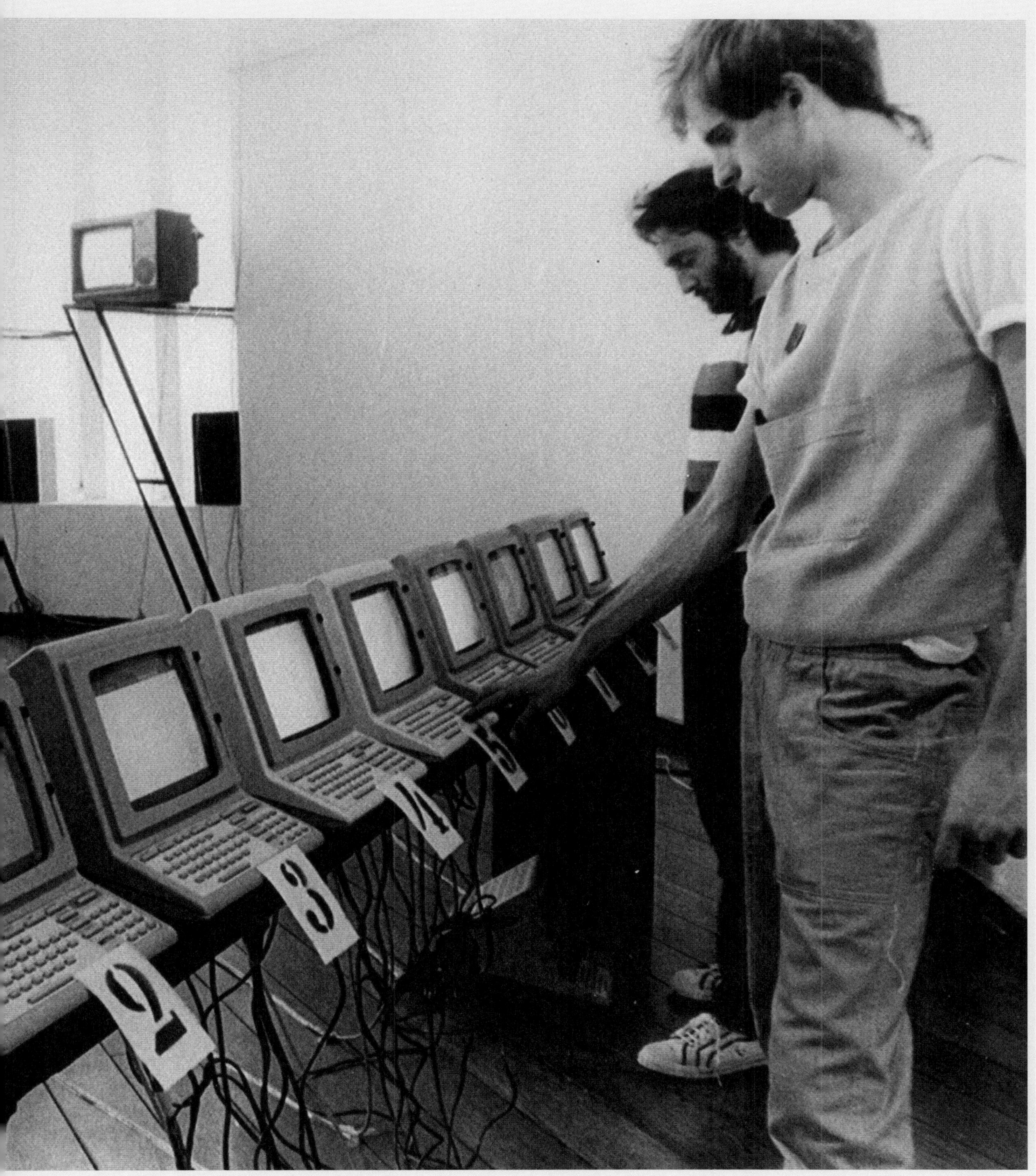
2
3
4
5

IMAGINARY

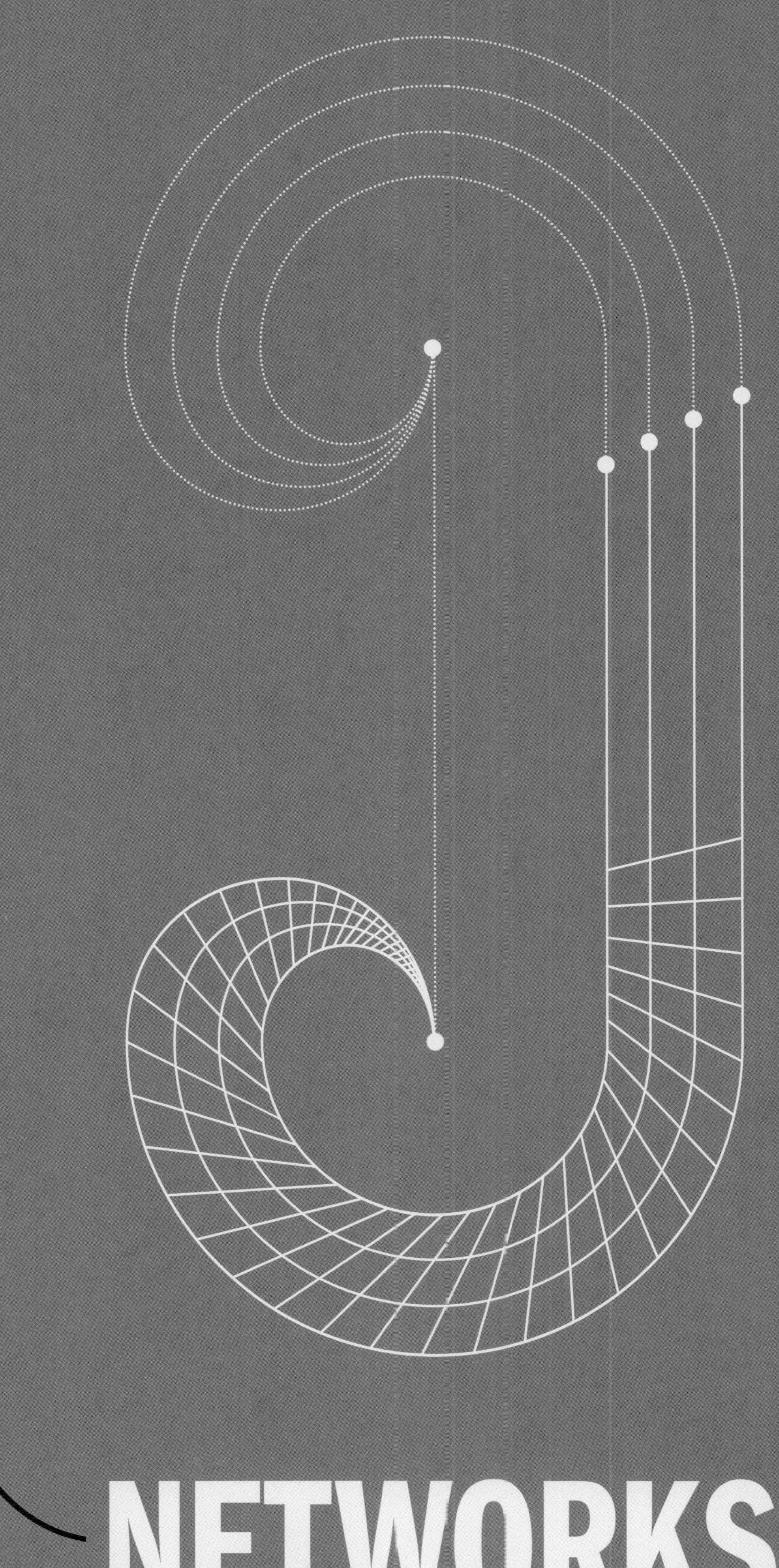

NETWORKS

IMAGINARY NETWORKS

In the opening pages of *The Book of Imaginary Media*, Eric Kluitenberg writes that "like communities, all media are partly real and partly imagined. Without either actual or imaginary characteristics, media cannot function." This book ends, then, with the invitation to explore the imaginary networks that have existed in, through, and around real networks, on a spectrum ranging from the bizarre to something just-this-side of the possible. What if networks involved the dead? Directly connected one brain to another? Used pigeons as part of a suite of internet protocols? Harnessed the quirks of quantum entanglement and produced faster-than-light communication [61]? This section uses the dystopic world of Neal Stephenson's fictional metaverse [63] (in which the wealthy appear in higher resolution in this world's network of virtual realities) as motivation to instead pave the way for alternative networks that are part and parcel of a collectively owned future no longer solely determined by ever-accelerating global accumulation of capital.

Necromancy [53]

COUNTRY OF ORIGIN Present-day Egypt and Iraq (ancient Egypt and Babylonia)
CREATOR(S) Unknown
YEAR CONCEIVED Age of Antiquity (3000–400 BCE)

BASIC INFRASTRUCTURE/MATERIALS May include hallucinogenic plants, specially prepared rooms or altars, stones, mirrors, crystals, child or adult mediums

RELATED Telepathy [56], cosmic internet [66]

DESCRIPTION Necromancy is a type of ritual magic involving the conjuring of and/or communication with demons, angels, fairies, and (more rarely in the long history of the practice) deceased humans and non-human animals, in order to divine the future, answer pressing questions, or bring the dead back to life. It is, therefore, a kind of one-to-one network between the living and the dead. Necromancy was first used in ancient Egypt, Babylonia, Greece, and Rome. With the arrival of Latin translations from the Iberian Peninsula, it also experienced a revival in twelfth and thirteenth century Europe. It continues today in many parts of the world, and involves diverse techniques including inscriptions (such as the drawing of circles on the ground), signs (such as the pentagram), spells, and actions involving a wide range of objects, from swords, jugs, and candles to stones, mirrors, crystals, and so-called child mediums.

SOURCES Frank Klassen, "Necromancy," *Routledge History of Medieval Magic*, Sophie Page and Catherine Rider, eds. (New York and London, 2019); Sebastià Giralt, "Medieval Necromancy: The Art of Controlling Demons," Sciencia.cat (October 4, 2017)

Pasilalinic-Sympathetic Compass [54]

COUNTRY OF ORIGIN France
CREATOR(S) Jacques-Toussaint Benoît
YEAR CONCEIVED 1850

BASIC INFRASTRUCTURE/MATERIALS Wooden beams, zinc bowls, cloth, copper-sulphate solution, glue, snails

RELATED Radiotelegraphy [16.1], telepathy [56]

DESCRIPTION The pasilalinic-sympathetic compass, also referred to as "snail telegraph," was created by French occultist Jacques-Toussaint Benoît, possibly with the assistance of someone named Monsieur Biat-Chrétien (an individual whose existence has not yet been proven), to demonstrate that snails are capable of instantaneously and wirelessly transmitting messages to each other across any distance. Benoît's theory was that in the course of mating, snails exchange so-called "sympathetic fluids," which create a lifelong telepathic bond and also enable them to communicate with each other. He believed he could induce snails to transmit messages faster and more reliably than wired telegraph by placing a snail on top of a letter and then prodding it with an electric charge, after which the snail would transmit the letter to another snail placed at some distance. The pasilalinic-sympathetic compass itself consisted of

[53] Portrait of “Edward Kelly a magician, raising the ghost of a person lately deceased, in the church yard of Walton-le-Dale, Lancaster” from roughly 1740.

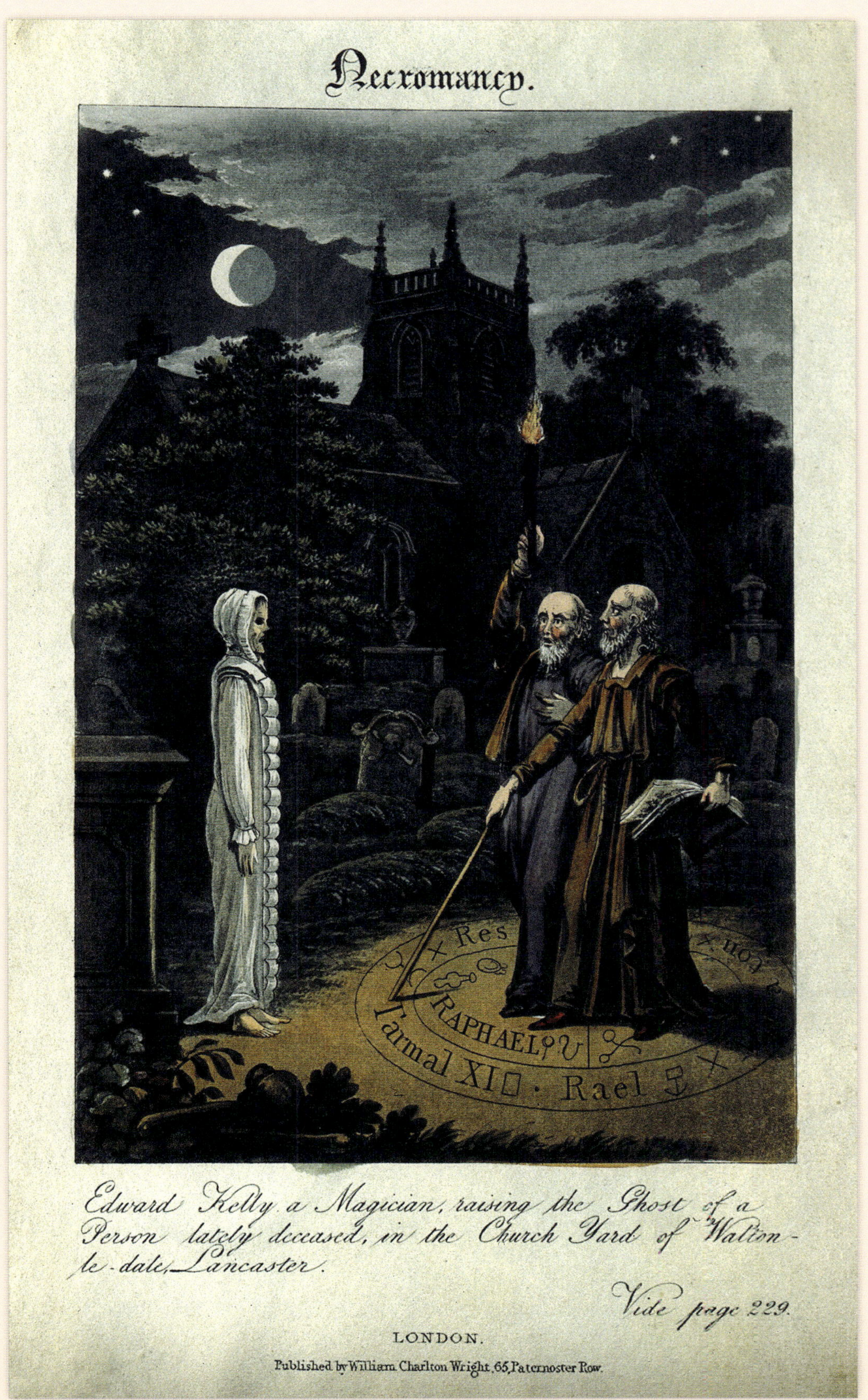

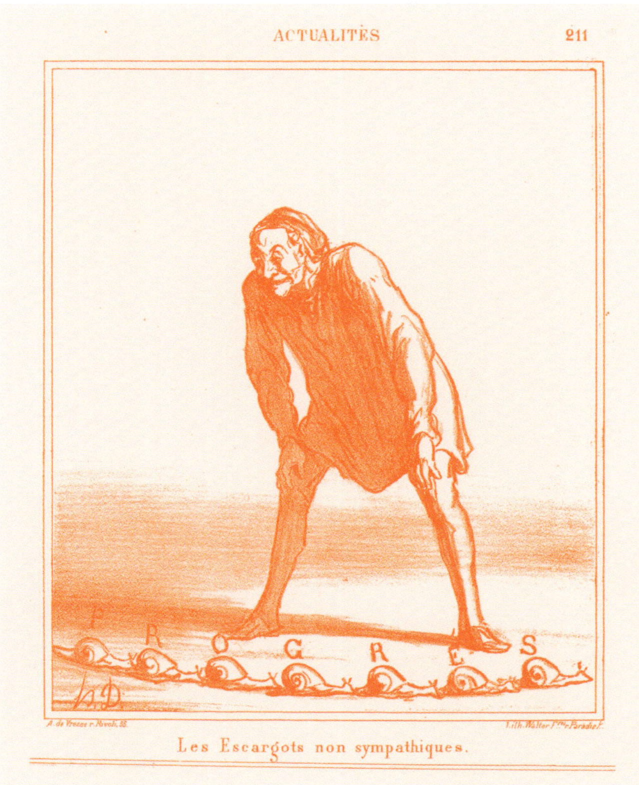

Satirical cartoon by Honoré Daumier (1869) captioned "Progrès. Les Escargots non sympathiques." ("Progress. Unsympathetic Snails.")

twenty-four different wooden structures containing a zinc bowl, cloth that had been soaked in copper sulphate, and a snail glued to the bottom of the bowl. Benoît unsuccessfully demonstrated the snail telegraph to a journalist from *La Presse*, Jules Allix, in October 1850.

SOURCES "Sympathetic Snail Compass," *Chambers Edinburgh Journal* 372 (February 15, 1851); Sabine Baring-Gold, *Historic Oddities and Strange Events* (Methuen,1889); Alex Butterworth, *The World That Never Was: A True Story of Dreamers, Schemers, Anarchists, and Secret Agents* (Vintage Books, 2011)

Telephonoscope [55]

COUNTRY OF ORIGIN UK
CREATOR(S) George du Maurier
YEAR CONCEIVED 1878

BASIC INFRASTRUCTURE/MATERIALS Two telephones, two camera obscuras

RELATED Telephone [34], telautograph [36], videophone [38], cable television [48]

DESCRIPTION On December 9, 1878, the British weekly magazine *Punch* published George Du Maurier's cartoon titled "Edison's Telephonoscope." The communication device (which was imagined as conveying two-way sound and one-way vision) was described as an "electric camera obscura." But, given that the cartoon depicted an older couple in London communicating with their daughter in the Crown colony of Ceylon (now Sri Lanka)—a daughter who is, in Doron Galili's words, "standing next to a badminton court surrounded by other colonialists and a native woman"—it's clear that the telephonoscope also represents the nineteenth-century desire for a technology that could globally transmit colonialist and bourgeois ideals.

SOURCES Doron Galili, *Seeing by Electricity* (Duke University Press, 2020); Ivy Roberts, "'Edison's Telephonoscope': The Visual Telephone and the Satire of Electric Light Mania," *Early Popular Visual Culture* 15:1 (2017)

Telepathy [56]

COUNTRY OF ORIGIN UK
CREATOR(S) W. F. Barrett, C. C. Massey, Rev. W. Stainton Moses, Frank Podmore, Edmund Gurney, Fredric W. H. Myers
YEAR CONCEIVED 1882

BASIC INFRASTRUCTURE/MATERIALS Two or more conscious and sentient beings

RELATED Necromancy [53], pasilalinic-sympathetic compass [54], cosmic internet [66]

DESCRIPTION Telepathy (also known as mind reading, thought reading, and brain-to-brain communication) is the purported human and/or nonhuman animal ability to transmit information directly to another without either speech or gesture and without the use of any physical media other than body/mind. The term "telepathy" was coined by W.F. Barrett, C.C. Massey, Rev. W. Stainton Moses, Frank Podmore, Edmund Gurney, and Fredric W. H. Myers, who wrote in 1882, "we venture to introduce the words *Telaesthesia* and *Telepathy* to cover all cases of impression received at a distance." In the late nineteenth century, various mentalists and magicians claimed telepathic abilities could be demonstrated as a result of finely honed skills for reading individuals' ideomotor and muscular movements; otherwise, nearly all attempts to prove the existence of telepathy in the twentieth century failed. In 2014, however, a team of researchers successfully demonstrated "Conscious Brain-to-Brain

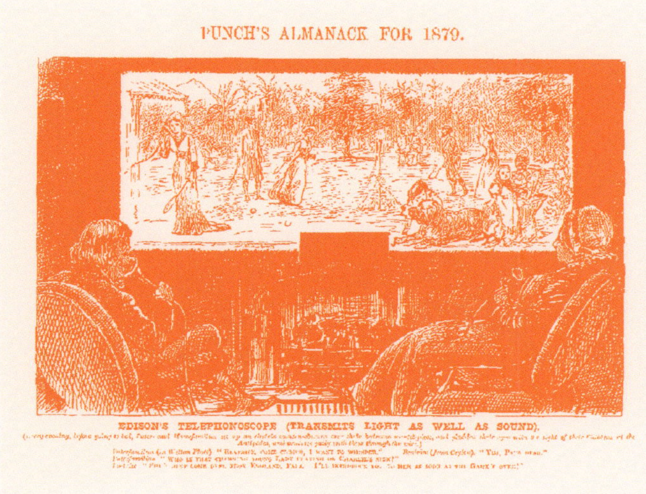

"Edison's Telephonoscope (Transmits Light as Well as Sound)" by George Du Maurier, published in *Punch's Almanack* for 1879.

Communication in Humans Using Non-Invasive Technologies" (described below); some interpret this as proof of the existence of the possibility of telepathic communication.

EXPERIMENTS In 1992, Brazilian artist Mario Ramiro and Japanese artist Morio Labonete Nishimura created *Entre o Norte e o Sul* (*Between North and South*), a telepathic telecommunications experience connecting Greece and Finland using the surrounding scenery. The work consisted of an installation in each country—a wood structure in the forest next to Lake Pitäjärvi (Finland) and a stone structure on a rock formation on the island of Amorgós (Greece). As Ramiro describes it, the antennas "were made of glass tubes containing water from the Rhine River and strands of the artists' hair connected by gold leaves. Nishimura's 'antenna' was placed at the center of his piece, inside a carved lotus flower, the image he chose for his transmission to Greece. Ramiro's 'antenna' was positioned atop a large stone, upon which he drew the image of a burning sword. The artists sketched the images they 'received' and later compared the drawings, some of which resembled the intended transmission."

While they didn't set out to prove telepathy as such, an international team of researchers demonstrated successful brain-to-brain communication in 2014. Corinne Iozzio describes the process as follows: "First, the team had to establish binary-code equivalents of letters . . . Then, with EEG (electroencephalography) sensors attached to the scalp, the sender moved either his hands or feet to indicate a 1 or a 0. The code then passed to the recipient over email. On the other end, the receiver was blindfolded with a transcranial magnetic stimulation (TMS) system on his head . . . The TMS headset stimulated the recipient's brain, causing him to see quick flashes of light. A flash was equivalent to a '1' and a blank was a '0.' From there, the code was translated back into text. It took about 70 minutes to relay the message."

SOURCES W.F. Barrett, et al., "Second Report on Thought-Transference," *Proceedings of the Society for Psychical Research* (December 1882); Janet Oppenheim, *The Other World: Spiritualism and Psychical Research in English, 1850–1914* (Cambridge University Press, 1985); Roger Luckhurst, *The Invention of Telepathy, 1870–1901* (Oxford University Press, 2002); Carles Grau et al., "Conscious Brain-to-Brain Communication in Humans Using Non-Invasive Technologies," *PLoS One* 9:8 (August 19, 2014); Corinne Iozzio, "Scientists Prove That Telepathic Communication Is Within Reach," *Smithsonian* (October 2, 2014); Mario Ramiro, "Between Form and Force: Connecting Architectonic, Telematic and Thermal Spaces," *Leonardo* 31:4 (1998)

Ley Lines [57]

COUNTRY OF ORIGIN England
CREATOR(S) Alfred Watkins
YEAR CONCEIVED 1921

BASIC INFRASTRUCTURE/MATERIALS May include any combination of prehistoric sites, ancient churches and crosses, moats and fords

RELATED Cosmic internet [66]

DESCRIPTION In 1921, businessperson and amateur archaeologist Alfred Watkins hypothesized that straight lines embedded in the British landscape formed what we would today call "a network" connecting prehistoric sites, ancient churches and crosses, moats, and fords. Watkins believed that these lines were likely created in the late Neolithic period by the alignment of hilltop sighting points, and that objects on the lines were thus actually vantage and/or signaling points. He dubbed these lines "leys" in reference to the ancient Saxon word for "cleared land" and stipulated that the existence of a line is proven by the alignment of any four objects listed above "with a hill peak at one end, and with bits of old tracks and antiquarian objects on the line." While archaeologists have consistently denied the existence of ley lines, enthusiasts since the 1970s have claimed the lines form a network of earth energies.

[56]

Illustration of brain-to-brain communication in a 2014 study by Grau et al.

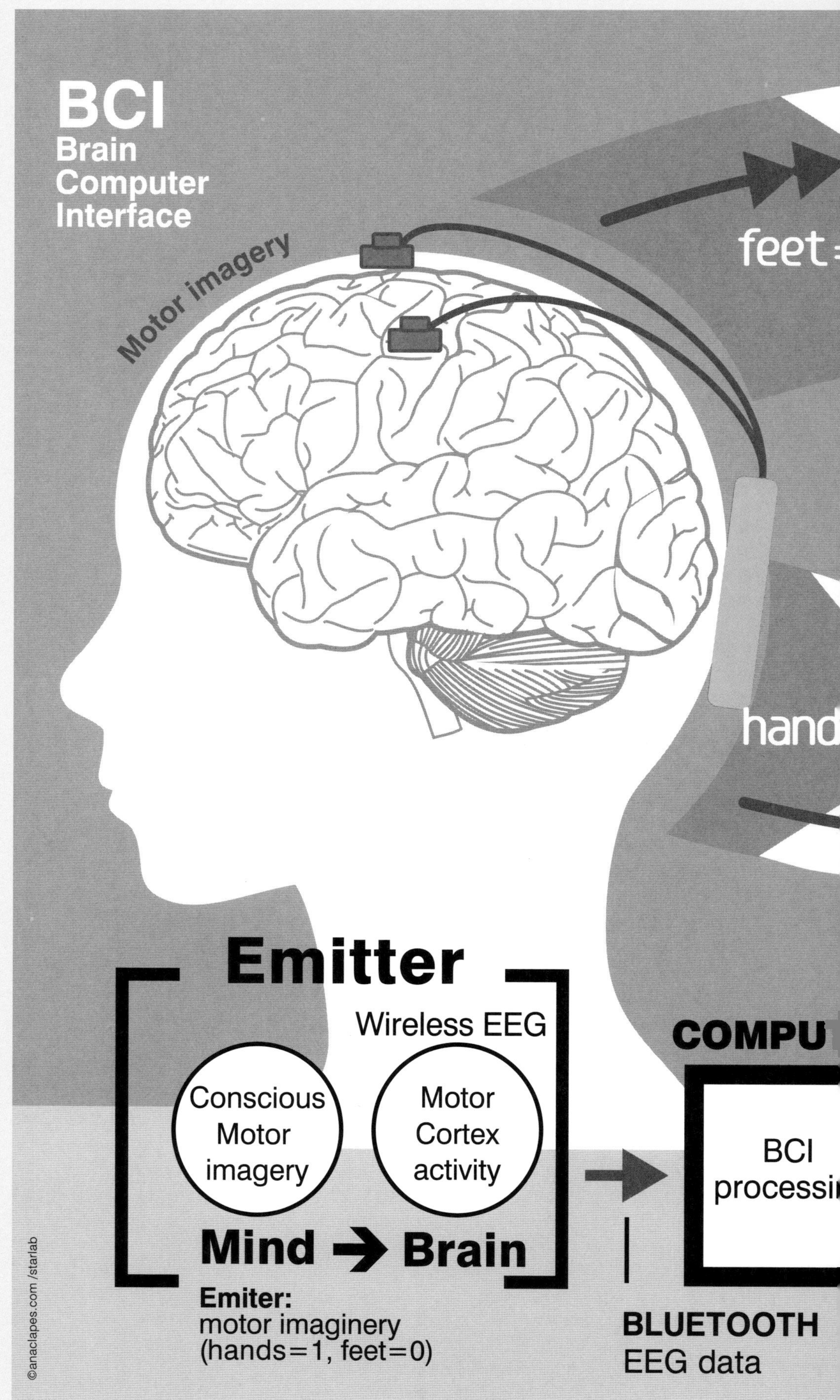

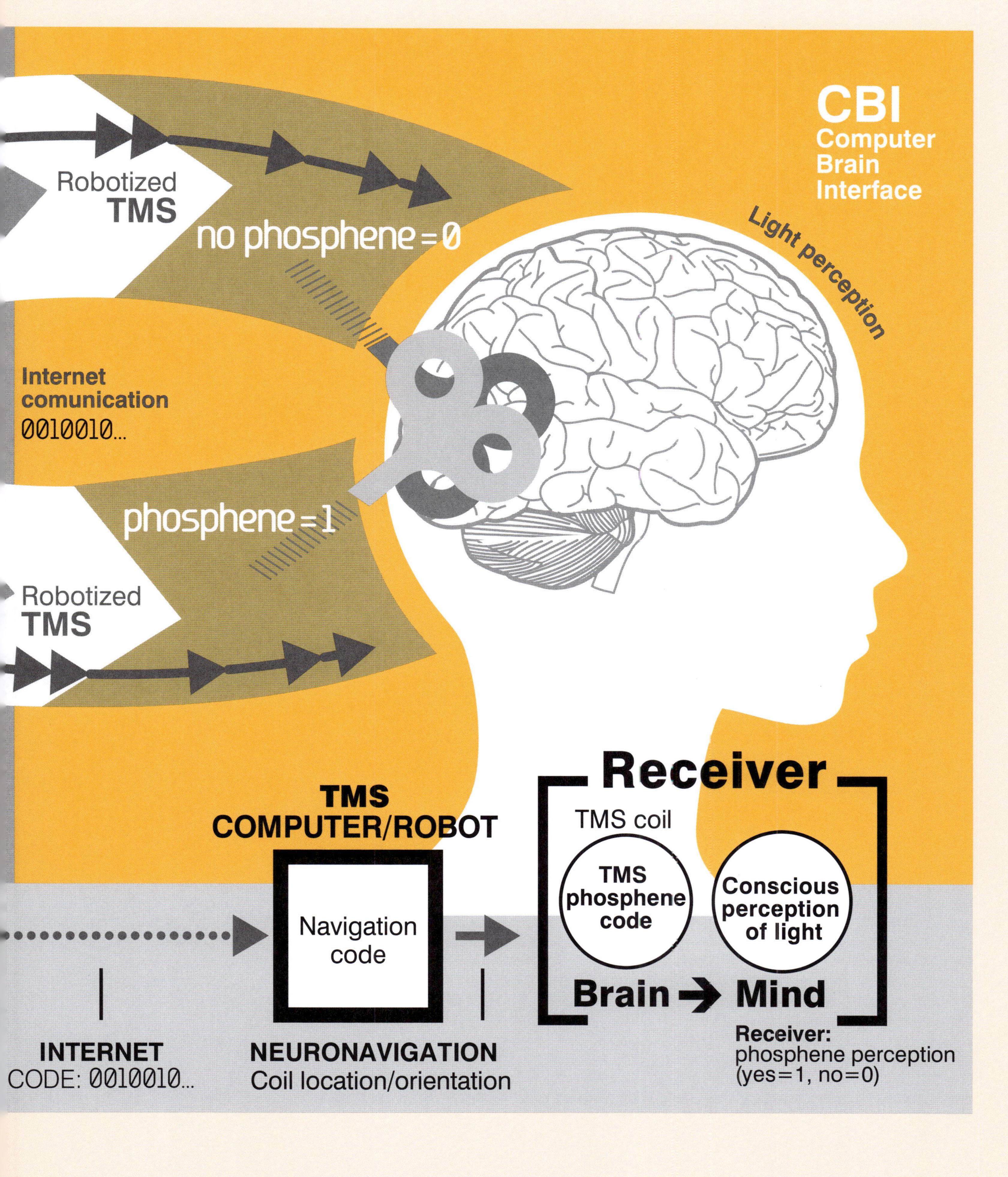

CBI
Computer
Brain
Interface
Light perception
Robotized
TMS
no phosphene = 0
Internet
comunication
0010010...
phosphene = 1
Robotized
TMS
TMS
COMPUTER/ROBOT
Navigation
code
Receiver
TMS coil
TMS
phosphene
code
Conscious
perception
of light
Brain → Mind
INTERNET
CODE: 0010010...
NEURONAVIGATION
Coil location/orientation
Receiver:
phosphene perception
(yes=1, no=0)

[57] A map of eight ley lines in Wales, from Alfred Watkins's *Early British Trackways* (1922).

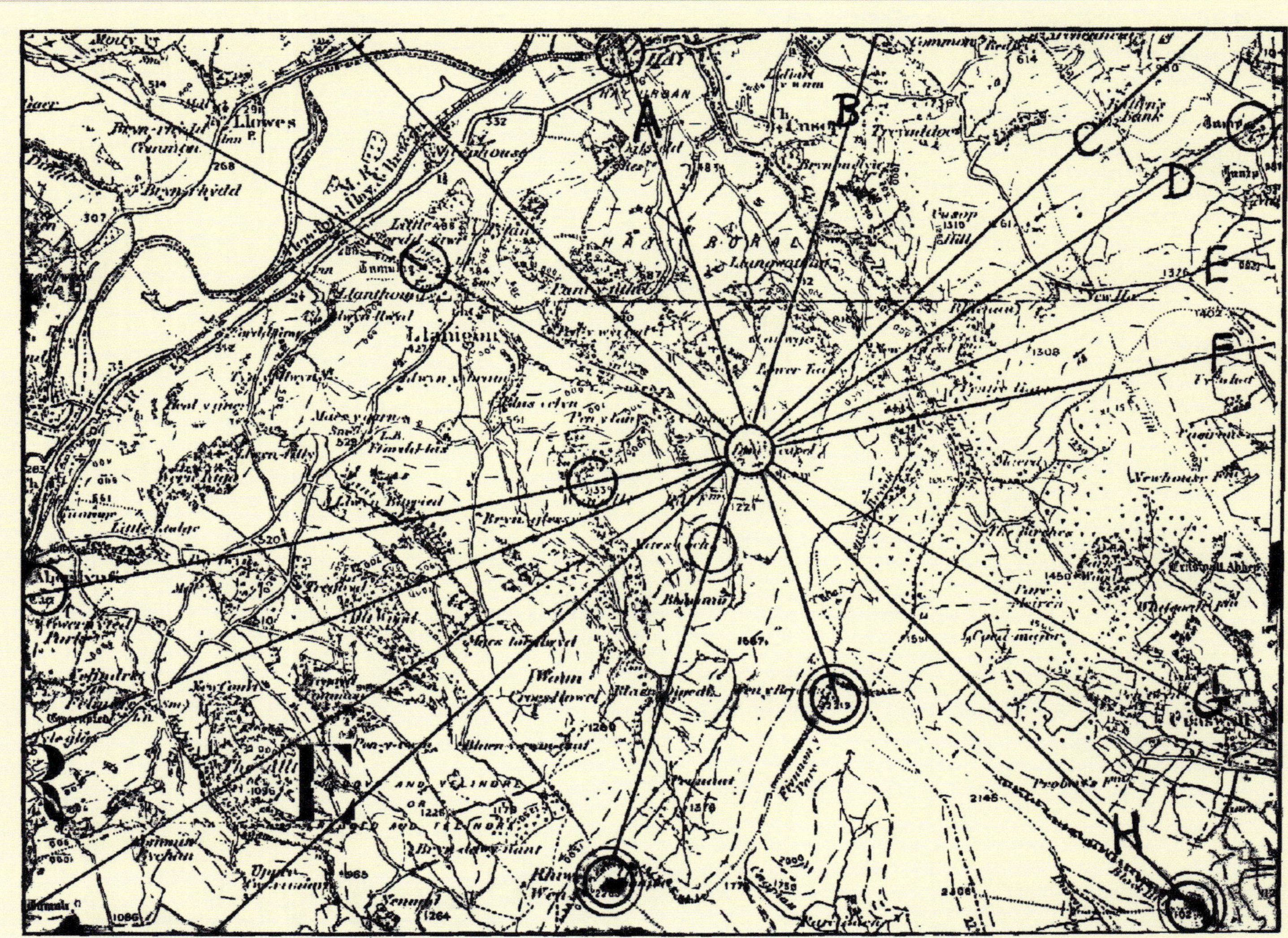

PLATE XX. MAP OF EIGHT LEYS THROUGH CAPEL-Y-TAIR-YWEN, HAY. SEE CONTENTS TABLE
(Based upon the Ordnance Survey with the sanction of the Controller of H.M. Stationery Office).

SOURCES Alfred Watkins, *Early British Trackways* (Watkins Meter Co., 1922); "Ley Line," in *Encyclopedic Dictionary of Archaeology* (Springer, 2021); Patricia D. Netzley, "Ley Lines," in *The Greenhaven Encyclopedia of Paranormal Phenomena* (Greenhaven Press, 2006); Michael Charlesworth, "Photography, the Index, and the Nonexistent: Alfred Watkins' Discovery (or Invention) of the Notorious Ley-lines of British Archaeology," *Visual Resources* 26:2 (2010)

The Mundaneum [58]

COUNTRY OF ORIGIN Belgium
CREATOR(S) Paul Otlet
YEAR CONCEIVED 1934

BASIC INFRASTRUCTURE/MATERIALS Index cards and/or documents; filing cabinets; any combination of media including telephone, radio, microfilm, phonograph/record, television; multimedia workstation

RELATED Radio broadcast [17], book [43], library [42], telephone [34], cable television [48], world brain [59], memex [60], Project Xanadu [62]

DESCRIPTION The Mundaneum was initially called Le Palais Mondiale and was co-conceived with Paul Otlet and Henri La Fontaine in 1910. Le Palais Mondiale took the form of numerous rooms in a government building in Brussels filled with index cards and vertical filing cabinets that provided an "encyclopedic survey of human knowledge, as an enormous intellectual warehouse of books, documents, catalogues and scientific objects." By 1934, after the project had been renamed the Mundaneum, Otlet moved on to envisioning a "réseau mondial" or worldwide network that would allow people to access information in numerous Mundaneums as well as millions of other linked documents, images, audio, and film via workstations. As Otlet describes it, the workstation "no longer carries any books. Instead there is a screen connected to a telephone." Over there, in a great building, are all the books and related material, with all the space necessary for cataloging and registering them . . . From there one could call up a page on screen to read the answer to questions posed by telephone. The screen could be double, quadruple or [decuple] if there are multiple texts to show simultaneously; there would be an audio speaker if needed for additional material to complement the text." Each station would also allow individuals to upload files and communicate with others wirelessly.

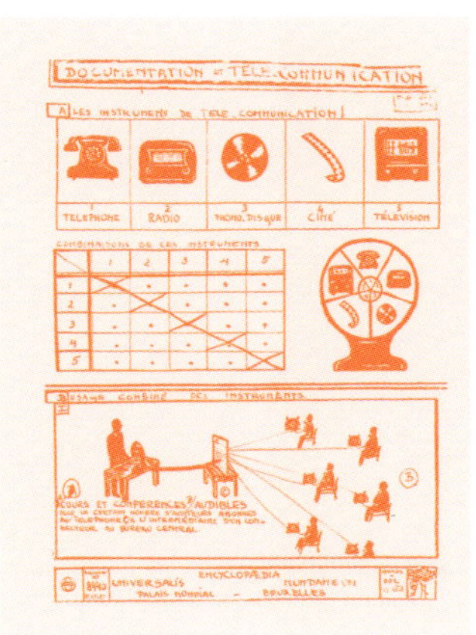

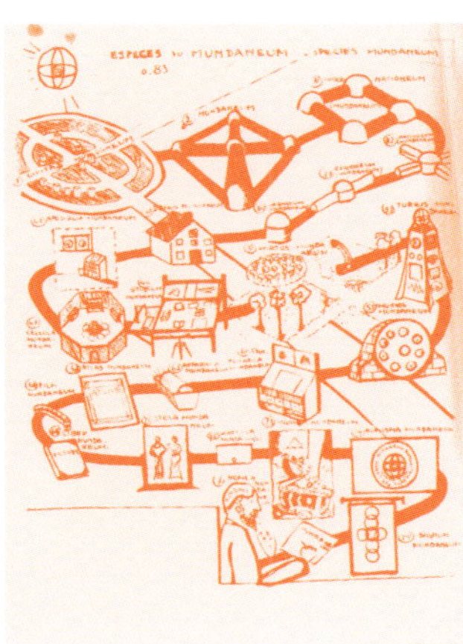

Otlet's drawings illustrating (left) the five media by which information would be drawn from and then shared with users via telephone, and (right) his vision for a global network of linked Mundaneums.

SOURCES Union of International Associations, "The Union of International Associations: A World Center," *International Organisation and Dissemination of Knowledge: Selected Essays of Paul Otlet*, trans. and ed. by W. B. Rayward (Elsevier, 1990); Alex Wright, *Cataloging the World: Paul Otlet and the Birth of the Information Age* (Oxford University Press, 2014); W. Boyd Rayward, "From the Index Card to the World City: Knowledge Organization and Visualization in the Work and Ideas of Paul Otlet," *Classification & Visualization-Interfaces to Knowledge: Proceedings of the International UDC Seminar* (October 24–25, 2013)

World Brain [59]

COUNTRY OF ORIGIN UK
CREATOR(S) H.G. Wells
YEAR CONCEIVED 1938

BASIC INFRASTRUCTURE/MATERIALS Any combination of printed material, radio, photography, microfilm

RELATED Book [43], library [42], mundaneum [58], memex [60], Project Xanadu [62]

DESCRIPTION In his 1938 collection of essays and lectures *World Brain*, H.G. Wells describes his vision of a "new social organ, a new institution" that he calls alternately a World Brain or World Encyclopaedia that would bring together "all the scattered and ineffective mental wealth of our world into something like a common understanding, and into effective reaction upon our vulgar everyday political, social and economic life." The new organization he proposed would, he believed, prevent "transatlantic misunderstandings" by way of this "common interpretation of

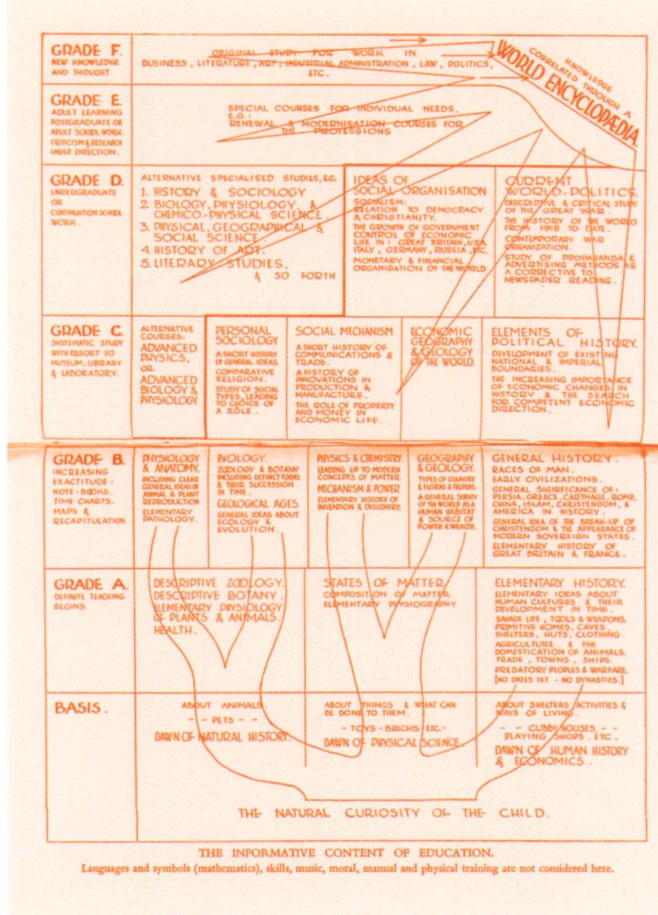

An illustration of the pedagogical value of a "World Encyclopedia," from H.G. Wells's book *World Brain* (1938).

reality" and it would include materials from museums, art galleries, libraries, atlases, and surveys. The process of bringing together these materials and documenting them would involve radio, photography, and especially microfilm so that "any student, in any part of the world, will be able to sit with his projector in his own study at his or her convenience to examine any book, any document, in an exact replica." The content of the World Brain itself would consist of expert-approved selections, extracts, and quotations that would, in W. Boyd Rayward's words, be "carefully collated and edited and critically presented. It would be not a miscellany, but a concentration, a clarification and a synthesis." It is often touted as an influence on the late-twentieth century World Wide Web and Wikipedia, even though both platforms have been shown to have highly unreliable methods for creating, editing, and moderating content.

SOURCES H.G. Wells, *World Brain* (Doubleday, 1938); W. Boyd Rayward, "H.G. Wells' Idea of a World Brain," *Journal of the American Society for Information Science* (January 1999)

Memex [60]

COUNTRY OF ORIGIN USA
CREATOR(S) Vannevar Bush
YEAR CONCEIVED 1945

INFRASTRUCTURE/MATERIALS Desk, screens, keyboard, buttons and levers, printed matter, microphotography

RELATED Book [43], library [42], mundaneum [58], world brain [59], Project Xanadu [62]

DESCRIPTION In July 1945, reflecting on what lay in store for scientists in the aftermath of World War II, Vannevar Bush described his vision of the Memex: a device he had been designing since the 1930s that could solve the ever-increasing production and specialization of scientific literature by allowing individuals to make, store, and consult any piece of information in almost any medium of the time. He also made clear that the problem at the time was not necessarily technological but rather organizational, and that the arrangement of records alphabetically or numerically does not allow the scientist to easily find the information they are seeking. Thus, the Memex mechanizes the organization of information by association—which is, for Bush, the way "the mind [already] works," "in accordance with some intricate web of trails carried by the cells of the brain." In terms of its physical design, the Memex would be a desk with screens, a keyboard, buttons, and levers for creating and accessing (at variable speeds) what he called "trails" through the user's collection of books,

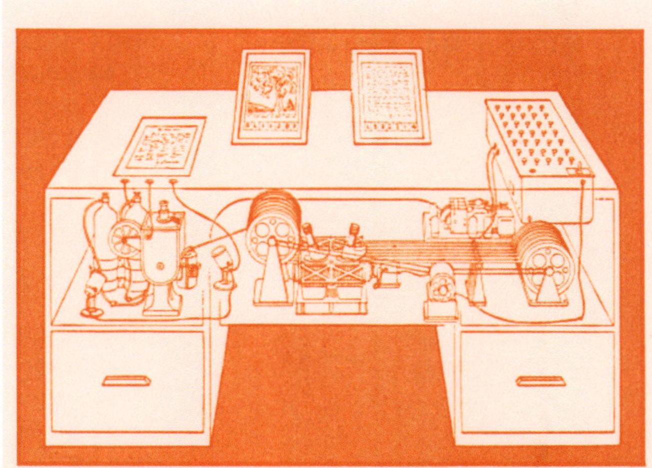

MEMEX in the form of a desk would instantly bring files and material on any subject to the operator's fingertips. Slanting translucent viewing screens magnify supermicrofilm filed by code numbers. At left is a mechanism which automatically photographs longhand notes, pictures and letters, then files them in the desk for future reference.

Illustration by Alfred D. Crimi in *Life* magazine (September 1945).

newspapers, periodicals, photographs, etc., which in turn would be stored on microfilm. While the Memex was never built, these references to associational thinking represented in terms of webs and trails certainly influenced Theodor Nelson's notion of hypertext as well as the creation of the World Wide Web.

SOURCES Vannevar Bush, "As We May Think," *The Atlantic* (July 1945); Belinda Barnet, *Memory Machines: The Evolution of Hypertext* (Anthem Press, 2013)

Faster-Than-Light Communication Networks [61]

COUNTRY OF ORIGIN USA
CREATOR(S) Albert Einstein
YEAR CONCEIVED 1947

BASIC INFRASTRUCTURE/MATERIALS Unknown

RELATED Infrared Communication [9], ultraviolet communication [13], laser communication [14], visible light communication [15], necromancy [53], telepathy [56], pandoran neural network [65], cosmic internet [66]

DESCRIPTION Sometimes referred to as "superliminal communication," faster-than-light (FTL) communication refers to a well-trodden (even amongst physicists) but still purely hypothetical situation in which information travels faster than the speed of light. In a letter to Max Born in 1947, Albert Einstein referred to the possibility of what he called "spooky action at a distance" that could result from quantum entanglement; the latter occurs when two particles are able to share information with each other even if they are physically separate. Although Einstein dismissed the idea in the same letter to Born "because the theory cannot be reconciled with the idea that physics should represent a reality in time and space," in recent decades physicists have in fact demonstrated that "spooky action at a distance" does actually occur. The unanswered question, however, is whether *information* can also travel faster than 186,282 miles per second (the speed of light in a vacuum).

EXPERIMENTS In Ursula LeGuin's novel *Rocannon's World* (1966) she coined the term "ansible" (a contracted form of "answerable") to refer to a particular kind of FTL communication device that can send and receive messages instantaneously, even when sending/receiving across star systems. She continued to develop the device in subsequent novels such as *The Left Hand of Darkness* (1969), in which she wrote that the ansible "doesn't involve radio waves, or any form of energy. The principle it works on, the constant of simultaneity, is analogous in some ways to gravity . . . One point has to be fixed, on a planet of certain mass, but the other end is portable." Other science fiction writers such as Isaac Asimov and Vernor Vinge have adopted the terms "ultrawave" and "hyperwave" to refer to other forms of FTL communication.

SOURCES Albert Einstein, *The Born-Einstein Letters 1916–1955: Friendship, Politics and Physics in Uncertain Times* (Macmillan Press, 2014); Jake Parks, "Quantum Entanglement Loophole Quashed by Quasar Light," *Astronomy Magazine* (August 23, 2018); Gabriel Popkin, "China's Quantum Satellite Achieves 'Spooky Action' at Record Distance," *Science* (June 15, 2017); Ursula LeGuin, *Rocannon's World* (Ace Books, 1966); Ursula LeGuin, *The Left Hand of Darkness* (Ace Books, 1969)

Project Xanadu [62]

COUNTRY OF ORIGIN USA
CREATOR(S) Theodor H. Nelson
YEAR CONCEIVED 1960

BASIC INFRASTRUCTURE/MATERIALS Personal computer, screen, keyboard, Xanadu software, wired or wireless internet connection

RELATED Book [43], library [42], mundaneum [58], world brain [59], memex [60]

DESCRIPTION Throughout its lifetime, Project Xanadu has been variously a proposal, software, imagined online network for linked documents, and Web-based demo. It is, then, in some sense only partly an imaginary network. Though implementations in Algol and Fortran existed in the early 1970s, with beta versions released in the 1980s, 1990s, and 2000s, a final version has not (yet) been released. At the heart of Project Xanadu is hypertext—an idea that Theodor Nelson came up with in 1960, coined in 1963, and wrote about in 1965. Nelson then chose "Xanadu" (named after the beautiful imagined landscape in Samuel Taylor Coleridge's poem "Kubla Khan") in 1967 to describe the project as a whole. As Belinda Barnet puts it, it should have been "like the Web, but much better: no links would ever be broken, no documents would ever be lost, and copyright and ownership would be scrupulously preserved." Certain incarnations of Xanadu

[62] Two illustrations from Theodor Nelson's *Literary Machines*, depicting Nelson's vision of linked personal computers and servers running Xanadu software, connecting even users in space.

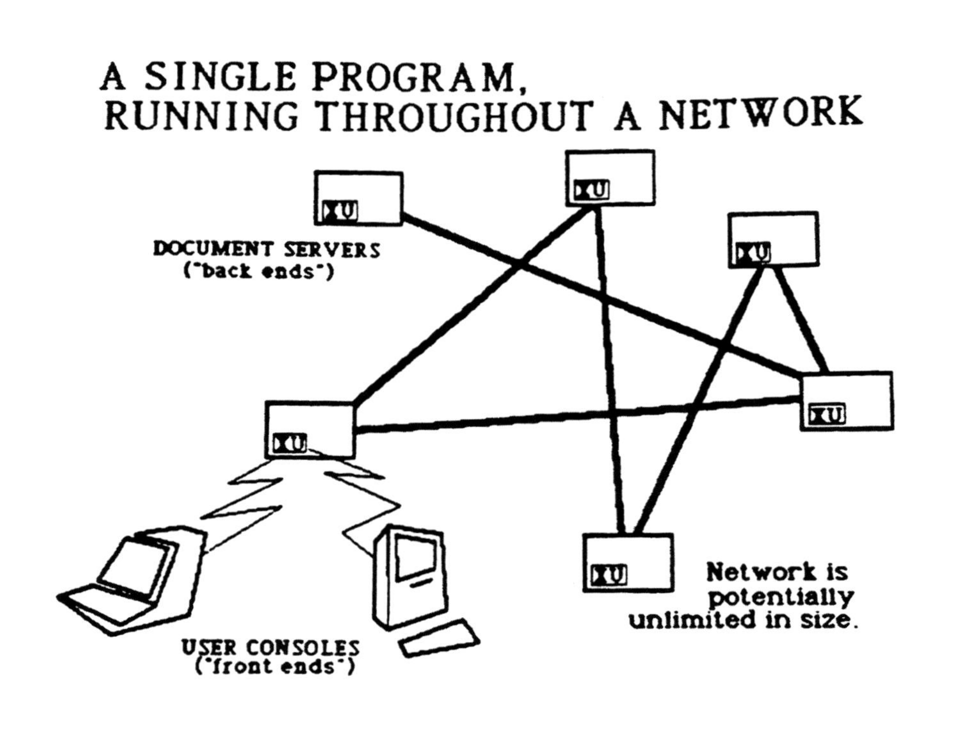

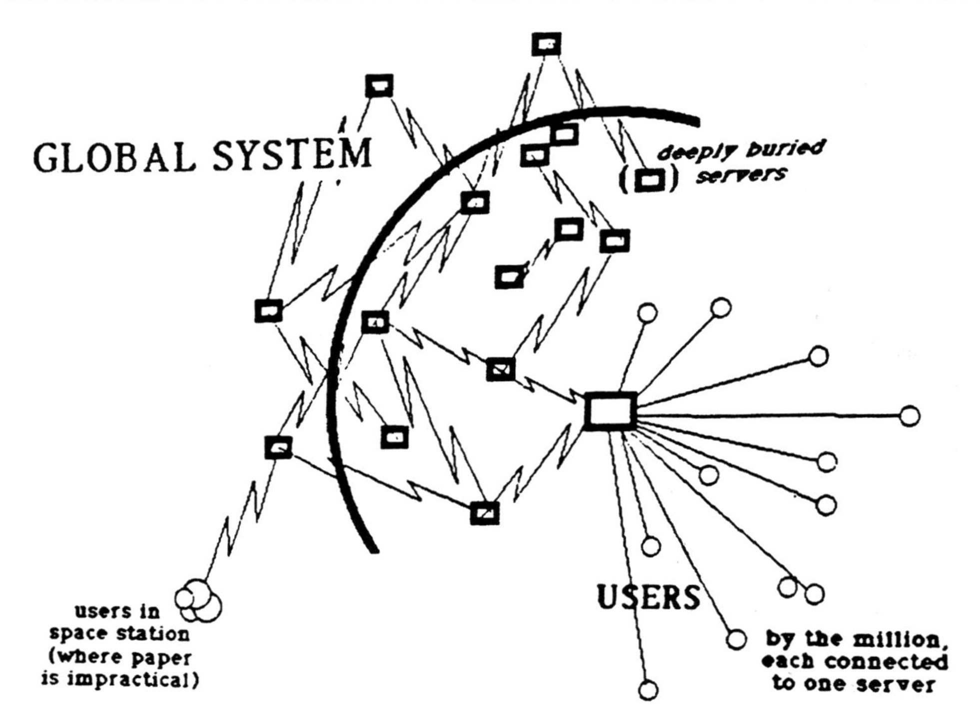

involved Xanadocs (the basic document unit within the system), Xanalinks (the linking mechanism that also provided for self-repairing links), and Xanadu servers running both locally and globally.

SOURCES Project Xanadu website; T.H. Nelson, "Complex Information Processing: A File Structure for the Complex, the Changing and the Indeterminate," *ACM '65: Proceedings of the 1965 20th National Conference* (August 1965); Theodor H. Nelson, *Computer Lib / Dream Machines* (1974); Theodor H. Nelson, *Literary Machines* (1980), Belinda Barnet, *Memory Machines: The Evolution of Hypertext* (Anthem Press, 2013)

Metaverse [63]

COUNTRY OF ORIGIN USA
CREATOR(S) Neal Stephenson
YEAR CONCEIVED 1992

BASIC INFRASTRUCTURE/MATERIALS Internet, goggles, earphones

RELATED N/A

DESCRIPTION Neal Stephenson coined the term "metaverse" in his 1992 cyberpunk novel *Snow Crash*. Set in the twenty-first century, the metaverse Stephenson envisions in his novel is—like the contemporary metaverse that's currently being driven in part by the rebranding of Facebook as "Meta"—an internet of connected virtual spaces that users enter; but unlike the rhetoric surrounding the real world metaverse that's slowly coming into view, Stephenson explicitly positions his fictional Metaverse as a dystopic escape from an even more dystopic bleak world of global economic collapse and corporate monopolies (ironically, not unlike the actual twenty-first-century world). In the novel, users access the Metaverse with goggles, earphones, and virtual avatars of themselves and, in extreme circumstances, they also experience real-world consequences (including addictions and brain damage) for their actions in the virtual world.

SOURCES Neal Stephenson, *Snow Crash* (Bantam Books, 1992)

The Clacks [64]

COUNTRY OF ORIGIN UK
CREATOR(S) Terry Pratchett
YEAR CONCEIVED 1999

BASIC INFRASTRUCTURE/MATERIALS Wood panels, pulleys, shutters, lamps

RELATED Hydraulic semaphore [6], optical telegraph [8], signal lamp [10], pony express [44.4], email letter [44.6]

DESCRIPTION In Terry Pratchett's fantasy novel *The Fifth Elephant*, the twenty-fourth book in the Discworld series, he introduces a network of semaphore communication towers called Clacks that he continued to develop throughout the series. Standing at roughly three stories tall and made of wood panels, pulleys, shutters, and lamps for nighttime transmission, the towers in the network closely resemble telegraph towers. No doubt as an echo of the Pony Express [44.4], the network was managed by the Grand Trunk Company. As the network expands across the series, it starts to incorporate technologies and terminologies from other time periods; for example, operators develop a system of punch cards to automate message transmission; in *Going Postal*, Pratchett also refers to messages sent over the network as "c-mail."

SOURCES Terry Pratchett, *The Fifth Elephant* (Doubleday, 1999); Terry Pratchett, *Going Postal* (Doubleday, 2004)

Pandoran Neural Network [65]

COUNTRY OF ORIGIN USA
CREATOR(S) James Cameron
YEAR CONCEIVED 2009

BASIC INFRASTRUCTURE/MATERIALS Living organisms

RELATED Necromancy [53], telepathy [56], faster-than-light communication network [61], cosmic internet [66]

DESCRIPTION In James Cameron's 2009 movie *Avatar*, Pandora is an exoplanetary moon on which all the flora and fauna, including the humanoid inhabitants the Na'vi, communicate through a neural network they access at certain hubs using an extension of their nervous system called a "queue." The network itself is a collection of electro-chemical connections between Pandoran trees. While it is partly used by the Na'vi to domesticate certain species, it also provides access to their own memories as well as those of their ancestors.

SOURCES James Cameron, *Avatar* (20th Century Studios, 2009)

Cosmic Internet [66]

COUNTRY OF ORIGIN Norway
CREATOR(S) Jenny Hval
YEAR CONCEIVED 2020

BASIC INFRASTRUCTURE/MATERIALS Living and/or dead organisms

RELATED Necromancy [53], pasilalinic-sympathetic compass [54], telepathy [56], ley lines [57], faster-than-light communication networks [61], pandoran neural network [65]

DESCRIPTION As it drifts across time periods and musings about capitalism, patriarchy, communes, witches, and the power of anger, hatred, the body, and filth, Jenny Hval's novel *Girls Against God* also dips into imagining another internet. Early in the book she describes it as a form of "intimacy through the body's waste and secretions. A self-constructed network between bodies." By the last third of the novel, the network has become "the cosmic internet"—an invisible, intangible network that exists entirely outside of capitalism and which, in fact, functions best when all connections to the actual internet are severed, a network that seems to exist in the ether, linking bodies, plants, life-forms, and even the dead. She writes, "the cosmic internet communicates through noise . . . It creates confusion, poor connections, pixelated images and digital one-way streets." And further, the two characters in the novel "agree that in the long run, when it trusts us, the web will evolve into a fleshy peer-to-peer network, where a small part of your flesh is always seeding."

SOURCES Jenny Hval (trans. Marjam Idriss), *Girls Against God* (Verso Books, 2020)

[66] Jenny Hval's typographic exploration of the cosmic internet in her novel *Girls Against God*.

(like the internet before the internet)
((like the internet during the internet))
(((or underground internet)))
((((deep web))))
(((((deep tissue)))))
((((((deep web and deep tissue))))))
(((((((ecstatic deep-webian intimacy)))))))
((((((((primitive language))))))))
(((((((((I want to understand this language)))))))))
((((((((((insist that this language signifies))))))))))
(((((((((((until it signifies)))))))))))
((((((((((((for you too))))))))))))

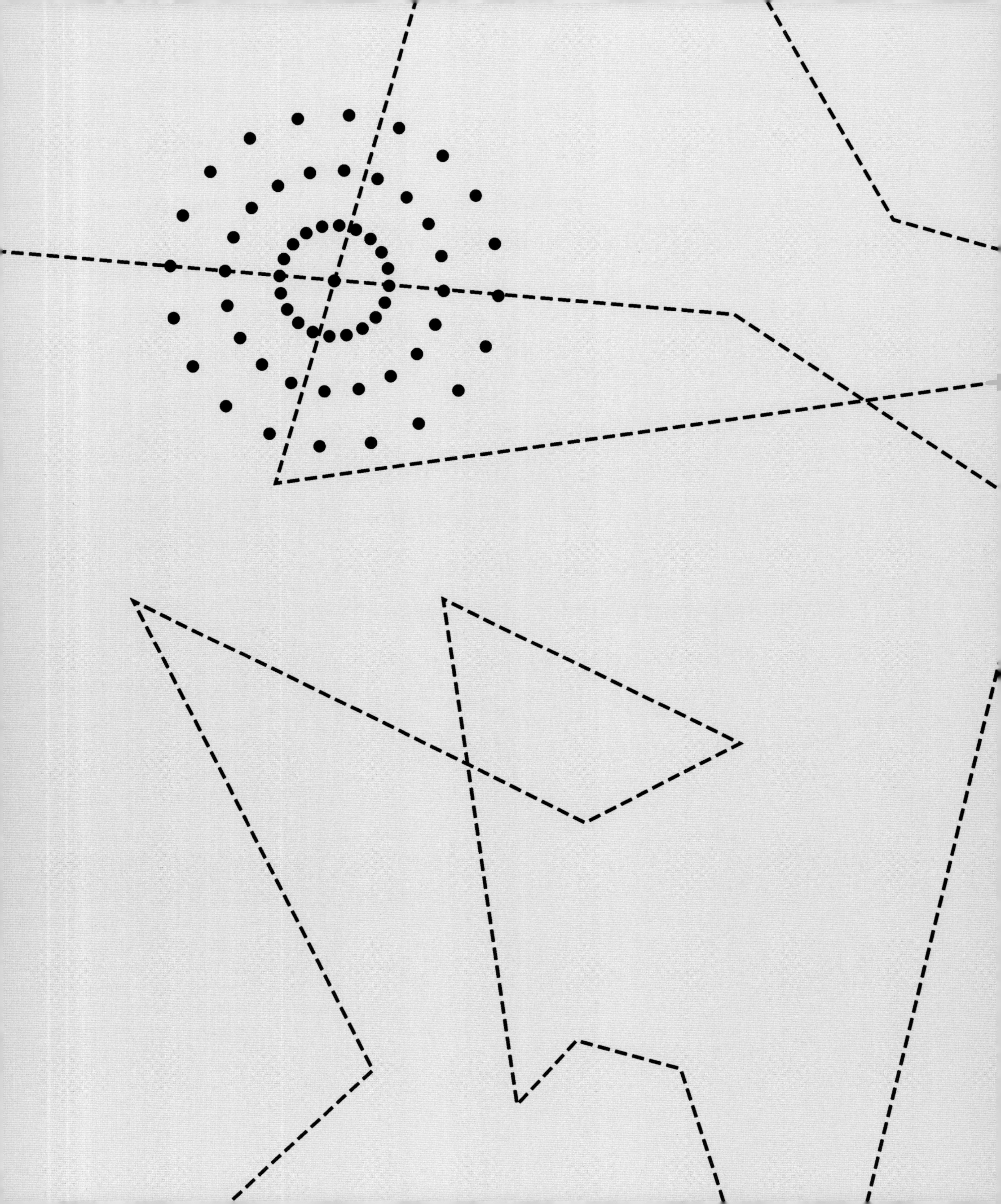

AFTERWORD
Howard Rheingold

When I wrote my book *Net Smart*, about the five essential social media literacies, I included "network literacy" (the other literacies are "attention," "crap detection," "participation," and "collaboration")—and I included network literacy when I taught this material at Berkeley and Stanford. Why do I consider awareness of networks that important? Because, as Manuel Castells asserted, we don't really live in the information age. We live in the age of networks—and the internet is only (an important) part of the network environment.

Life on earth is a network we call ecology—and organisms emerged when cells began to link up in networks. Our bodies are grown and maintained by networks of cells, processes, electrochemical messages. Our brains function through networks of neurons. While our communities consist of people who mostly know one another, each one of us has a social network unique to us but linked to others; our family and closest friends are part of our social network, but so is the person we buy a cup of coffee from every morning, our regular grocery store clerk, our kid's teacher.

Most important, the configuration of the networks we live in can determine how we behave: in what are known as small-world networks, each person on earth is connected to every other person through no more than six links; in an organization's network of networks, people who can bridge between otherwise sparsely connected networks can be arbiters of innovation ("structural holes"). Indeed, a number of "laws of networks" have arisen in regard to communication networks. While one fax machine was useless, the value of the network of fax machines increased proportionally to the square of the number of nodes—Metcalfe's law. If the nodes are people rather than fax machines, and the people are able to form groups (think of eBay for buying and selling), Reed's law states that the value of a network increases exponentially with the number of potential subgroups that users can establish within the network, in contrast to Metcalfe's law. Networks and the ways they are configured have enormous potential impact on the way humans communicate, cooperate, and collaborate—in ways both beneficial and harmful. Political control, political organizing, and political power in particular are also affected by the forms and uses of networks.

The field of social network analysis has revealed that the characteristic shapes and flows within human networks also can influence the behavior of the humans within those networks. In some networks, gossip and disease can spread rapidly; in other networks, with different topologies, these phenomena spread more slowly. Other research has shown that people whose networks include smokers are more likely to smoke—and are more likely even if their friends' networks include smokers that they don't know directly.

But the networks we are most familiar with are far from the only ones. Now that colleges are beginning to teach network theory, there is room for an expansion of network science to include the myriad lesser-known networks, from technological networks such as the electrical grid or the internet (a network of networks) to water networks, the barbed wire telegraph, even imaginary networks. That's where Lori Emerson's *Other Networks: A Radical Technology Sourcebook* comes in. So far, network science has demonstrated the depth of networks in our lives. *Other Networks* reveals the breadth of networks that enmesh our world. In addition, this sourcebook illuminates alternatives to the ways we relate to networks, pointing the way to more participative, communitarian, and democratic methods of gaining and wielding wealth and power—in contrast to the passive, hierarchical, corporatized format that currently feels ascendant.

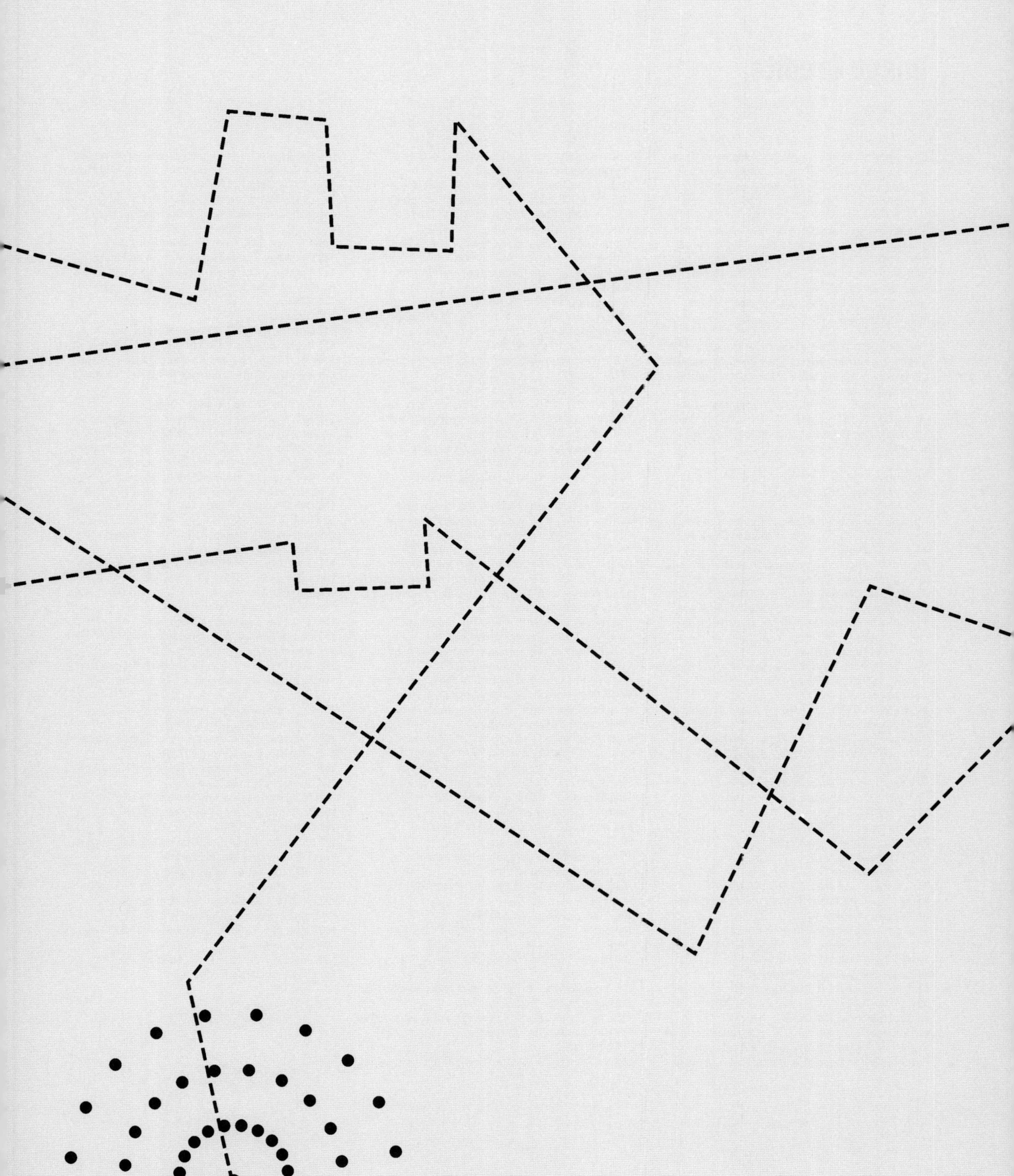

Image Credits

Every effort has been made to trace or contact all copyright holders. The publishers would be pleased to rectify any omissions or errors brought to their notice at the earliest opportunity.

The following images are believed to be in the public domain. If any rights have been inadvertently overlooked, please contact us so we may rectify the situation: pp. 20, 25 (top), 31, 48, 51, 55 (top), 80, 86–87, 89–90, 95–98, 100–101, 104–105, 107 (top), 111, 121, 126, 133, 136, 138–139, 143, 146, 149, 169, 179–181, 184, 198.

The following images are of unknown copyright status. Extensive research has been conducted to identify the copyright holder without success. If any rights have been inadvertently overlooked, please contact us so we may rectify the situation: pp. 15, 32–33, 55 (bottom), 58, 88, 91, 108 (top), 117 (top), 132, 135, 140, 154–155, 158, 167 (top), 170–171.

The following images are courtesy of AT&T: 70–71, 110, 114–116, 137 (top).

xiii Top: Image courtesy Kac Studios and Eduardo Kac; bottom: Photograph by Gwenn Thomas.
4 Top: Paul Baran, *On Distributed Communications: I. Introduction to Distributed Communications Networks*. Santa Monica, CA: RAND Corporation, 1964. https://www.rand.org/pubs/research_memoranda/RM3420.html. Also available in print form; bottom: Bettmann via Getty Images.
10 Courtesy Zulumathabo Zulu.
11 Julien Meyer, *Whistled Languages: A Worldwide Inquiry on Human Whistled Speech*, 19, 2015, Springer Nature.
14 The Print Collector / Alamy Stock Photo.
16–17 Harper Stereograph Collection, Boston Public Library, Print Department (Creative Commons, Attribution 2.0 Generic [CC BY 2.0]).
18 Classic Image / Alamy Stock Photo.
21 Used by permission of Charles Bernstein, for Hannah Weiner, in trust.
22 CC BY-SA 3.0.
23 Courtesy Eric Voskull (beforemario.com).
25 Bottom: ilbusca via iStock.
29 CC BY-SA 3.0.
30 Courtesy Anna Friz.
35 Courtesy Gerry Walsh / KB6OOC.
36–37 Copyright © 2008 Frank Dörenberg (www.hellschreiber.com).
38 Courtesy Siemens Historical Institute, Copyright © Siemens AG, Munich/Berlin.
40–41 Photo © Katie Paterson.
43 Copyright ©ARTSAT Project.
44 Courtesy Bruce R. MacKinnon.
45 E.P. Joe Tozer.
46 Courtesy Silvia Neuhaus.
49 Courtesy Lucy Helton. The piece was created as part of the Land Art Agency's Sustainable Futures: Outer Space residency and was made with support from Lui Gough, the Land Art Agency, and Lisa Ruth Rand.
52 Johnson Publishing Company Archive. Courtesy J. Paul Getty Trust and Smithsonian National Museum of African American History and Culture. Made possible by the Ford Foundation, J. Paul Getty Trust, John D. and Catherine T. MacArthur Foundation, The Andrew W. Mellon Foundation, and Smithsonian Institution.
53 Courtesy Brannon Dorsey.
54 P. A. Forsyth, "The Principles of JANET—A Meteor-Burst Communication System." *Proceedings of the IRE*, vol. 45, no. 12. Copyright © IEEE. Reprinted with permission from *Proceedings of the IRE*.
56–57 Courtesy Open Space Arts Society.
59 Courtesy of MIT Lincoln Laboratory, Lexington, Massachusetts. Reprinted with permission.
60 Photo by David R. Jones.
62 Courtesy Computer History Museum.
63 Courtesy Tetsuo Kogawa.
64 Courtesy Floris Vanhoof. Reprinted by permission.
67 Courtesy Danja Vasiliev / k0a1a.net, 2009.
68 Alex Ogle / AFP via Getty Images.
69 Courtesy Bill Fontana.
74–75 Courtesy Liza Béar, co-producer Send/Receive Satellite Network, 1977. More information: Liza Béar, "Send/Receive: Some History," https://sendreceivesatellitenetwork.blogspot.com (June 30, 2020) and Julie Ingram, "An Enchanted Evening: A Q&A with Video Pioneers Liza Béar and Milly Iatrou," *Huffington Post*, https://www.huffpost.com/entry/an-enchanted-evening-a-qa_b_3628853 (December 6, 2017).
81 Courtesy Granger Historical Picture Archive.
82–83 Copyright © Science Museum Group.
85 *Subtle Distress* (2007), © Kristen Haring, photograph by Frank Kleinbach.
92–93 Left: John Giorno with Dial-A-Poem, 1970. Copyright © 1970 Courtesy of the Giorno Poetry Systems Archive; right: *The Village Voice* Jan 16–Mar 30,1969 (from 1969-1970 Publicity Scrapbook 4). Copyright © 1969 Courtesy of the Giorno Poetry Systems Archive.
94 Courtesy Computer History Museum.
99 Courtesy MIT Libraries.
102 Source: Museum Foundation Post and Telecommunications.
103 Copyright © Tom Klinkowstein, all rights reserved.
106 Courtesy Marisa González.
107 Bottom: Courtesy Associated Press / AP Photo.
108 Bottom: Image courtesy Kac Studios and Eduardo Kac.
109 Image courtesy Kac Studios and Eduardo Kac.
112 Courtesy Kit Galloway.
113 Copyright © Ponton / Van Gogh TV, 1992–1993.
117 Bottom: Adolfsson Martin, *The New Yorker*, © Condé Nast.
118 N.E. Thing Co., *Telexed Triangle*, 1969, offset lithograph, ink and foil and paper seal on paper, 45.5 × 60.9 cm. Collection of the Morris and Helen Belkin Art Gallery, University of British Columbia, Purchased with the support of the Canada Council for the Arts Acquisition Assistance program and the Morris and Helen Belkin Foundation, 1999. Photo: Howard Ursuliak.
122 Courtesy Philip Peters and David Rueter.
127 Courtesy Christian Swinehart / https://samizdat.co/cyoa.
128–129 Courtesy of Charleston County Public Library.
131 Copyright © Ray Johnson Estate, New York.
137 Bottom: Weiner, Morris "Moe" to Sylvia Weiner, 25 December 1943. Autographed letter, signed (The Gilder Lehrman Institute of American History, GLC09414.0880).
141 Courtesy Jung Kwang-Il 정광일.
142 Top: CC BY-SA 4.0; bottom: *Commercial Broadcasting Pioneer: The WEAF Experiment, 1922–1926* by William Peck Banning, Cambridge, Mass.: Harvard University Press, Copyright © 1946 by the President and Fellows of Harvard College. Used by permission. All rights reserved.
144–145 Copyright © 2024 Chris Burden / licensed by The Chris Burden Estate and Artists Rights Society (ARS), New York. Photo: G. Beydler. Courtesy Gagosian.
147 Courtesy of McCasland Maps and Spatial Data, Oklahoma State University Library.
148 Copyright © Schweizer Radio und Fernsehen (SRF). Reprinted by permission.
150 Courtesy DeeDee Halleck.
151 Photo: Evangelos Dousmanis, courtesy Experimental Television Center.
152 Courtesy the Jaime Davidovich Foundation / Institute for Studies on Latin American Art (ISLAA)
157 Courtesy of Zbigniew Stachniak.
159 Copyright © Julian Oliver, 2015.
160 Courtesy of the University of Illinois Archives (RS 7/13/810, Box 3, April 1971 Report).
161 Courtesy of Trustees of Dartmouth College.
162 Courtesy Kit Galloway.
163–165 Courtesy Computer History Museum.
167 Bottom: Courtesy Per-Olof Jernberg.
168 Copyright © BBC Archive.
172 Image courtesy Kac Studios and Eduardo Kac.
174–175 Courtesy Mario Ramiro.
182–183 Illustration by Ana Clapés. Courtesy Giulio Ruffini.
185 Copyright © Collections Mundaneum, Belgium. Used by permission.
186 Top and bottom: Courtesy Computer History Museum.
188 Courtesy Theodor Holm Nelson.
191 Courtesy Verso Books.

Acknowledgments

I wish I could count the number of people who have been incredible, constant, enthusiastic supporters over the years it's taken to write this book. All the same, the most important people I want to thank are Ben Robertson, who always surprises me with his unwavering belief in me and his willingness to listen to me recount over dinner, for years, the weird facts about networks I learned every day; Mark Iosifescu, who approached me about doing this project with Anthology Editions in 2020 and then, somewhat miraculously, trusted in the value and the worth of *Other Networks* and provided genuinely invaluable vision, insights, encouragement, and help with the project; Jesse Pollock and the entire Anthology Editions team, including the brilliant cover designer Robert Beatty and equally brilliant book designer Ella Gold; and libi striegl, who has been a stalwart companion, friend, and irreplaceable educator on this *Other Networks* journey, patiently and repeatedly walking me through the basics of the electromagnetic spectrum, electricity, and electronics (among many other things) and even encouraging this recovering poet and literary scholar to get her ham radio license. I like to joke that libi teaches me about how things work in the world—things I should have learned in middle school and high school and somehow never did until now.

When Twitter existed, I also leaned heavily on the smart and helpful people there who were always willing to help me find sources and explain the workings of anything from pneumatic tubes [4] to telegraphs. Doron Galili is one of these exceptional people I met on Twitter who continues to amaze me with his willingness to share his deep knowledge of the history of television, among many other things. Thankfully, the usefulness of Twitter has been replaced by Mastodon, where my gratitude continues for all the people who follow #othernetworks and share their enthusiasm for learning (mostly in their spare time) and for the untapped potential of alternative networks. In a way, this book is a tribute to all the amateur historians, tinkerers, radio enthusiasts, and Wikipedia contributors whose deep know-how, research, and generosity I couldn't have done without.

Some particular people and entities I want to make sure I name, even though naming hardly stands in for the deep thanks I owe each and every one of you: Roel Abbing, Penny Ahlstrad and the wonderful staff at the Computer History Museum, William Aspray, Gabriele Balbi, Brett Ian Balogh, Fred Beavers, Robert Bernecky, August Black, Andrew Brandt, Nick Briz, Hank Bull and the equally wonderful staff at Western Front, Stephen Cass, Rob Cruickshank, CU Boulder Interlibrary Loan library staff and reference librarians, Kevin Curran, Mike Dank, John Day, Lydia and Rich Dissly, Paul Dourish, Michael Driscoll, Steve Dunbar, Christina Dunbar-Hester, John Durno, Benjamin Gaulon, Bernard Dionysius Geoghegan, Jason Gladstone, Sarah Grant, Dave Hartzell, Cheryl Higashida, Janice Ho, Tatyana Huseynova, Douglas Kahn, Laura Hyunjhee Kim, Chris J. King, Matthew Kirschenbaum, Tom Klinkowstein, William Kuskin, Mirco Lange, Caroline Langhill, Patrick Lichty, Ramesh Mallipeddi, Judy Malloy, Will Mari, Meredith Martin, Jeffrey Mathias, Shannon Mattern, John McVey, Darija Medic, Roger Moore, Open Space Archives, Élika Ortega, Jussi Parikka, Rick Prelinger, Erik Radio, Paul Riismandel, Sonia Robles, Andrew Russell, Jerome Saint-Clair, Edward Shanken, Iain Sharp, Josh Shepperd, Rory Solomon, Alan Sondheim, Zbigniew Stachniak, Melanie Swalwell, Ezra Teboul, @topleftbrick, Sean Tudor, Biyi Wen, Darren Wershler, Norman White, Rowan Wilken, Geoffrey Winthrop-Young, and Hannah Zeavin.

First published in the United States of America
in 2025 by Anthology Editions

87 Guernsey Street
Brooklyn, NY 11222

anthologyeditions.com

Editor: Mark Iosifescu
Editorial Assistants: Erin Youngblood, Jana Horn, Donna Allen, Rainer Turim
Designer: Ella Gold
Retouching: Alex Tults
Cover and Section Headings: Robert Beatty
Art Direction: Jesse Pollock
Sales and Marketing: Casey Whalen
Proofreader: Chris Peterson

First Edition, Second Printing
ARC 109
Printed in Turkey
ISBN: 978-1-944860-65-3

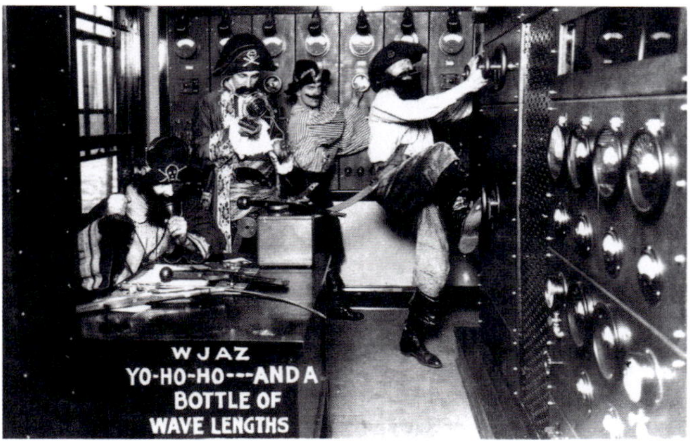

[18]

Uncredited photograph of engineering staff at WJAZ in Chicago—a radio station that, in 1926, began transmitting on an unassigned frequency. The original caption reads: "THE CREW OF A 'PIRATE STATION' IN ACTION: Undaunted by the disciplinary action of Uncle Sam in bringing suit against WJAZ on the ground that it has been pirating wave-lengths other than those assigned to it, the outlaw Chicago station recently—and appropriately—broadcast the operetta 'The Pirate.'"

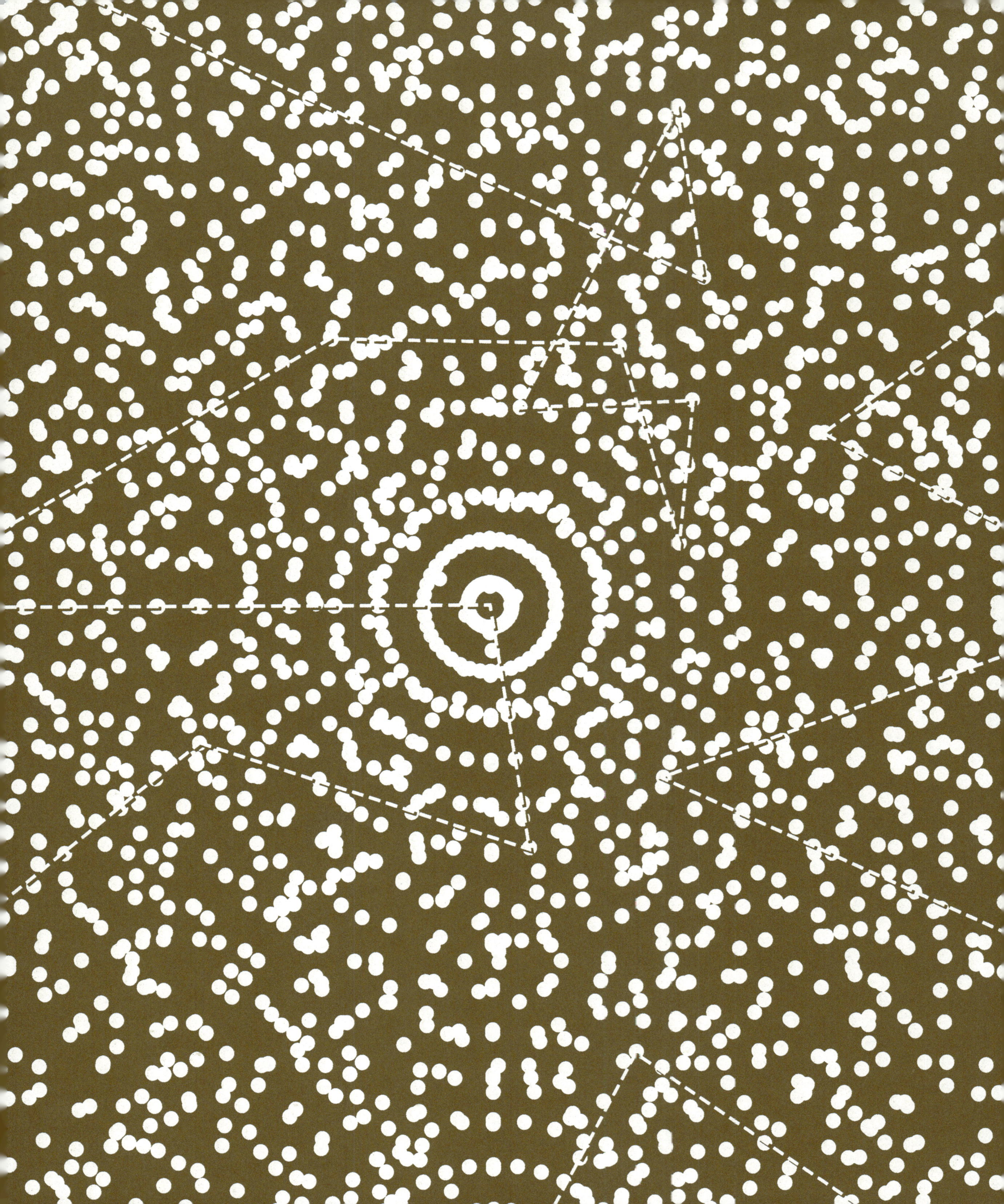

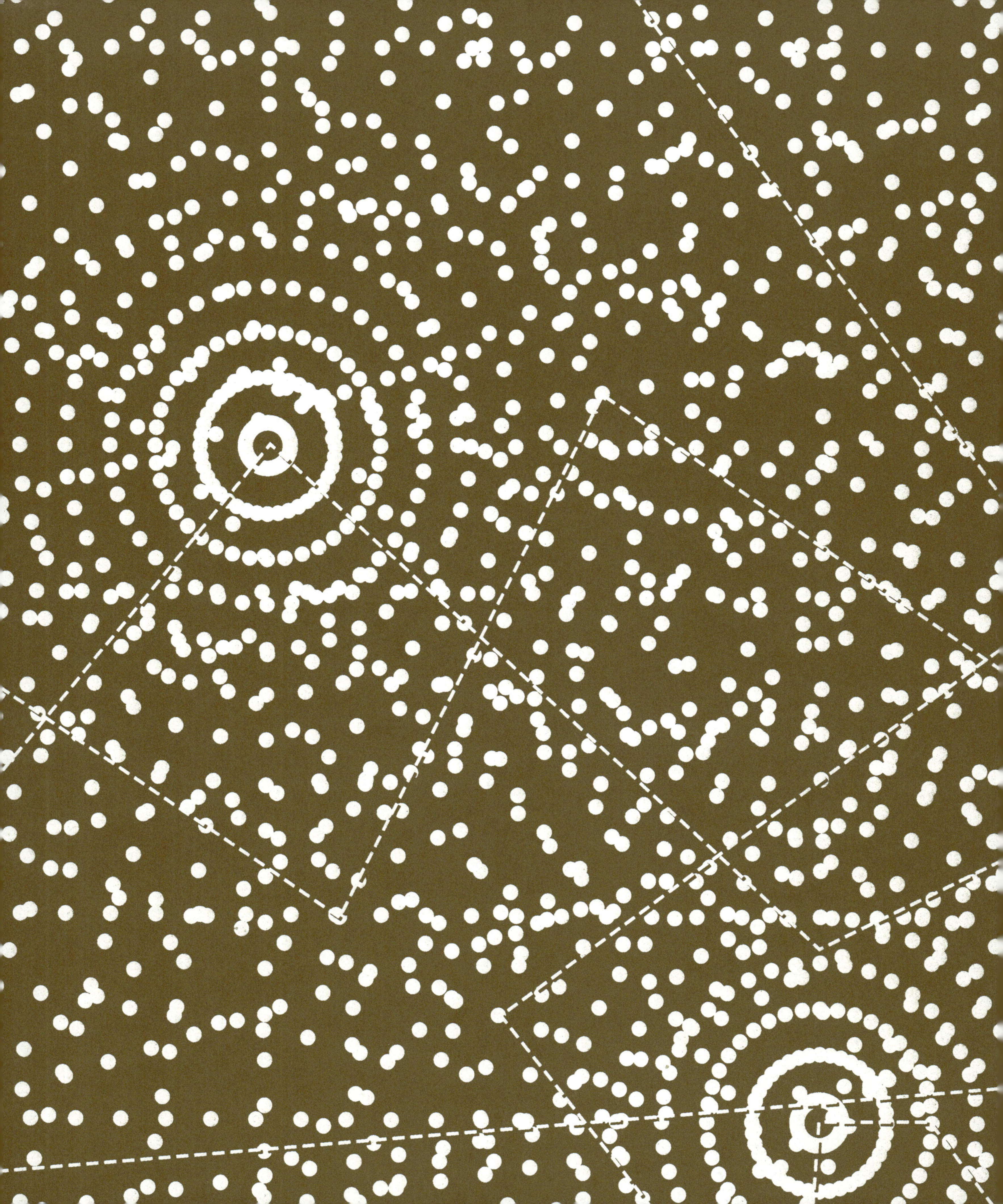